PEM Fuel Cell Modeling and Simulation Using MATLAB®

PEM Fuel Cell Modeling and Simulation Using MATLAB®

Colleen Spiegel

AMSTERDAM • BOSTON • HEIDELBERG • LONDON
NEW YORK • OXFORD • PARIS • SAN DIEGO
SAN FRANCISCO • SINGAPORE • SYDNEY • TOKYO

ELSEVIER   Academic Press is an imprint of Elsevier

Academic Press is an imprint of Elsevier
30 Corporate Drive, Suite 400, Burlington, MA 01803, USA
525 B Street, Suite 1900, San Diego, California 92101-4495, USA
84 Theobald's Road, London WC1X 8RR, UK

Library of Congress Cataloging-in-Publication Data
Spiegel, Colleen.
  PEM fuel cell modeling and simulation using Matlab / Colleen Spiegel.
    p. cm.
  Includes index.
  ISBN 978-0-12-374259-9
  1. Proton exchange membrane fuel cells–Design and construction. 2. Proton exchange membrane fuel cells–Computer simulation. 3. Fuel cells–Design and construction. 4. MATLAB. I. Title.
  TK2931.S66 2008
  621.31'2429–dc22

                                                                    2008006621

British Library Cataloguing-in-Publication Data
A catalogue record for this book is available from the British Library.

ISBN: 978-0-12-374259-9

For information on all Academic Press publications
visit our Web site at www.books.elsevier.com

Transferred to Digital Printing, 2010

Printed and bound in Great Britain by
CPI Antony Rowe, Chippenham and Eastbourne

Contents

Acknowledgments

To my husband, Brian, who always encourages me to pursue all of my dreams.

To my son, Howard, who had to endure many days and nights wondering what his mother is doing in that office. Maybe this text will be of use to you one day?

To my parents, Chris and Shirley, and my in-laws, Mark and Susan (pseudo parents), for helping to watch Howard while I completed this text.

To my aunt, Maureen, who helped me to become the person that I am.

To Dr. Shekhar Bhansali, who encouraged me to pursue my passion.

To all of the other friends, scientists, engineers, and colleagues that had a positive influence in my life.

Thank you very much. Your influence has had a positive impact on my life.

Acknowledgments

To my husband, Brian, who always encourages me to pursue all of my dreams.

To my son, Howard, who had to endure many days and nights wondering what his mother is doing in that office. Maybe this text will be of use to you one day.

To my parents, Clark and Shirley, and my in-laws, Mark and Susan, again for their time helping to watch Howard while I completed this text.

To my aunt, Maureen, who helped me to become the person that I am.

To Dr. Shekhar Banerjee, who encouraged me to pursue my passion.

To all of the other friends, relatives, mentors, and influences that had a positive influence on my life.

And to my parents. Your influence has had a positive impact on my life.

CHAPTER 1

An Introduction to Fuel Cells

1.1 Introduction

Fuel cells are set to become the power source of the future. The interest in fuel cells has increased during the past decade due to the fact that the use of fossil fuels for power has resulted in many negative consequences. Some of these include severe pollution, extensive mining of the world's resources, and political control and domination of countries that have extensive resources. A new power source is needed that is energy efficient, has low pollutant emissions, and has an unlimited supply of fuel. Fuel cells are now closer to commercialization than ever, and they have the ability to fulfill all of the global power needs while meeting the efficacy and environmental expectations.

Polymer electrolyte membrane (PEM) fuel cells are the most popular type of fuel cell, and traditionally use hydrogen as the fuel. PEM fuel cells also have many other fuel options, which range from hydrogen to ethanol to biomass-derived materials. These fuels can either be directly fed into the fuel cell, or sent to a reformer to extract pure hydrogen, which is then directly fed to the fuel cell.

There are only 30 additional years left of the supply of fossil fuels for energy use. Changing the fuel infrastructure is going to be costly, but steps should be taken now to ensure that the new infrastructure is implemented when needed. Since it is impossible to convert to a new economy overnight, the change must begin slowly and must be motivated by national governments and large corporations. Instead of using fossil fuels directly, they can be used as a "transitional" fuel to provide hydrogen that can be fed directly into the fuel cells. After the transition to the new economy has begun, hydrogen can then be obtained from cleaner sources, such as biomass, nuclear energy, and water. This chapter discusses fuel cell basics and introduces the modeling of fuel cells with the following topics:

- What is a PEM fuel cell?
- Why do we need fuel cells?
- The history of fuel cells
- Mathematical models in the literature
- Creating mathematical models

These introductory fuel cell topics are discussed to help the reader to appreciate the relevance that fuel cell modeling has in addressing the global power needs.

1.2   What Is a Fuel Cell?

A fuel cell consists of a negatively charged electrode (anode), a positively charged electrode (cathode), and an electrolyte membrane. Hydrogen is oxidized on the anode and oxygen is reduced on the cathode. Protons are transported from the anode to the cathode through the electrolyte membrane, and the electrons are carried to the cathode over the external circuit. In nature, molecules cannot stay in an ionic state, therefore they immediately recombine with other molecules in order to return to the neutral state. Hydrogen protons in fuel cells stay in the ionic state by traveling from molecule to molecule through the use of special materials. The protons travel through a polymer membrane made of persulfonic acid groups with a Teflon backbone. The electrons are attracted to conductive materials and travel to the load when needed. On the cathode, oxygen reacts with protons and electrons, forming water and producing heat. Both the anode and cathode contain a catalyst to speed up the electrochemical processes, as shown in Figure 1-1.

A typical PEM fuel cell (proton exchange membrane fuel cell) has the following reactions:

Anode: H_2 (g) $\rightarrow$ $2H^+$ (aq) + $2e^-$
Cathode: $^1/_2O_2$ (g) + $2H^+$ (aq) + $2e^-$ $\rightarrow$ H_2O (l)
Overall: H_2 (g) + $^1/_2O_2$ (g) $\rightarrow$ H_2O (l) + electric energy + waste heat

Reactants are transported by diffusion and/or convection to the catalyzed electrode surfaces where the electrochemical reactions take place. The water and waste heat generated by the fuel cell must be continuously removed and may present critical issues for PEM fuel cells.

The basic PEM fuel cell stack consists of a proton exchange membrane (PEM), catalyst and gas diffusion layers, flow field plates, gaskets and end plates as shown in Table 1-1: The actual fuel cell layers are the PEM, gas diffusion and catalyst layers. These layers are "sandwiched" together using various processes, and are called the membrane electrode assembly (MEA). A stack with many cells has MEAs "Sandwiched" between bipolar flow field plates and only one set of end plates.

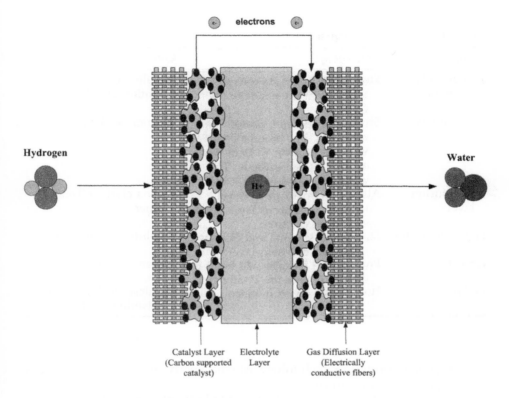

FIGURE 1-1. A single PEM fuel cell configuration.

Some advantages of fuel cell systems are as follows:

- Fuel cells have the potential for a high operating efficiency.
- There are many types of fuel sources, and methods of supplying fuel to a fuel cell.
- Fuel cells have a highly scalable design.
- Fuel cells produce no pollutants.
- Fuel cells are low maintenance because they have no moving parts.
- Fuel cells do not need to be recharged, and they provide power instantly when supplied with fuel.

Some limitations common to all fuel cell systems include the following:

- Fuel cells are currently costly due to the need for materials with specific properties. There is an issue with finding low-cost replacements. This includes the need for platinum and Nafion material.

TABLE 1-1
Basic PEM Fuel Cell Components

Component	Description	Common Types
Proton exchange membrane	Enables hydrogen protons to travel from the anode to the cathode.	Persulfonic acid membrane (Nafion 112, 115, 117)
Catalyst layers	Breaks the fuel into protons and electrons. The protons combine with the oxidant to form water at the fuel cell cathode. The electrons travel to the load.	Platinum/carbon catalyst
Gas diffusion layers	Allows fuel/oxidant to travel through the porous layer, while collecting electrons	Carbon cloth or Toray paper
Flow field plates	Distributes the fuel and oxidant to the gas diffusion layer	Graphite, stainless steel
Gaskets	Prevent fuel leakage, and helps to distribute pressure evenly	Silicon, Teflon
End plates	Holds stack layers in place	Stainless steel, graphite, polyethylene, PVC

- Fuel reformation technology can be costly and heavy and needs power in order to run.
- If another fuel besides hydrogen is fed into the fuel cell, the performance gradually decreases over time due to catalyst degradation and electrolyte poisoning.

1.3 Why Do We Need Fuel Cells?

Power traditionally relies upon fossil fuels, which have several limitations: (1) they produce large amounts of pollutants, (2) they are of limited supply, and (3) they cause global conflict between regions. Fuel cells can power anything from a house to a car to a cellular phone. They are especially advantageous for applications that are energy-limited. For example, power for portable devices is limited, therefore, constant recharging is necessary to keep a device working.

Table 1-2 compares the weight, energy, and volume of batteries to a typical PEM fuel cell. As shown in the Table 1-1, the fuel cell system can provide a similar energy output to batteries with a much smaller system weight and volume. This is especially advantageous for portable power system. Future markets for fuel cells include the portable, transportation and stationary sectors (basically every sector!). Each market needs fuel cells for varying reasons, as described in sections 1.3.1 to 1.3.3.

TABLE 1-2
General Fuel Cell Comparison with Other Power Sources

	Weight	*Energy*	*Volume*
Fuel cell	9.5 lb	2190 Whr	4.0 L
Zinc-air cell	18.5 lb	2620 Whr	9.0 L
Other battery types	24 lb	2200 Whr	9.5 L

1.3.1 Portable Sector

One of the major future markets for fuel cells is the portable sector. There are numerous portable devices that would use fuel cells in order to power the device for longer amounts of time. Some of these devices include laptops, cell phones, video recorders, ipods, etc. Fuel cells will power a device as long as there is fuel supplied to it. The current trend in electronics is the convergence of devices, and the limiting factor of these devices is the amount of power required. Therefore, power devices that can supply greater power for a longer period of time will allow the development of new, multifunctional devices. The military also has a need for high-power, long-term devices for soldiers' equipment. Fuel cells can easily be manufactured with greater power and less weight for military applications. Other military advantages include silent operation and low heat signatures.

1.3.2 Transportation Market

The transportation market will benefit from fuel cells because fossil fuels will continue to become scarce, and because of this, there will be inevitable price increases. Legislation is becoming stricter about controlling environmental emissions. There are certain parts of countries that are passing laws to further reduce emissions and to sell a certain number of zero emission vehicles annually. Fuel cell vehicles allow a new range of power use in smaller vehicles and have the ability to be more fuel efficient than vehicles that are powered by other fuels.

1.3.3 Stationary Sector

Large stationary fuel cells can produce enough electricity to power a house or business. These fuel cells may also make enough power to sell back to the grid. This fuel cell type is especially advantageous for businesses and residences where no electricity is available. Fuel cell generators are also more reliable than other generator types. This can benefit companies by saving money when power goes down for a short time.

1.4 History of Fuel Cells

William Grove is credited with inventing the first fuel cell in 1839[1]. Fuel cells were not investigated much during the 1800s and most of the 1900s. Extensive fuel cell research began during the 1960s at NASA. During the past decade, fuel cells have been extensively researched and are finally nearing commercialization.

A summary of fuel cell history is given in Figure 1-2.

The process of using electricity to break water into hydrogen and oxygen (electrolysis) was first described in 1800 by William Nicholson and Anthony Carlisle[2]. William Grove invented the first fuel cell in 1839, using the idea from Nicholson and Carlisle to "recompose water." He accomplished this by combining electrodes in a series circuit, with separate platinum electrodes in oxygen and hydrogen submerged in a dilute sulfuric acid electrolyte solution. The gas battery, or "Grove cell," generated 12 amps of current at about 1.8 volts[3]. Some of the other individuals who contributed to the invention of fuel cells are summarized as follows:

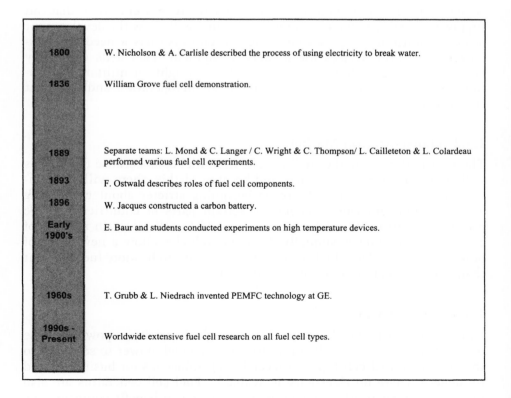

FIGURE 1-2. The history of fuel cells.

- Friedrich Wilhelm Ostwald (1853–1932), one of the founders of physical chemistry, provided a large portion of the theoretical understanding of how fuel cells operate. In 1893, Ostwald experimentally determined the roles of many fuel cell components[4].
- Ludwig Mond (1839–1909) was a chemist who spent most of his career developing soda manufacturing and nickel refining. In 1889, Mond and his assistant Carl Langer performed numerous experiments using a coal-derived gas. They used electrodes made of thin, perforated platinum and had many difficulties with liquid electrolytes. They achieved 6 amps per square foot (the area of the electrode) at 0.73 volt[5].
- Charles R. Alder Wright (1844–1894) and C. Thompson developed a similar fuel cell around the same time. They had difficulties in preventing gases from leaking from one chamber to another. This and other causes prevented the battery from reaching voltages as high as 1 volt. They thought that if they had more funding, they could create a better, robust cell that could provide adequate electricity for many applications[6].
- Louis Paul Cailleteton (1832–1913) and Louis Joseph Colardeau (France) came to a similar conclusion but thought the process was not practical due to needing "precious metals." In addition, many papers were published during this time saying that coal was so inexpensive that a new system with a higher efficiency would not decrease the prices of electricity drastically[7].
- William W. Jacques (1855–1932) constructed a "carbon battery" in 1896. Air was injected into an alkali electrolyte to react with a carbon electrode. He thought he was achieving an efficiency of 82% but actually obtained only an 8% efficiency[8].
- Emil Baur and students (1873–1944) (Switzerland) conducted many experiments on different types of fuel cells during the early 1900s. Their work included high-temperature devices and a unit that used a solid electrolyte of clay and metal oxides[9].
- Thomas Grubb and Leonard Niedrach invented PEM fuel cell technology at General Electric in the early 1960s. GE developed a small fuel cell for the U.S. Navy's Bureau of Ships (Electronics Division) and the U.S. Army Signal Corps. The fuel cell was fueled by hydrogen generated by mixing water and lithium hydride. It was compact, but the platinum catalysts were expensive.

NASA initially researched PEM fuel cell technology for Project *Gemini* in the early U.S. space program. Batteries were used for the preceding Project *Mercury* missions, but Project *Apollo* required a power source that would last a longer amount of time. Unfortunately, the first PEM cells developed had repeated difficulties with the internal cell contamination and leakage of oxygen through the membrane. GE redesigned their fuel cell, and the new model performed adequately for the rest of the *Gemini* flights. The designers of Project *Apollo* and the Space Shuttle ultimately chose to use alkali fuel cells.

GE continued to work on PEM fuel cells in the 1970s, and designed PEM water electrolysis technology, which led to the U.S. Navy Oxygen Generating Plant. The British Royal Navy used PEM fuel cells in the early 1980s for their submarine fleet, and during the past decade, PEM fuel cells have been researched extensively by commercial companies for transportation, stationary, and portable power markets.

Based upon the research, development, and advances made during the past century, technical barriers are being resolved by a world network of scientists. Fuel cells have been used for over 40 years in the space program, and the commercialization of fuel cell technology is rapidly approaching.

1.5 Mathematical Models in the Literature

Fuel cell modeling is helpful for fuel cell developers because it can lead to fuel cell design improvements, as well as cheaper, better, and more efficient fuel cells. The model must be robust and accurate and be able to provide solutions to fuel cell problems quickly. A good model should predict fuel cell performance under a wide range of fuel cell operating conditions. Even a modest fuel cell model will have large predictive power. A few important parameters to include in a fuel cell model are the cell, fuel and oxidant temperatures, the fuel or oxidant pressures, the cell potential, and the weight fraction of each reactant. Some of the parameters that must be solved for in a mathematical model are shown in Figure 1-3.

The necessary improvements for fuel cell performance and operation demand better design, materials, and optimization. These issues can only be addressed if realistic mathematical process models are available. There are many published models for PEM fuel cells in the literature, but it is often a daunting task for a newcomer to the field to begin understanding the complexity of the current models. Table 1-3 shows a summary of equations or characteristics of fuel cell models presented in recent publications.

The first column of Table 1-3 shows the number of dimensions the models have in the literature. Most models in the early 1990s were one dimensional, models in the late 1990s to early 2000s were two dimensional, and more recently there have been a few three dimensional models for certain fuel cell components. The second column specifies that the model can be dynamic or steady-state. Most published models have steady-state voltage characteristics and concentration profiles. The next column of Table 1-3 presents the types of electrode kinetic expressions used. Simple

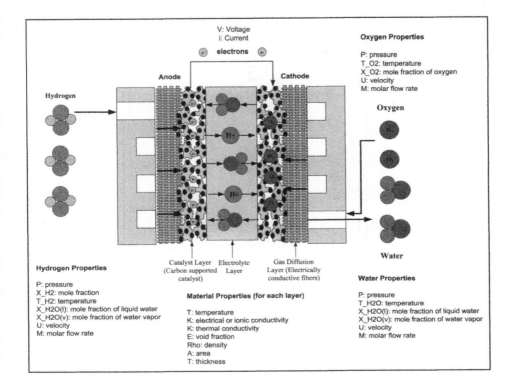

FIGURE 1-3. Parameters that must be solved for in a mathematical model.

Tafel-type expressions are often employed. Certain papers use Butler-Volmer–type expressions, and a few other models use more realistic, complex multistep reaction kinetics for the electrochemical reactions. The next column compares the phases used for the anode and cathode structures. It is well known that there are two phases (liquid and gas) that coexist under a variety of operating conditions. Inside the cathode structure, water may condense and block the way for fresh oxygen to reach the catalyst layer.

An important feature of each model is the mass transport descriptions of the anode, cathode, and electrolyte. Several mass transport models are used. Simple Fick diffusion models and effective Fick diffusion models typically use experimentally determined effective transport coefficients instead of Fick diffusivities, and do not account for convective flow contributions. Therefore, many models use Nernst-Planck mass transport expressions that combine Fick's diffusion with convective flow. The convective flow is typically calculated from Darcy's law using different formulations of the hydraulic permeability coefficient. Some models use Schlogl's formulations for convective flow instead of Darcy's law, which also accounts

TABLE 1-3
Comparison of Recent Mathematical Models

No. of Dimensions	Dyn/SS	Anode and Cathode Kinetics	Anode and Cathode Phase	Mass Transport (Anode and Cathode)	Mass Transport (Electrolyte)	Membrane Swelling	Energy Balance
one dimension, two dimension, or three dimension	Dynamic or steady-state	Tafel-type expressions, Butler-Volmer, complex kinetics equations	Gas, liquid, combination of gas and liquid	Effective Fick's diffusion, Nernst-Planck, Nernst-Planck + Schlogl, Maxwell-Stefan	Nernst-Planck + Schlogl, Nernst-Planck + drag coefficient, Maxwell-Stefan, Effective Fick's diffusion	Empirical or thermodynamic models	Isothermal or full energy balance

for electroosmotic flow, and can be used for mass transport inside the PEM. A very simple method of incorporating electroosmotic flow in the membrane is by applying the drag coefficient model, which assumes a proportion of water and fuel flow to proton flow. Another popular type of mass transport description is the Maxwell-Stefan formulation for multicomponent mixtures. This has been used for gas-phase transport in many models, but this equation would be better used for liquid–vapor–phase mass transport. Very few models use this equation for both phases. Surface diffusion models and models derived from irreversible thermodynamics are seldom used. Mass transport models that use effective transport coefficients and drag coefficients usually only yield a good approximation to experimental data under a limited range of operating conditions.

The second to last column of Table 1-3 shows that the swelling of polymer membranes is modeled through empirical or thermodynamic models for PEM fuel cells. Most models assume a fully hydrated PEM. In certain cases, the water uptake is described by an empirical correlation, and in other cases a thermodynamic model is used based upon the change of Gibbs free energy inside the PEM based upon water content.

The last column notes whether the published model includes energy balances. Most models assume an isothermal cell operation and therefore have no energy balances included. However, including energy balance equations is an important parameter in fuel cell models because the temperature affects the catalyst reactions and water management in the fuel cell.

A model is only as accurate as its assumptions allow it to be. The assumptions need to be well understood in order to understand the model's limitations and to accurately interpret its results. Common assumptions used in fuel cell modeling are:

- Ideal gas properties
- Incompressible flow
- Laminar flow
- Isotropic and homogeneous electrolyte, electrode, and bipolar material structures
- A negligible ohmic potential drop in components
- Mass and energy transport is modeled from a macroperspective using volume-averaged conservation equations

The concepts presented in this chapter can be applied to all fuel cell types, regardless of the fuel cell geometry. Even simple fuel cell models will provide tremendous insight into determining why a fuel cell system performs well or poorly. The physical phenomenon that occurs in a fuel cell can be represented by the solution of the equations presented throughout the book, especially in Chapters 2 through 11.

1.6 Creating Mathematical Models

The basic steps used for creating a mathematical model are the same regardless of the system being modeled. The details vary somewhat from method to method, but an understanding of the common steps, combined with the required method, provides a framework in which the results from almost any method can be interpreted and understood. The basic steps of the model-building process are:

1. Model selection
2. Model fitting
3. Model validation

These three basic steps are used iteratively until an appropriate model for the data has been developed. In the model selection step, plots of the data, process knowledge, and assumptions about the process are used to determine the form of the model to be fit to the data. Then, using the selected model and data, an appropriate model-fitting method is used to estimate the unknown parameters in the model. When the parameter estimates have been made, the model is then carefully assessed to see if the underlying assumptions of the analysis appear reasonable. If the assumptions seem valid, the model can be used to answer the scientific or engineering questions that initiated the modeling effort. If the model validation identifies problems with the current model, however, then the modeling process is repeated using information from the model validation step to select and/or fit an improved model.

1.6.1 A Variation on the Basic Steps

The three basic steps of process modeling assume that the data have already been collected and that the same data set can be used to fit all of the candidate models. Although this is often the case in model-building situations, one variation on the basic model-building sequence comes up when additional data are needed to fit a newly hypothesized model based on a model fit to the initial data. In this case, two additional steps—experimental design and data collection—can be added to the basic sequence between model selection and model-fitting. Figure 1-4 shows the basic model-fitting sequence with the integration of the related data collection steps into the model-building process.

Considering the model selection and fitting before collecting the initial data is also a good idea. Without data in hand, a hypothesis about what the data will look like is needed in order to guess what the initial model should be. Hypothesizing the outcome of an experiment is not always possible, but efforts made in the earliest stages of a project often maximize the efficiency of the whole model-building process and result in the best possible models for the process. The remainder of this book is devoted to the background theory and modeling of the fuel cell layers to help one to better understand the fuel cell system, and to create an accurate overall fuel cell model.

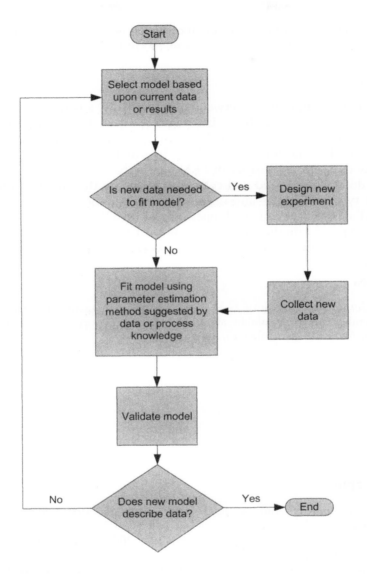

FIGURE 1-4. Model-building sequence.

Chapter Summary

While the fuel cell is a unique and fascinating system, accurate system selection, design, and modeling for prediction of performance are needed to obtain optimal performance and design. In order to make strides in performance, cost, and reliability, one must possess an interdisciplinary understanding of electrochemistry, materials, manufacturing, and mass and heat transfer. The remaining chapters in this book will provide the necessary bases in these areas in order to create models to provide the required fuel cell information and predictions in order to improve fuel cell designs.

Problems

- Describe the differences between a fuel cell and a battery.
- William Grove is usually credited with the invention of the fuel cell. In what way does his gaseous voltaic battery represent the first fuel cell?
- Describe the functions of each layer in a fuel cell stack.
- Briefly describe the history of PEM fuel cells.
- Why does society need fuel cells, and what can they be used for?
- What type of kinetics and mass transport equations would you use to model a PEM fuel cell stack operating at 80 °C?
- What type of kinetics and mass transport equations would you use to model an air-breathing PEM fuel cell stack operating at 22 °C?

Endnotes

[1] *Collecting the History of Fuel Cells.* Smithsonian Institution. Last updated December 2005. Available at: http://americanhistory.si.edu/fuelcells/index .htm. Accessed September 15, 2006.
[2] Ibid.
[3] Ibid.
[4] Ibid.
[5] Ibid.
[6] Ibid.
[7] Ibid.
[8] Ibid.
[9] Ibid.

Bibliography

Appleby, A., and F. Foulkes. *Fuel Cell Handbook.* 1989. New York: Van Nostrand Reinhold.
Barbir, F. *PEM Fuel Cells: Theory and Practice.* 2005. Burlington, MA: Elsevier Academic Press.
Energy Research Center of the Netherlands (ECN). *Fuel Cells.* Last updated September 18, 2006. Available at: http://www.ecn.nl/en/h2sf/additional/ fuel-cells-explained/.
Li, X. *Principles of Fuel Cells.* 2006. New York: Taylor & Francis Group.
Mikkola, M. Experimental studies on polymer electrolyte membrane fuel cell stacks. Helsinki University of Technology, Department of Engineering, Physics and Mathematics, master's thesis, 2001.
O'Hayre, R., S.-W. Cha, W. Colella, and F.B. Prinz. *Fuel Cell Fundamentals.* 2006. New York: John Wiley & Sons.
Stone, C., and A.E. Morrison. From curiosity to power to change the world. *Solid State Ionics.* 2002, Vol. 152–153, pp. 1–13.

CHAPTER 2

Fuel Cell Thermodynamics

2.1 Introduction

Thermodynamics is the study of energy changing from one state to another. The predictions that can be made using thermodynamic equations are essential for understanding and modeling fuel cell performance since fuel cells transform chemical energy into electrical energy. Basic thermodynamic concepts allow one to predict states of the fuel cell system, such as potential, temperature, pressure, volume, and moles in a fuel cell. The specific topics to be covered in this chapter are:

- Enthalpy
- Specific heats
- Entropy
- Free energy change of a chemical reaction
- Fuel cell reversible and net output voltage
- Theoretical fuel cell efficiency

The first few concepts relate to reacting systems in fuel cell analysis: absolute enthalpy, specific heat, entropy, and Gibbs free energy. The absolute enthalpy includes both chemical and sensible thermal energy[1]. Chemical energy or the enthalpy of formation (h_f) is associated with the energy of the chemical bonds, and sensible thermal energy (Δh_s) is the enthalpy difference between the given and reference state. The next important property is specific heat, which is a measure of the amount of heat energy required to increase the temperature of a substance by $1\,°C$ (or another temperature interval). Entropy is another important concept, which is a measure of the quantity of heat that shows the possibility of conversion into work. Gibbs free energy is the amount of useful work that can be obtained from an isothermal, isobaric system when the system changes from one set of steady-state conditions to another. The maximum fuel cell performance is then examined through the reversible voltage. The net output voltage is the actual fuel cell voltage after activation, ohmic, and concentration losses (which are explained in more detail in Chapters 3–5). Finally, the maximum efficiency of fuel cells is examined and demonstrated.

2.2 Enthalpy

When analyzing thermodynamic systems, the sum of the internal energy (U) and the product of pressure (P) and volume (V) appears so frequently that it has been termed "enthalpy" (H), and is denoted as[2]:

$$H = U + pV \tag{2-1}$$

The values for the internal energy and enthalpy can be obtained from thermodynamic tables when the temperature and pressure are known. When dealing with two-phase liquid-vapor mixtures (as in fuel cells), the specific internal energy and specific enthalpy can be calculated by Equations 2-1 and 2-3, respectively:

$$u = (1 - x)u_f + xu_g = u_f + x(u_g - u_f) \tag{2-2}$$

$$h = (1 - x)h_f + xh_g = h_f + x(h_g - h_f) \tag{2-3}$$

Sometimes the increase in enthalpy during vaporization $(h_g - h_f)$ is calculated under the heading h_{fg}. The use of Equations 2-1 and 2-2 is illustrated in Example 2-1.

EXAMPLE 2-1: Calculating the Enthalpy of Water

Water is in a state with a pressure of 1 psi and a temperature of 100 °C.

(a) Calculate the enthalpy.
   From Appendix H, $v = 1.696$ m³/kg and $u = 2506.7$ kJ/kg, then:

$$H = U + pV$$

$$H = 2506.7\frac{kJ}{kg} + \left(6894.76\frac{N}{m^2}\right)\left(1.696\frac{m^3}{kg}\right)\left(\frac{1kJ}{10^3\,N\cdot m}\right)$$

$$H = 2506.7 + 11.69 = 2518.4\frac{kJ}{kg}$$

Using MATLAB to solve:

%%

% EXAMPLE 2-1: Calculating the Enthalpy of Water

% UnitSystem SI

%%

% Inputs

V = 1.696;   % Specific Volume (m³/kg)
U = 2506.7;   % Specific Internal Energy (kJ/kg)
% Conversions
% 1 psi = 6894.76 N/m^2
% 1 kJ = 1000 N*m
P = 6894.76/1000;   % Conversion to get to the correct units
H = U + p*V

(b) If the water is heated to 80°C, calculate the enthalpy with a specific internal energy of 400 kJ/kg:

Looking at Appendix H, the given internal energy value falls between u_f and u_g at 80°C, therefore, the state is a two-phase liquid-vapor mixture. The quality of the mixture can be found from Equation 2-2 as follows:

$$x = \frac{u - u_f}{u_g - u} = \frac{400 - 334.86}{2482.2 - 400} = \frac{65.14}{2082.2} = 0.0313$$

With the values from Appendix H:

$$h = (1 - x)h_f + xh_g$$

$$h = (1 - 0.0313)*334.91 + 0.0313*2643.7$$

$$h = 324.43 + 82.75 = 407.14 \frac{kJ}{kg}$$

% Inputs

U = 400; % Specific Internal Energy (kJ/kg)
Uf = 334.86;
Ug = 2482.2;
Hf = 334.91;
Hg = 2643.7;
X = (u - uf)/(ug - u)
H = ((1 - x)*hf) + (x*hg)

The tables for specific internal energy and enthalpy data are given in Appendices H–J. The values of u, h, and s were not obtained directly from measurements but were calculated from other data that were more easily obtained from experiments. When using these properties with energy balances (see Chapter 6), the differences between the values at each state are more important than the actual values.

The enthalpy of formation for a substance is the amount of heat absorbed or released when one mole of the substance is formed from its elemental substances at the reference state. The enthalpy of substances in their naturally occurring state is defined as zero at the reference state (reference state is typically referred to as $T_{ref} = 25\,°C$ and $P_{ref} = 1$ atm). For example, hydrogen and oxygen at reference state are diatomic molecules (H_2 and O_2) and, therefore, the enthalpy of formation for H_2 and O_2 at $T_{ref} = 25\,°C$ and $P_{ref} = 1$ atm is equal to zero. The enthalpy of formation is typically determined by laboratory measurements, and can be found in various thermodynamic tables. Appendix B lists some values for the most common fuel cell substances.

2.3 Specific Heats

Another property that is important in thermodynamics and the study of fuel cells are the specific heats—which are useful when using the ideal gas model, which is introduced in Equation 2-12. The specific heats (or heat capacities) can be defined for pure, compressible substances as the partial derivatives of u(T,v) and h(T,p):

$$c_v = \frac{\partial u}{\partial T}\bigg|_v \qquad (2\text{-}4)$$

$$c_p = \frac{\partial h}{\partial T}\bigg|_p \qquad (2\text{-}5)$$

where v and p are the variables that are held constant when differentiating. c_v is a function of v and T, and c_p is a function of T and p. The specific heat is available in many thermodynamic property tables, and Appendices C, D, and E show the specific heats for some of the common fuel cell substances. These values are obtained through spectrographic measurements, or through exacting property measurements. The ratio of the heat capacities is called the specific heat ratio (k), and can be defined by:

$$k = \frac{c_p}{c_v} \qquad (2\text{-}6)$$

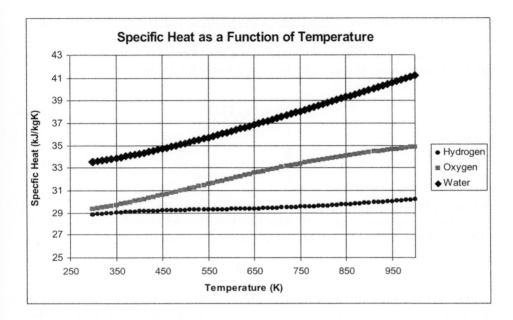

FIGURE 2-1. Specific heat values for hydrogen, oxygen, and water as a function of temperature.

The values c_p, c_v, and k are all very usual in thermodynamics and energy balances (see Chapter 6 because they relate the temperature change of a system to the amount of energy added by heat transfer). The specific heat values for hydrogen, oxygen, and water as a function of temperature are shown in Figure 2-1.

The values for v, u, and h can be obtained at liquid states using saturated liquid data. In order to simplify evaluations involving liquids and solids, the specific volume is often assumed to be constant and the specific internal energy is assumed to vary only with temperature. When a substance is defined in this manner, it is called "incompressible." Because the values for v, u, and h vary only slightly with changes in pressure at a fixed temperature, the following can be assumed for most engineering calculations:

$$v(T,p) \approx v_f(T) \tag{2-7}$$

$$u(T,p) \approx u_f(T) \tag{2-8}$$

Therefore, substituting these back into Equation 2-1, after deriving:

$$h(T,p) \approx u_f(T) + pv_f(T) \approx h_f(T) + v_f(T)[p - p_{sat}(T)] \tag{2-9}$$

When a substance is modeled as incompressible, the specific heats are assumed to be equal, $c_p = c_v$. The changes in specific internal energy and specific enthalpy between two states are a function of temperature and can be calculated using the specific heat at a constant temperature:

$$u_2 - u_1 = \int_{T_1}^{T_2} c_p(T)dT \tag{2-10}$$

$$h_2 - h_1 = \int_{T_1}^{T_2} c_p(T)dT + v(p_2 - p_1) \tag{2-11}$$

where $c_p(T)$ is the specific heat at a constant pressure.

The temperature/pressure/specific volume relationship for gases at many states can be given approximately by the ideal gas law:

$$pv = RT \tag{2-12}$$

where p is the pressure, v is the specific volume, T is the temperature, and R is the ideal gas constant. With $v = V/n$, the more familiar version of the equation can be expressed as:

$$pV = nRT \tag{2-13}$$

When the ideal gas model is used, the specific internal energy and specific enthalpy depend only upon temperature; therefore:

$$u_2(T_2) - u_1(T_1) = \int_{T_1}^{T_2} c_v(T)dT \tag{2-14}$$

$$h_2(T_2) - h_1(T_1) = \int_{T_1}^{T_2} c_p(T)dT \tag{2-15}$$

The relationship between the ideal gas–specific heats can be expressed by:

$$c_p(T) = c_v(T) + R \tag{2-16}$$

There are several alternative specific heat equations that are available in the literature. One that is easy to integrate is the polynomial form:

$$\frac{c_p}{R} = \alpha + \beta T + \gamma T^2 + \delta T^3 + \varepsilon T^4 \tag{2-17}$$

The values for α, β, γ, δ, and ε are listed in many thermodynamics texts, and can be found on the NIST website. Specific heats are painstakingly obtained from property measurements. Using the ideal gas tables, enthalpy can be obtained from:

$$\Delta h_T = h_{298.15} + \int_{298.15}^{T} \Delta c_p dT \qquad (2\text{-}18)$$

where $h_{298.15}$ is the enthalpy at a reference temperature. The specific heat is available in many thermodynamic property tables, and Appendices D through F show the specific heats for some of the common fuel cell substances. The average specific heat can be approximated as a linear function of temperature:

$$\overline{c_p} = c_p\left(\frac{T + T_{ref}}{2}\right) \qquad (2\text{-}19)$$

where $\overline{c_p}$ is the average specific heat at a constant pressure, T is the given temperature, and $T_{ref} = 25\,°C$.

The enthalpy of dry gas is

$$h_g = c_{pg}t \qquad (2\text{-}20)$$

where h_g is the enthalpy of dry gas, J/g, C_{pg} is the specific heat of gas, J/gK, and T is the temperature in °C.

When dealing with two-phase liquid-vapor mixtures, Equations 2-16 through 2-20 are useful. The enthalpy of the water vapor is

$$h_v = c_{pv}t + h_{fg} \qquad (2\text{-}21)$$

where h_{fg} is the heat of evaporation, 2500 J/g at 0 °C. Enthalpy of the moist gas is then:

$$h_{vg} = c_{pg}t + x(c_{pv}t + h_{fg}) \qquad (2\text{-}22)$$

and the unit is J/g dry gas.

The enthalpy of liquid water is

$$h_w = c_{pw}t \qquad (2\text{-}23)$$

If the gas contains both water and vapor, the enthalpy can be found by:

$$h_{vg} = c_{pg}t + x_v(c_{pv}t + h_{fg}) + x_w c_{pw}t \qquad (2\text{-}24)$$

where x_v is the water vapor content and x_w is the liquid water content. The total water content thus is

$$x = x_v + x_w \qquad (2\text{-}25)$$

Air going into the cell is typically humidified to prevent drying of the membrane near the cell inlet. Air enters the cell relatively dry. At lower temperatures, smaller amounts of water are required to humidify the cell than at higher temperatures. As the cell's air is heated up and the pressure increases, it needs more and more water.

EXAMPLE 2-2: Calculating the Enthalpy of H_2, O_2, and Water

Determine the absolute enthalpy of H_2, O_2, and water (H_2O) at the pressure of 1 atm at a temperature of (a) 25 °C using Appendix F, (b) 80 °C using Appendix F, and (c) 300 to 1000 K using increments of 50, using Equation 2-17. (d) Plot the hydrogen and oxygen enthalpy as a function of temperature. Calculate the absolute enthalpy for the vapor and liquid form if applicable.

(a) For T = 25 °C = 298 K, from Appendix F, the enthalpy of formation at 25 °C and 1 atm is $hf_{H2} = 0$, $hf_{O2} = 0$, $hf_{H2O(l)} = -285,826$ (J/mol), $hf_{H2O(g)} = -241,826$ (J/mol).

(b) For T = 80 °C = 353 K, the average temperature is (298 + 353)/2 = 325.5 K. From Appendix D, after interpolation and converting to a per mol basis:

$$c_{p,H2} = 14.3682 \frac{kJ}{kgK} \times 2.016 \frac{kg}{kmol} = 28.9663 \frac{kJ}{kmolK}$$

$$c_{p,O2} = 0.9231 \frac{kJ}{kgK} \times 31.999 \frac{kg}{kmol} = 29.5383 \frac{kJ}{kmolK}$$

$$c_{p,H2O(g)} = 33.8638 \frac{kJ}{kmolK}$$

For liquid water from Appendix E:

$$c_{p,H2O(l)} = 4.1821 \frac{kJ}{kgK} \times 18.015 \frac{kg}{kmol} = 75.3403 \frac{kJ}{kmolK}$$

Using MATLAB, the inputs are:

```
%%%%%%%%%%%%%%%%%%%%%%%%%%%%%%%%%%%%%%%%%%
% EXAMPLE 2-2:  Calculating the Enthalpy of H2, O2, and Water
% UnitSystem SI
%%%%%%%%%%%%%%%%%%%%%%%%%%%%%%%%%%%%%%%%%%
% Inputs

T = 353;                     % Temperature (K)
T_ref = 298;                 % Reference Temperature (K)
T_av = (T + T_ref)/2;        % Average Temperature (K)
m_H2 = 2.016;                % Moles of Hydrogen
m_O2 = 31.999;               % Moles of Oxygen
m_H2O = 18.015;              % Moles of H2O
hf_H2 = 0;                   % Enthalpy at standard state
hf_O2 = 0;                   % Enthalpy at standard state
hf_H2Ol = −285826;           % Enthalpy at standard state of water in liquid phase(J/mol)
hf_H2Og = −241826;           % Enthalpy at standard state of water in gas phase(J/mol)

%%%%%%%%%%%%%%%%%%%%%%%%%%%%%%%%%%%%%%%%%%
```

One method of obtaining the interpolated specific heat values is using MATLAB's "interp1" function. The columns in the tables in Appendices D and E can be put into MATLAB as vectors, and the "interp1" function will interpolate between data points to get the required value. This is shown in MATLAB by:

```
% Interpolate values from Appendix D

T_table = [250 300 350 400 450 500 550 600 650 700 750 800 900 1000];
Cp_H2_table = [14.051 14.307 14.427 14.476 14.501 14.513 14.530 14.546 14.571
    14.604 14.645 14.695 . . . 14.822 14.983];
Cp_O2_table = [0.913 0.918 0.928 0.941 0.956 0.972 0.988 1.003 1.017 1.031 1.043
    1.054 1.074 1.090];
Cp_H2Og_table = [33.324 33.669 34.051 34.467 34.914 35.390 35.891 36.415
    36.960 37.523 38.100 . . . 38.690 39.895 41.118];

% From Appendix E

T_table2 = 273:20:453;
Cp_H2Ol_table = [4.2178 4.1818 4.1784 4.1843 4.1964 4.2161 4.250 4.283 4.342
    4.417];

% Calculate specific heats (KJ/kgK)

Cp_H21 = interp1(T_table,Cp_H2_table,T_av);
Cp_O21 = interp1(T_table,Cp_O2_table,T_av);
Cp_H2Og1 = interp1(T_table,Cp_H2Og_table,T_av);
Cp_H2Ol1 = interp1(T_table2,Cp_H2Ol_table,T_av);
```

% Convert to a per mole basis

```
Cp_H2 = Cp_H21*m_H2;
Cp_O2 = Cp_O21*m_O2;
Cp_H2Og = Cp_H2Og1;
Cp_H2Ol = Cp_H2Ol1*m_H2O;
```

The absolute enthalpy is determined as follows:

$$h_{H2} = h_{f,H2} + c_{p,H2}(T - T_{ref}) = 0 + 28.954 \times (353 - 298) = 1593.1 \frac{J}{mol}$$

$$h_{O2} = h_{f,O2} + c_{p,O2}(T - T_{ref}) = 0 + 29.535 \times (353 - 298) = 1624.6 \frac{J}{mol}$$

$$h_{H2O(l)} = h_{f,H2O(l)} + c_{p,H2O(l)}(T - T_{ref}) = -285,826 + 75.321 \times (353 - 298)$$

$$= -281,680.0 \frac{J}{mol}$$

$$h_{H2O(g)} = h_{f,H2O(g)} + c_{p,H2O(g)}(T - T_{ref}) = -241,826 + 33.845 \times (353 - 298)$$

$$= -239,960.0 \frac{J}{mol}$$

% Determine absolute enthalpy

```
h_H2 = hf_H2 + Cp_H2*(T – T_ref)
h_O2 = hf_O2 + Cp_O2*(T – T_ref)
h_H2Ol = hf_H2Ol + Cp_H2Ol*(T – T_ref)
h_H2Og = hf_H2Og + Cp_H2Og*(T – T_ref)
```

(c) Inserting Equation 2-17 into Equation 2-18, and integrating with respect to temperature, yields:

$$h_2 - h_1 = R \int_{T_1}^{T_2} \alpha + \beta T + \gamma T^2 + \delta T^3 + \varepsilon T^4$$

$$h_2 - h_1 = R \left[\alpha(T_2 - T_1) + \frac{\beta}{2}(T_2^2 - T_1^2) + \frac{\gamma}{3}(T_2^3 - T_1^3) + \frac{\delta}{4}(T_2^4 - T_1^4) + \frac{\varepsilon}{5}(T_2^5 - T_1^5) \right]$$

One very useful feature of using a programming language to obtain the solution to problems is the ability to perform numerous calculations simultaneously. The solution to (c) can be made simple in MATLAB using "loops." Loops enable the programmer to repeat a calculation for a number of inputted values. In this particular calculation, the enthalpies will be found for a temperature range of 300 to 1000 K, with increments of 50.

Using MATLAB, the inputs are:

```
%%%%%%%%%%%%%%%%%%%%%%%%%%%%%%%%%%%%%%%%
% EXAMPLE 2-2: Calculating the Enthalpy of H2, O2,
and Water
% UnitSystem SI
%%%%%%%%%%%%%%%%%%%%%%%%%%%%%%%%%%%%%%%%
% Inputs
R = 8.314;                % Ideal Gas Constant
T = 300:50:1000;          % Temperature range from 300 K to 1000 K with increments
                            of 50
```

Create the temperature loop:

```
% Create Temperature Loop
I = 0;                    % Initialization of loop variable
for T = 300:50:1000;      % Temperature range from 300 K to 1200 K with
                            increments of 50
I = I + 1;                % Loop variable
```

The enthalpy calculations are in the temperature loop:

```
%%%%%%%%%%%%%%%%%%%%%%%%%%%%%%%%%%%%%%%%
% Enthalpy Calculations
%%%%%%%%%%%%%%%%%%%%%%%%%%%%%%%%%%%%%%%%
% Hydrogen enthalpy calculations
hf_h = 0;                      % Hydrogen enthalpy at the standard state
hs_h = R * (((3.057 * T) + ((1/2) * 2.677E-3 * T^2) − ((1/3) * 5.810E-6 * T^3) +
      ((1/4) * 5.521E-9 * T^4) − ((1/5) * 1.812E-..12 * T^5)) − (3.057 * 298 +
      ((1/2) * 2.677E-3 * 298^2) − ((1/3) * 5.810E-6 * 298^3) + ((1/4) * 5.521E-9 * 298^4) −
      ((1/5) * 1.812E-12 * 298^5)));
H_hydrogen = (hf_h + hs_h);    % Enthalpy of Hydrogen
%%%%%%%%%%%%%%%%%%%%%%%%%%%%%%%%%%%%%%%%
% Oxygen enthalpy calculations
hf_o = 0;                      % Oxygen enthalpy at the standard state
hs_o = R * (((3.626 * T) − ((1/2) * 1.878E-3 * T^2) + ((1/3) * 7.055E-6 * T^3) −
      ((1/4) * 6.764E-9 * T^4) + ((1/5) * 2.156E-12 * T^5)) − (3.626 * 298 −
      ((1/2) * 1.878E-3 * 298^2) + ((1/3) * 7.055E-6 * 298^3) − ((1/4) * 6.764E-9 * 298^4) +
      ((1/5) * 2.156E-12 * 298^5)));
H_oxygen = (hf_o + hs_o);      % Enthalpy of Oxygen
```

```
%%%%%%%%%%%%%%%%%%%%%%%%%%%%%%%%%%%%%%%%
% Water enthalpy calculations

hf_w = -241820;              % Water enthalpy at the standard state
hs_w = R*(((4.070*T) - ((1/2)*1.108E-3*T^2) + ((1/3)*4.152E-6*T^3) -
   ((1/4)*2.964E-9*T^4) + ((1/5)*0.807E-12*T^5)) - (4.070*298) -
   ((1/2)*1.108E-3*298^2) + ((1/3)*4.152E-6*298^3) - ((1/4)*2.964E-9*298^4) +
   ((1/5)*0.807E-12*298^5));
H_water = (hf_w + hs_w)   % Enthalpy of Water

%%%%%%%%%%%%%%%%%%%%%%%%%%%%%%%%%%%%%%%%
```

In order to save the new calculated values at each temperature, new variables have to be created as follows:

```
%%%%%%%%%%%%%%%%%%%%%%%%%%%%%%%%%%%%%%%%
% Create new variables to save the new calculated values for
hydrogen, oxygen, and water enthalpy at each temperature
increment

Hydrogen_Enthalpy(i) = H_hydrogen;
Oxygen_Enthalpy(i) = H_oxygen;
Water_Enthalpy(i) = H_water;
Temperature(i) = T;
end % End Loop
```

   (d) The following MATLAB code can be used to plot the hydrogen and oxygen enthalpy as a function of temperature:

```
figure1 = figure('Color',[1 1 1]);
hdlp = plot(Temperature,Hydrogen_Enthalpy,Temperature,Oxygen_Enthalpy);
title('Hydrogen and Oxygen Enthalpies','FontSize',12,'FontWeight','Bold')
xlabel('Temperature (K)','FontSize',12,'FontWeight','Bold');
ylabel('Hydrogen and Oxygen Enthalpies (KJ/kg)','FontSize',12,'FontWeight','Bold');
set(hdlp,'LineWidth',1.5);
grid on;
```

The hydrogen and oxygen enthalpies as a function of temperature are plotted in Figure 2-2. Another simple method of accomplishing this is MATLAB is through airways and matrices. There will be many examples throughout this book that will use this method.

   Example 2-2 shows that as the temperature increases, the specific heat at constant temperature increases very slowly. Specific heat is a very weak function of temperature. The increase in specific heat is the smallest for H_2, which is the smallest molecule in the periodic table compared with all of the other elements.

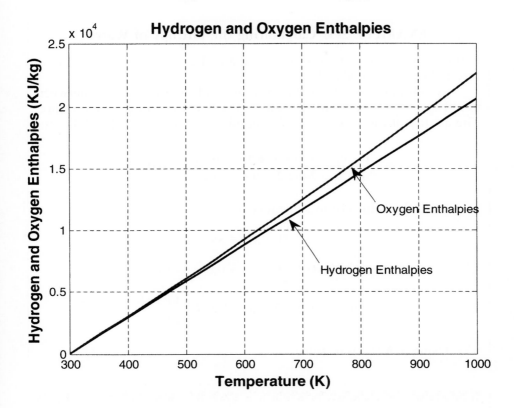

FIGURE 2-2. Hydrogen and oxygen enthalpies as a function of temperature.

2.4 Entropy

Entropy can be defined loosely as the amount of "disorder" in a system, and can be expressed as:

$$S2 - S1 = \left[\int_1^2 \frac{\delta Q}{T} \right]_{int\,rev} \tag{2-26}$$

This is valid for any reversible process that links two states. Entropy is calculated in the same way that enthalpy was calculated—using the properties v, u, and h.

When dealing with two-phase liquid-vapor mixtures as in fuel cells, the specific entropy can be calculated in the same manner as enthalpy:

$$s = (1 - x)s_f + xs_g = s_f + x(s_g - s_f) \tag{2-27}$$

EXAMPLE 2-3: Calculating the Entropy of Water

If the water is heated to 80°C, calculate the entropy with a specific internal energy of 400 kJ/kg:

    Looking at Appendix B, the given internal energy value falls between u_f and u_g at 80°C; therefore, the state is a two-phase liquid-vapor mixture. The quality of the mixture can be found from Equation 2-2 as follows:

$$x = \frac{u - u_f}{u_g - u} = \frac{400 - 334.86}{2482.2 - 400} = \frac{65.14}{2082.2} = 0.0313$$

With the values from Appendix B:

$$s = (1 - x)s_f + xs_g$$

$$s = (1 - 0.0313) * 1.0753 + 0.0313 * 7.6122$$

$$h = 1.0417 + 0.2381 = 1.2798 \frac{kJ}{kgK}$$

Using MATLAB to solve:

```
%%%%%%%%%%%%%%%%%%%%%%%%%%%%%%%%%%%%%%%%
% EXAMPLE 2-3: Calculating the Entropy of Water
% UnitSystem SI
%%%%%%%%%%%%%%%%%%%%%%%%%%%%%%%%%%%%%%%%
% Inputs (From Appendix B)
U = 400;    % Specific Internal Energy (kJ/kg)
Uf = 334.86;
Ug = 2482.2;
Sf = 1.0753;
Sg = 7.6122;
%%%%%%%%%%%%%%%%%%%%%%%%%%%%%%%%%%%%%%%%
% Calculate the mole fraction of the mixture
X = (u − uf)/(ug − u)
% Calculate the enthalpy
S = ((1 − x) * sf) + (x * sg)
```

Like enthalpy, the values for v, u, and h vary only slightly with changes in pressure at a fixed temperature; therefore, the following can be assumed for most engineering calculations:

$$s(T,p) \approx s_f(T) \qquad (2\text{-}28)$$

When a pure, compressible system undergoes an internally reversible process in the absence of gravity and overall system motion, an energy balance can be written as:

$$\delta Q_{int\ rev} = dU + \delta W_{int\ rev} \qquad (2\text{-}29)$$

In a simple compressible system, the work can be defined as:

$$\delta W_{int\ rev} = p \cdot dV \qquad (2\text{-}30)$$

Substituting equations 2-30 into 2-29, one obtains:

$$TdS = dU - pdV \qquad (2\text{-}31)$$

Another useful equation can be obtained by substituting equation 2-1:

$$TdS = dH - Vdp \qquad (2\text{-}32)$$

Although these equations have been obtained by considering an internally reversible process, the entropy change calculated by these equations is valid for the change in any process of the system, reversible or irreversible, between two equilibrium states.

When the ideal gas model is used, the specific entropy depends only upon temperature and can be derived from Equations 2-31 and 2-32, therefore:

$$s_2(T_2, v_2) - s_1(T_1, v_1) = \int_{T_1}^{T_2} c_v(T)\frac{dT}{T} + R \ln\frac{v_2}{v_1} \qquad (2\text{-}33)$$

$$s_2(T_2, p_2) - s_1(T_1, p_1) = \int_{T_1}^{T_2} c_p(T)\frac{dT}{T} + R \ln\frac{p_2}{p_1} \qquad (2\text{-}34)$$

Like enthalpy, the entropy can be obtained from using the ideal gas tables:

$$\Delta s_T = s_{298.15} + \int_{298.15}^{T} \Delta c_p dT \qquad (2\text{-}35)$$

where $s_{298.15}$ is the entropy at a reference temperature.

EXAMPLE 2-4: Calculating the Entropy of H₂, O₂, and Water

Determine the absolute entropy of H_2, O_2, and water (H_2O) at the pressure of 1 atm at a temperature of 300 to 1000 K using increments of 50, using Equation 2-17. Plot the hydrogen and oxygen enthalpy as a function of temperature. Calculate the absolute enthalpy for the vapor and liquid form if applicable.

(a) As seen previously, the specific heat can be calculated by:

$$\frac{c_p}{R} = \alpha + \beta T + \gamma T^2 + \delta T^3 + \varepsilon T^4$$

Inserting Equation 2-17 into Equation 2-34 and integrating with respect to temperature yields:

$$s_2 - s_1 = R \int_{T_1}^{T_2} \left(\frac{\alpha + \beta T + \gamma T^2 + \delta T^3 + \varepsilon T^4}{T} \right) dT,$$

$$s_2 - s_1 = R \left[\alpha \ln \frac{T_2}{T_1} + \beta(T_2 - T_1) + \frac{\gamma}{2}(T_2^2 - T_1^2) + \frac{\delta}{3}(T_2^3 - T_1^3) + \frac{\varepsilon}{4}(T_2^4 - T_1^4) \right]$$

As shown previously in Example 2-2, a very useful feature of using a programming language to obtain the solution to problems is the ability to perform numerous calculations simultaneously. This solution is again obtained in MATLAB using "loops." Loops enable the programmer to repeat a calculation for a number of inputted values. In this particular calculation, the enthalpies will be found for a temperature range of 300 to 1000 K, with increments of 50.

Using MATLAB, the inputs are:

```
%%%%%%%%%%%%%%%%%%%%%%%%%%%%%%%%%%%%%%%%%
% EXAMPLE 2-4:  Calculating the Entropy of H2, O2, and Water
% UnitSystem SI
%%%%%%%%%%%%%%%%%%%%%%%%%%%%%%%%%%%%%%%%%
% Inputs

R = 8.314          % Ideal Gas Constant
T = 300:50:1 000   % Temperature range from 300 K to 1000 K with increments
                     of 50
```

Create the temperature loop:

% Create Temperature Loop

```
i = 0;                    % Initialization of loop variable
for T = 300:50:1000;      % Temperature range from 300 K to 1200 K with
                          increments of 50
i = i + 1;                % Loop variable
```

The enthalpy calculations are in the temperature loop:

```
%%%%%%%%%%%%%%%%%%%%%%%%%%%%%%%%%%%%%%%%%
```

% Enthalpy Calculations

```
%%%%%%%%%%%%%%%%%%%%%%%%%%%%%%%%%%%%%%%%%
```

% hydrogen entropy calculations

```
sf_h = 130.57;
st_h = log(24.42) + 22.26E-3*(T - 298) - 24.2E-6*(T^2 - 298^2) + 15.3E-9*
   (T^3 - 298^3) - 3.78E-12*(T^4 - 298^4);
S_hydrogen = (sf_h + st_h);
```

```
%%%%%%%%%%%%%%%%%%%%%%%%%%%%%%%%%%%%%%%%%
```

% oxygen entropy calculations

```
sf_o = 205.03;
st_o = log(30.15) - 15.6E-3*(T - 298) + 29.33E-6*(T^2 - 298^2) - 18.7E-9*
   (T^3 - 298^3) + 4.48E-12*(T^4 - 298^4);
S_oxygen = (sf_o + st_o);
```

```
%%%%%%%%%%%%%%%%%%%%%%%%%%%%%%%%%%%%%%%%%
```

% water entropy calculations

```
sf_w = 188.72;
st_w = log(33.84) - 9.216E-3*(T-298) + 17.26E-6*(T^2-298^2) - 8.21E-9*(T^3-
   298^3) + 1.67E-12*(T^4-298^4);
S_water = (sf_w + st_w);
```

```
%%%%%%%%%%%%%%%%%%%%%%%%%%%%%%%%%%%%%%%%%
```

In order to save the new calculated values at each temperature, new variables have to be created as follows:

```
%%%%%%%%%%%%%%%%%%%%%%%%%%%%%%%%%%%%%%%%%
```

% Create new variables to save the new calculated values for hydrogen, oxygen and water entropy at each temperature increment

```
Hydrogen_Entropy(i) = S_hydrogen;
Oxygen_Entropy(i) = S_oxygen;
Water_Entropy(i) = S_water;
```

```
Temperature(i) = T;
end % End Loop
```

(b) The following MATLAB code can be used to plot the hydrogen
and oxygen enthalpy as a function of temperature:

```
figure1 = figure('Color',[1 1 1]);
hdlp = plot(Temperature,Hydrogen_Entropy,Temperature,Oxygen_Entropy,
    Temperature,Water_Entropy);
title('Hydrogen, Oxygen and Water Entropies','FontSize',12,'FontWeight','Bold')
xlabel('Temperature (K)','FontSize',12,'FontWeight','Bold');
ylabel('Entropies (KJ/kgK)','FontSize',12,'FontWeight','Bold');
set(hdlp,'LineWidth',1.5);
grid on;
```

Figure 2-3 shows the plot of hydrogen and oxygen entropies as a function
of temperature.

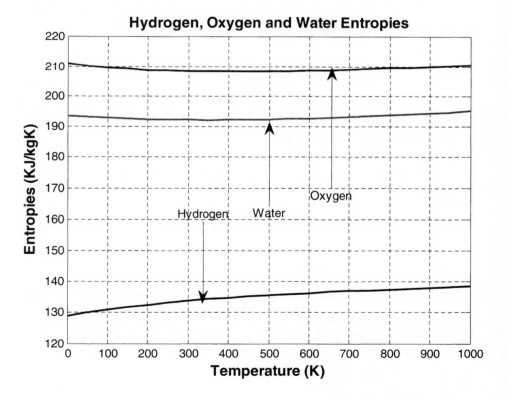

FIGURE 2-3. Hydrogen and oxygen entropies as a function of temperature.

2.5 Free Energy Change of a Chemical Reaction

The conversion of the free energy change associated with a chemical reaction directly into electrical energy is the electrochemical energy conversion. This free energy change is a measure of the maximum electrical work (W_{elec}) a system can perform at a constant temperature and pressure from the reaction. This is given by the negative change in Gibbs free energy change (ΔG) for the process, and can be expressed in molar quantities as:

$$W_{elec} = -\Delta G \qquad (2\text{-}36)$$

The Gibbs free energy is the energy required for a system at a constant temperature with a negligible volume, minus any energy transferred to the environment due to heat flux. This equation is valid at any constant temperature and pressure for most fuel cell systems. From the second law of thermodynamics, the change in free energy, or maximum useful work, can be obtained when a "perfect" fuel cell operating irreversibly is dependent upon temperature. Thus, W_{elec}, the electrical power output, is

$$W_{elec} = \Delta G = \Delta H - T\Delta S \qquad (2\text{-}37)$$

where G is the Gibbs free energy, H is the heat content (enthalpy of formation), T is the absolute temperature, and S is entropy. The Gibbs free energy will be equal to the enthalpy if the change in entropy is zero. As can be seen by the definition of the Gibbs free energy function, it is a linearly decreasing function with temperature, but the trend is complicated by the temperature dependence of the enthalpy and entropy terms. The $T\Delta S$ term grows faster with increasing T than the ΔH term because it has a stronger dependence upon temperature, as it is of the form $T*\ln(T/T)$, whereas the enthalpy is simply $(T - T)$.

The potential of a system to perform electrical work by a charge, Q (coulombs), through an electrical potential difference, E in volts, is[3]:

$$W_{elec} = EQ \qquad (2\text{-}38)$$

If the charge is assumed to be carried out by electrons:

$$Q = nF \qquad (2\text{-}39)$$

where n is the number of moles of electrons transferred and F is the Faraday constant (96,485 coulombs per mole of electrons). Combining the last three equations to calculate the maximum reversible voltage provided by the cell:

$$\Delta G = -nFE_r \qquad (2\text{-}40)$$

where n is the number of moles of electrons transferred per mol of fuel consumed, F is Faraday's constant, and E_r is the standard reversible potential.

The relationship between voltage and temperature is derived by taking the free energy, linearizing about the standard conditions of 25 °C, and assuming that the enthalpy change ΔH does not change with temperature:

$$E_r = -\frac{\Delta G_{rxn}}{nF} = -\frac{\Delta H - T\Delta S}{nF}$$

$$\Delta E_r = \left(\frac{dE}{dT}\right)(T - 25) = \frac{\Delta S}{nF}(T - 25) \qquad (2\text{-}41)$$

where E_r is the standard-state reversible voltage, and ΔG_{rxn} is the standard free energy change for the reaction. The change in entropy is negative; therefore, the open circuit voltage output decreases with increasing temperature. The fuel cell is theoretically more efficient at low temperatures. However, mass transport and ionic conduction are faster at higher temperatures, and this more than offsets the drop in open-circuit voltage[4].

For any chemical reaction

$$jA + kB \rightarrow mC + nD$$

The change in Gibbs free energy between the products and reactants is

$$\Delta G = mG_c + nG_D - jG_A - kG_B \qquad (2\text{-}42)$$

In the case of a hydrogen–oxygen fuel cell under standard-state conditions:

$$H_2(g) + \frac{1}{2}O_2(g) \rightarrow H_2O(l)$$

$$(\Delta H = -285.8\,kJ/mol;\ \Delta G = -237.3\,kJ/mol)$$

$$E_{H_2/O_2} = \frac{-237.3\,kJ/mol}{2\,mol * 96,485\,C/mol} = 1.229\,V$$

At standard temperature and pressure, this is the highest voltage obtainable from a hydrogen–oxygen fuel cell. Most fuel cell reactions have voltages in the 0.8- to 1.5-V range. To obtain higher voltages, several cells have to be connected together in series.

Fuel cells can operate at any pressure, and often it is advantageous to operate the fuel cell at pressures above atmospheric. The typical range for fuel cells is atmospheric pressure to 6 to 7 bars. The change in Gibbs free energy as related to pressure can be written as:

$$dG = V_m dP \qquad (2\text{-}43)$$

where V_m is the molar volume (m^3/mol), and P is the pressure in Pascals. For an ideal gas[5]:

$$PV_m = RT \qquad (2\text{-}44)$$

Therefore:

$$dG = RT \frac{dP}{P} \qquad (2\text{-}45)$$

After integration:

$$G = G_0 + RT \ln\left(\frac{P}{P_0}\right) \qquad (2\text{-}46)$$

where G_0 is the Gibbs free energy at standard pressure and temperature (25 °C and 1 atm), and P_0 is the standard pressure (1 atm), which is a form of the Nernst equation.

If Equation 2-42 is substituted into the Equation 2-46[6]:

$$G = G_0 + RT \ln \frac{\left(\dfrac{P_C}{P_0}\right)^m \left(\dfrac{P_D}{P_0}\right)^n}{\left(\dfrac{P_A}{P_0}\right)^j \left(\dfrac{P_B}{P_0}\right)^k} \qquad (2\text{-}47)$$

where P is the partial pressure of the reactant or product species, and P_0 is the reference pressure.

For the hydrogen–oxygen fuel cell reaction, the Nernst equation becomes

$$G = G_0 + RT \ln\left(\frac{P_{H_2} P_{O_2}^{0.5}}{P_{H_2O}}\right) \qquad (2\text{-}48)$$

Therefore, the cell potential as a function of temperature and pressure is:

$$E_{T,P} = \left(\frac{\Delta H}{nF} - \frac{T\Delta S}{nF}\right) + RT \ln\left(\frac{P_{H_2} P_{O_2}^{0.5}}{P_{H_2O}}\right) \qquad (2\text{-}49)$$

For nonstandard conditions, the reversible voltage of the fuel cell may be calculated from the energy balance between the reactants and the products[7]. Equation 2-49 reduces to the common form of the Nernst equation:

$$E_t = E_r - \frac{RT}{nF} \ln\left[\prod_i \alpha_i^{v_i} \right] \tag{2-50}$$

where R is the ideal gas constant, T is the temperature, a_i is the activity of species i, v_i is the stoichiometric coefficient of species i, and E_r is the standard-state reversible voltage, which is a function of temperature and pressure.

The hydrogen–oxygen fuel cell reaction is written as follows using the Nernst equation:

$$E = E_r - \frac{RT}{2F} \ln \frac{a_{H_2O}}{\alpha_{H_2} \alpha_{O_2}^{1/2}} \tag{2-51}$$

where E is the actual cell voltage, E_r is the standard-state reversible voltage, R is the universal gas constant, T is the absolute temperature, N is the number of electrons consumed in the reaction, and F is Faraday's constant. If the fuel cell is operating under 100 °C, the activity of water can be set to 1 because liquid water is assumed. At a pressure of 1.00 atm absolute (as it is at sea level on a normal day), with an effective concentration of 1.00 mol of H^+ per liter of the acid electrolyte, the ratio of $1.00^{1/2} : 1.00 = 1$, and $\ln 1 = 0$. Therefore, $E = E_r$. The standard electrode potential is that which is realized when the products and reactants are in their standard states.

At standard temperature and pressure, the theoretical potential of a hydrogen–air fuel cell can be calculated as follows:

$$E = 1.229 - \frac{8.314(J/(mol*K))*298.15}{2*96,485(C/mol)} \ln \frac{1}{1*0.21^{1/2}} = 1.219\,V$$

The potential between the oxygen cathode where the reduction occurs and the hydrogen anode at which the oxidation occurs will be 1.229 V at standard conditions with no current flowing. When a load connects the two electrodes, the current will flow as long as there is hydrogen and oxygen gas to react. If the current is small, the efficiency of the cell (measured in voltages) could be greater than 0.9 V, with an efficiency greater than 90%. This efficiency is much higher than the most complex heat engines such as steam engines or internal combustion engines, which can only reach a maximum 60% thermal efficiency.

Figure 2-4 illustrates the decrease in Nernst voltage with increasing temperature. By analyzing the Nernst equation, one can see why this trend occurs:

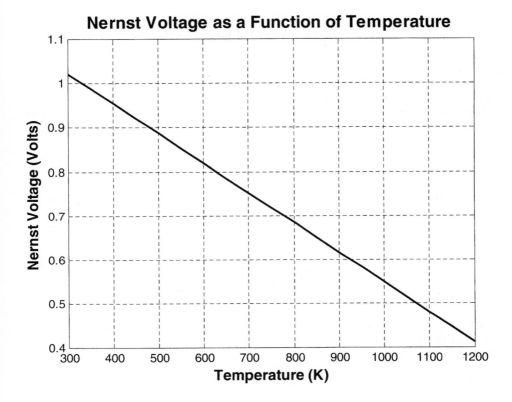

FIGURE 2-4. Nernst voltage as a function of temperature.

$$E_{OCV} = \frac{-\Delta G}{nF} + \frac{R_u T}{nF} \ln\left[\frac{a^{v_a} a^{v_b}}{a^{v_c}}\right] = \frac{-\Delta G}{nF} + \frac{R_u T}{nF} \ln\left[\frac{\left(\frac{\gamma P_a}{P_o}\right)_{H_2}\left(\frac{\gamma P_c}{P_o}\right)^{1/2}_{O_2}}{1}\right]$$

$$= \frac{-\Delta G}{nF} + \frac{R_u T}{nF} \ln[1] = \frac{-\Delta G}{nF}$$

The pressure dependence is nullified because both the anode and cathode are at 1 atm for our simple example. Further, the activity of the water is set to the relative humidity at the reaction site, which is unity because the water is being created at the cathode catalyst layer and does not limit the reaction in any way.

By assuming the gases are ideal (the activities of the gases are equal to their partial pressures, and the activity of the water phase is equal to unity), Equation 2-50 can be written as:

$$E_t = E_r - \frac{RT}{nF} \ln \left[\prod_i \left(\frac{p_i}{p_0} \right)^{v_i} \right] \tag{2-52}$$

where p_i is the partial pressure of species i, and p_0 is the reference pressure. For ideal gases or an estimate for a nonideal gas, partial pressure of species A, $p_A^{\star}$ can be expressed as a product of total pressure P_A and molar fraction x_A of the species:

$$p_A^{\star} = x_A P_A \tag{2-53}$$

If the molar fraction for the fuel is unknown, it can be estimated by taking the average of the inlet and outlet conditions[8]:

$$x_A = \frac{1 - x_{C,Anode}}{1 + \left(\dfrac{x_{Anode}}{2} \right) \left(1 + \left(\dfrac{\zeta_A}{(\zeta_{A-1})} \right) \right)} \tag{2-54}$$

where ζ_A is the stoichiometric flow rate, x_{Anode} is the molar ratio of species $2:1$ in dry gas, and $x_{C,Anode}$ is:

$$x_{C,Anode} = \frac{P_{sat}}{P_A} \tag{2-55}$$

The molar fractions are simply ratios of the saturation pressure (P_{sat}) at the certain fuel cell temperature to the anode and cathode pressures.

The water saturation temperature is a function of cell operating temperature. For a PEM hydrogen–oxygen fuel cell, P_{sat} can be calculated using[9]:

$$\log_{10} P_{sat} = -2.1794 + 0.02953 * T - 9.1837 \times 10^{-5} * T^2 + 1.4454 \times 10^{-7} * T^3 \tag{2-56}$$

where T is the cell operating temperature in °C.

If the current is large, the cell voltage falls fairly rapidly due to various nonequilibrium effects. The simplest of these effects is the voltage drop due to the internal resistance of the cell itself. According to Ohm's law, the voltage drop is equal to the resistance times the current flowing. At maximum current density of 1 amp/cm², the cell can drop 0.5 V.

EXAMPLE 2-5: Calculating Reversible Cell Potential

Determine the reversible cell potential as a function of temperature at 40 °C and 1 atm. Assume that $T_{ref} = 25$ °C. The reaction is

$$H_2 + \frac{1}{2}O_2 \rightarrow H_2O$$

The entropy for the fuel cell reaction at the standard reference temperature and pressure, and the reversible cell potential is:

$$\Delta G = \Delta H - T\Delta S$$

$$\Delta G = \left(h_{H2O(l)} - \left(h_{H2} + \frac{1}{2}h_{O2} \right) \right) - T * \left(s_{H2O(l)} - \left(s_{H2} + \frac{1}{2}s_{O2} \right) \right)$$

$$\Delta G = \left(-285{,}826\frac{J}{mol\_H_2O} - \left(0 + \frac{1}{2}*0 \right) \right) - T*\left(69.92\frac{J}{mol\_H_2O} - \right.$$
$$\left. \left(130.68\frac{J}{mol\_H_2O} + \frac{1}{2}*205.14\frac{J}{mol\_O2} \right) \right)$$

$$\Delta G = -285{,}826\frac{J}{molK} - (298\,K)*-163.25\frac{J}{molK} = -237{,}177.50\frac{J}{molK}$$

$$E_r = \frac{\Delta G}{nF} = \frac{-237{,}177.50\dfrac{J}{molK}}{\left(\dfrac{2mol\_e^-}{mol\_fuel}*96{,}487\dfrac{C}{mol\_e^-} \right)} = 1.229\frac{J}{C} = 1.229\,V$$

For liquid water at $T = 25$ °C and 1 atm:

$$\frac{\Delta s(T_{ref}, P)}{nF} - \frac{-165.25\,J/(mol\_fuel*K)}{2mol\_e^-/mol\_fuel \times 96{,}487\,C/mol\_e^-} = -0.8460 \times 10^{-3}\,V/K$$

Therefore, the desired expression is

$$E_r(T, P) = 1.229\,V - 0.8460 \times 10^{-3}\frac{V}{K}*(T - T_{ref})$$

For every degree of temperature increase, reversible cell potential is reduced by 0.8460 mV. At a temperature of 40 °C:

$$E_r(40,1) = 1.229\,V - 0.8460 \times 10^{-3}\frac{V}{K}*(40-25)K = 1.216\,V$$

The reversible potential is reduced from 1.229 V to 1.216 V when the temperature increases from 25 °C to 40 °C.

Using MATLAB to solve:

% EXAMPLE 2-5: Calculating the Reversible Cell Potential

% UnitSystem SI

%%%

% Inputs

```
hf_H2 = 0;              % Enthalpy of formation at standard state (J/mol)
hf_O2 = 0;              % Enthalpy of formation at standard state (J/mol)
hf_H2Ol = -285826;      % Enthalpy of formation at standard state (J/mol)
sf_H2 = 130.68;         % Entropy of formation at standard state (J/mol)
sf_O2 = 205.14;         % Entropy of formation at standard state (J/mol)
sf_H2Ol = 69.92;        % Entropy of formation at standard state (J/mol)
T_ref = 298;            % Reference temperature
T = 313;                % Given temperature
n = 2;                  % mol e- per mole fuel
F = 96487;              % Faraday's constant
S = -163.25;            % (J/mol fuel K) for liquid H2O
```

%%%

% Calculate Gibbs Free Energy

```
delH = hf_H2Ol - (hf_H2 + ((1/2)*hf_O2));
delS = sf_H2Ol - (sf_H2 + ((1/2)*sf_O2));
delG = delH - (T_ref*delS)
```

% Calculate reversible voltage

```
Er = delG/(n*F);
Sr = S/(n*F);
```

% Expression for reversible voltage

```
E = Er - Sr*(T - T_ref)
```

EXAMPLE 2-6: Calculating Reversible Cell Potential

Determine the reversible cell potential for the following reaction:

$$H_2(g) + \frac{1}{2}O_2(g) \rightarrow H_2O(l)$$

The molar fraction of H_2 in the fuel stream is 0.5 and the molar fraction of O_2 in the oxidant stream is 0.21. The remaining species are chemically inert.

Since the fuel cell operates at the standard temperature and pressure, $T = 25\,°C$ and P is 1 atm. Because the reactant streams are not pure, the reversible cell potential for the reaction is

$$E_r(T,\ P_i) = E_r(T,\ P) - \frac{RT}{nF}\ln K$$

The molar fraction of the reactants are $X_{H2} = 0.5$ and $X_{O2} = 0.21$, and K can be calculated as follows:

$$K = \prod_{i=1}^{N}(X_i)^{|vf-vi|/vf} = X_{H2}^{(0-1)/1}X_{O2}^{(0-1)/2/1} = X_{H2}^{-1}X_{O2}^{-1/2}$$

The reversible cell potential at standard pressure and temperature when pure H_2 and O_2 are used as reactants is 1.229 V.

$$E_r(T,\ P_i) = 1.229V -$$
$$\frac{8.314J/molK \times 298K}{2mol\_e^-/molfuel \times 96,487C/mol\_e^-}\ln 0.5^{-1} \times 0.21^{-1/2} = 1.210V$$

The cell potential is decreased as a result of dilute reactant products but not as drastically decreased as one would expect.

Using MATLAB to solve:

```
%%%%%%%%%%%%%%%%%%%%%%%%%%%%%%%%%%%%%%%%
% EXAMPLE 2-6: Calculating the Reversible Cell Potential
% UnitSystem SI
%%%%%%%%%%%%%%%%%%%%%%%%%%%%%%%%%%%%%%%%
% Inputs
R = 8.314;        % Ideal gas constant (J/molK)
T = 298;          % Temperature
N = 2;            % mol e- per mole fuel
F = 96487;        % Faraday's constant
Er = 1.229;       % Reversible cell potential (v)
x_H2 = 0.5;       % mole fraction of H2
x_O2 = 0.21;      % mole fraction of O2
%%%%%%%%%%%%%%%%%%%%%%%%%%%%%%%%%%%%%%%%
% Expression for reversible voltage
E = Er − (R*T/(n*F))*log((x_H2^(− 1)*(x_O2^(−1/2))))
```

EXAMPLE 2-7: Calculating Reversible Cell Potential

Determine the inlet and outlet Nernst potential for the following reaction:

$$H_2 + \frac{1}{2}O_2 \rightarrow H_2O(g)$$

at a temperature of 25 °C and 80 °C and a pressure of 1 atm. Assume the fuel is pure H_2 and O_2 from the air supplied to the cell. The H_2 utilization is zero, the oxygen utilization is 0.5, and the reactant product water is formed on the oxidant side.

The inlet and outlet Nernst potential can be calculated based upon the reactant composition at the cell inlet and outlet. The reversible cell potential at the cell inlet is

$$E_r(T, P_i) = E_r(T, P) - \frac{RT}{nF} \ln x_{O2,in}^{-1/2}$$

At the cell inlet: $X_{H2,in} = 1$, $X_{O2,in} = 0.21$, $X_{N2,in} = 0.79$, $P_{O2,in} + P_{N2,in} = 1$ atm.

From Example 5-5, E_r (25 °C, 1 atm) = 1.185 V.

Calculating ΔG at 353 K and 1 atm, it is

$$E_r = -\frac{\Delta G}{nF} = \frac{-228,170 \dfrac{J}{molK}}{\left(\dfrac{2mol\_e^-}{mol\_fuel} * 96,487 \dfrac{C}{mol\_e^-} \right)} = 1.1824 \frac{J}{C} = 1.1824V$$

The inlet Nernst potential is

$$E_r(25,1) = 1.185V - \frac{8.314J/molK \times 298K}{2mol\_e^-/molfuel \times 96,487C/mol\_e^-} \ln 0.21^{-1/2} = 1.175V$$

$$E_r(353,1) = 1.1824V - \frac{8.314J/molK \times 353K}{2mol\_e^-/molfuel \times 96,487C/mol\_e^-} \ln 0.21^{-1/2}$$

$$= 1.1705V$$

At the cell outlet, $X_{H2,out} = 1$, $X_{O2,out} = 0.095$, $P_{O2,out} + P_{H2O,out} + P_{N2,out} = P_{out} = P_{in} = 1$ atm.

The Nernst potential becomes

$$E_r(T, P_{O2,out}) = E_r(T, P) - \frac{RT}{nF} \ln X_{O2,out}^{-1/2}$$

$$E_r(25,1) = 1.185 - 0.012\,84 \times \ln 0.095^{1/2} = 1.170V$$

$$E_r(353,1) = 1.170\,5 - 0.015\,2 \times \ln 0.095^{1/2} = 1.1645V$$

Using MATLAB to solve:

```
%%%%%%%%%%%%%%%%%%%%%%%%%%%%%%%%%%%%%%%%%%
% EXAMPLE 2-7: Calculating the Reversible Cell Potential
% UnitSystem SI
%%%%%%%%%%%%%%%%%%%%%%%%%%%%%%%%%%%%%%%%%%
% Inputs
R = 8.314;              % Ideal gas constant (J/molK)
T = 298;                % Temperature
N = 2;                  % mol e- per mole fuel
F = 96487;              % Faraday's constant
x_O2in = 0.21;          % mole fraction of O2 at the inlet
x_O2out = 0.095;        % mole fraction of O2 at the outlet
hf_H2 = 0;              % Enthalpy of formation at standard state (J/mol)
hf_O2 = 0;              % Enthalpy of formation at standard state (J/mol)
hf_H2Ol = -285826;      % Enthalpy of formation at standard state (J/mol)
sf_H2 = 130.68;         % Entropy of formation at standard state (J/mol)
sf_O2 = 205.14;         % Entropy of formation at standard state (J/mol)
sf_H2Ol = 69.92;        % Entropy of formation at standard state (J/mol)
T_ref = 353;            % Reference temperature

%%%%%%%%%%%%%%%%%%%%%%%%%%%%%%%%%%%%%%%%%%
% Expression for reversible voltage
% Calculate Gibbs Free Energy
delH = hf_H2Ol - (hf_H2 + ((1/2)*hf_O2));
delS = sf_H2Ol - (sf_H2 + ((1/2)*sf_O2));
delG = delH - (T_ref*delS)

% Calculate reversible voltage
Er = - delG/(n*F);

% Expression for reversible voltage at the inlet
E_in = Er - (R*T_ref/(n*F))*log((x_O2in^(-1/2)))

% Expression for reversible voltage at the outlet
E_out = Er - (R*T_ref/(n*F))*log((x_O2out^(-1/2)))
```

2.6 Fuel Cell Reversible and Net Output Voltage

If the total energy based upon the higher heating value could be converted into electrical energy, then a theoretical potential of 1.48 V per cell could be obtained. The theoretical potential based upon the lower heating value is also shown. Because of the $T\Delta S$ limitation, the maximum theoretical potential of the cell is 1.229 V. This is the voltage that could be obtained if the free energy could be converted entirely to electrical energy without any losses.

The maximum electrical energy output, and the potential difference between the cathode and anode, is achieved when the fuel cell is operated under the thermodynamically reversible condition. This maximum possible cell potential is the reversible cell potential. The net output voltage of a fuel cell at a certain current density is the reversible cell potential minus the irreversible potential that is discussed in this section, and can be written as[10]:

$$V(i) = V_{rev} - V_{irrev} \tag{2-57}$$

where $V_{rev} = E_r$ is the maximum (reversible) voltage of the fuel cell, and V_{irrev} is the irreversible voltage loss (overpotential) occurring at the cell.

The actual work in the fuel cell is less than the maximum useful work because of other irreversibilities in the process. These irreversibilities (irreversible voltage loss) are the activation potential (v_{act}), ohmic overpotential (v_{ohmic}), and concentration overpotential (v_{conc}). This is shown by the following equation:

$$V_{irrev} = v_{act} + v_{ohmic} + v_{conc} \tag{2-58}$$

The variables in Equation 2-58 are discussed in more detail in Chapters 3 through 5. Chapter 3 covers fuel cell electrochemistry and discusses activation potential, Chapter 4 covers fuel cell charge transport and discusses ohmic overpotential, and Chapter 5 covers fuel cell mass transport and concentration overpotential. V_{irrev} in Equation 2-58 is substituted into Equation 2-57 to account for the irreversible voltage losses to obtain an accurate fuel cell net output voltage. Figure 2-5 illustrates the fuel cell voltage losses that need to be considered when designing fuel cells.

2.7 Theoretical Fuel Cell Efficiency

The efficiency of a chemical process must be evaluated differently than the conventional heat engine. Efficiency can be defined in two ways:

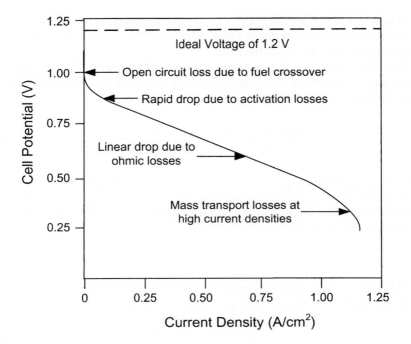

FIGURE 2-5. Hydrogen–oxygen fuel cell performance curve at equilibrium.

$\eta_{\Delta G} \equiv$ (actual useful work)/(maximum useful work) = (power $\times$ time)/ΔG

$\eta_{\Delta H} \equiv$ (actual useful work)/(maximum useful work) = (power $\times$ time)/ΔH

Since $\Delta G = \Delta H - T\,\Delta S$, $\eta_{\Delta H} < \eta_{\Delta G}$ for the same power output.

Efficiency of an ideal fuel cell based upon heat content ΔH is obtained by dividing maximum work out by the enthalpy input, so the fuel cell efficiency is

$$\eta_{\text{fuel\_cell}} = \Delta G/\Delta H \qquad (2\text{-}59)$$

Using the standard free energy and enthalpy given previously ($\Delta G = -237.2$ kJ/mol, $\Delta H = -285.8$ kJ/mol) shows that the maximum thermodynamic efficiency under standard conditions is 83%.

The fuel cell directly converts chemical energy into electrical energy. The maximum theoretical efficiency can be calculated using the following equation:

$$\eta_{\text{max}} = 1 - T * \Delta S/\Delta H \qquad (2\text{-}60)$$

Therefore, even an ideal fuel cell operating reversibly and isothermally will have an efficiency ranging from 60% to 90%. The heat quantity T ΔS is exchanged with the surroundings.

The efficiency is not a major function of device size. Energy consumed is measured in terms of the higher heating value of the fuel used. In other words, for hydrogen:

$$\eta = \frac{power_{out}}{power_{in}} = \frac{n_{electrons}FV_{output}}{n_{hydrogen}\Delta H_{HHV}} = \frac{2FV_{output}}{\Delta H_{HHV}} \tag{2-61}$$

where $n_{electrons}$ and $n_{hydrogen}$ are the flow rates in moles per second, F is Faraday's constant, V_{output} is the voltage of the cell output, and ΔH_{HHV} is −285.8 kJ/mol. The higher heating value enthalpy can be converted to an equivalent voltage of 1.481 V[11] so that

$$\eta = V_{output}/1.481V \tag{2-62}$$

This equivalent voltage concept is very useful in calculating efficiency and waste heat. The waste heat generated is simply

$$Q = n\Delta H_{HHV}(1 - \eta) \tag{2-63}$$

And the maximum efficiency is a thermodynamically limited 83%. (If the assumption that water stays in the liquid form is incorrect, the waste heat that must be rejected decreases because the vaporization of water cools the stack.) Heat generation is curved in great detail in Chapter 6.

2.7.1 Energy Efficiency

Fuel consumption rates can be calculated simply as a function of current density and Faraday's constant:

$$n_{A,reacted} = n_{B,reacted} = \frac{i}{2F} \tag{2-64}$$

The mass continuity of two reactants:

$$n_{A,in} = n_{A,reacted} + n_{A,out} \tag{2-65}$$

where $n_{A,in}$ is the molar flow rate to the fuel cell, and $n_{A,out}$ is the molar flow rate from the fuel cell. The energy efficiency of the fuel cell is

$$n_{Energy} = \frac{W_{FC}}{(n_{A,reacted} + n_{A,out}) \times HHV_A} \tag{2-66}$$

The fuel consumption rates are explained in greater detail in Chapters 4 and 9, which includes calculations for all fuel cell operating conditions, and greater detail about the PEM fuel cell catalyst layer.

Chapter Summary

The study of thermodynamics and its relation to fuel cells is very important for predicting fuel cell performance. The determination of fuel cell potential and efficiency depends heavily on the evaluation of thermodynamic properties. Some of the important properties explored in this chapter include the enthalpy, specific heat, entropy, Gibbs free energy, reversible voltage, net output voltage, and the fuel cell efficiency. These thermodynamic concepts allow one to predict states of the fuel cell system, such as potential, temperature, pressure, volume, and moles in a fuel cell. Learning and applying these concepts are the bases of all fuel cell modeling and analysis, and is essential for understanding the remainder of this book.

Problems

- Calculate the theoretical cell potential for a hydrogen–oxygen fuel cell operating at 50 °C with the reactant gases at 3 atm and 35 °C with liquid water as a product.
- Calculate and compare the differences in theoretical cell potential between three hydrogen–oxygen fuel cells: (1) operating at 25 °C and 1 atm, (2) operating at 50 °C and 2 atm, and (3) operating at 75 °C and 3 atm.
- Determine the inlet and outlet Nernst potential, as well as the associated Nernst loss for the following reaction:

$$H_2 + \frac{1}{2}O_2 \rightarrow H_2O(l)$$

  at a temperature of 80 °C and a pressure of 3 atm. Assume pure O_2 and H_2; the reaction product water is formed at the oxidant side.
- For the reaction in the above problem, what are the reversible cell efficiency and the cell current efficiency if the cell operates at a cell voltage of 0.65 V and a current of 0.7 A with a fuel flow of 4 mL/min?
- For the reaction given below, what is the amount of entropy generation, the amount of cell potential loss, and the amount of waste heat at 25 °C and 1 atm, if the cell operates at a cell voltage of 0.7 V?

$$H_2 + \frac{1}{2}O_2 \rightarrow H_2O(l)$$

Endnotes

[1] Li, X. *Principles of Fuel Cells*. 2006. New York: Taylor & Francis Group.
[2] Moran, M.J., and H.N. Shapiro. *Fundamentals of Engineering Thermodynamics*. 3rd ed. 1995. New York: John Wiley & Sons.
[3] O'Hayre, R., S.-W. Cha, W. Colella, and F.B. Prinz. *Fuel Cell Fundamentals*. 2006. New York: John Wiley & Sons.
[4] Barbir, F. *PEM Fuel Cells: Theory and Practice*. 2005. Burlington, MA: Elsevier Academic Press.
[5] Ibid.
[6] Ibid.
[7] Ay, M., A. Midilli, and I. Dincer. Thermodynamic modeling of a proton exchange membrane fuel cell. *Int. J. Energy*. 2006. Vol. 3, No. 1, pp. 16–44.
[8] Rowe, A., and X. Li. Mathematical modeling of proton exchange membrane fuel cells. *J. Power Sources*. 2001. Vol. 102, pp. 82–96.
[9] Springer et al. Polymer electrolyte fuel cell model. *J. Electrochem. Soc.* 1991. Vol. 138, No. 8, pp. 2334–2342.
[10] Hussain, M.M., J.J. Baschuk, X. Li, and I. Dincer. Thermodynamic analysis of a PEM fuel cell power system. *Int. J. Thermal Sci.* 2005. Vol. 44, pp. 903–911.
[11] Lin, B. Conceptual design and modeling of a fuel cell scooter for urban Asia. 1999. Princeton University, masters thesis.

Bibliography

Mench, M.M., C.-Y. Wang, and S.T. Tynell. *An Introduction to Fuel Cells and Related Transport Phenomena*. Department of Mechanical and Nuclear Engineering, Pennsylvania State University. Draft. Available at: http://mtrl1 .mne.psu.edu/Document/jtpoverview.pdf Accessed March 4, 2007.

Mench, M.M., Z.H. Wang, K. Bhatia, and C.Y. Wang. *Design of a Micro-Direct Methanol Fuel Cell*. 2001. Electrochemical Engine Center, Department of Mechanical and Nuclear Engineering, Pennsylvania State University, Pennsylvania.

Simon, W.E. *Transient Thermodynamic Analysis of a Fuel-Cell System*. June 1968. Houston, TX: NASA Technical Note. NASA TN D-4-4601, Manned Spacecraft Center.

Spiegel, C.S. *Designing & Building Fuel cells*. 2007. New York: McGraw-Hill.

Tester, J.W. and M. Modell. *Thermodynamics and its Applications*. 3rd ed. 1997. New Jersey: Prentice Hall.

CHAPTER 3

Fuel Cell Electrochemistry

3.1 Introduction

The thermodynamic concepts covered in Chapter 2 allow one to calculate the theoretical performance for a fuel cell reaction potential difference between the anode and cathode, and the fuel cell energy conversion efficiency. However, thermodynamics cannot provide information on how fast a reaction occurs in order to produce electric current, how reactants create products, how to predict the reaction rate to produce electric current in the cell, and how much energy loss occurs during the actual electrochemical reaction. This chapter covers the electrochemistry needed in order to predict or model basic electrode kinetics, activation overpotential, currents, and potentials in a fuel cell. The specific topics to be covered are:

- Basic electrokinetics concepts
- Charge transfer
- Activation polarization for charge transfer reations
- Electrode kinetics
- Voltage losses
- Internal currents and crossover currents

It is essential to understand the underlying reaction process occurring at the anode and cathode when modeling fuel cells. The electrochemical reactions control the rate of power generation, and are the cause of activation voltage losses. A lot of progress has been made in the area of electrode kinetics, but there is still a lot of work that needs to be developed in order to fully understand the actual complex anode and cathode kinetics. This chapter introduces the basics of electrochemical kinetics, and discusses activation polarization in detail.

3.2 Basic Electrokinetics Concepts

All electrochemical processes involve the transfer of electrons between an electrode and a chemical species with a change in Gibbs free energy. The

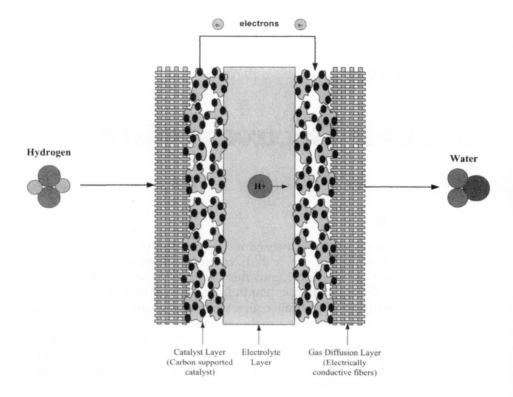

FIGURE 3-1. Fuel cell electrochemical reactions at the electrolyte and electrode.

electrochemical reaction occurs at the interface between the electrode and the electrolyte, as shown in Figure 3-1.

The overall PEM fuel cell reactions were introduced in Chapter 1.2. Although these reactions do not seem overly complicated, the actual reactions proceed through many steps and intermediate species. For example, for the anodic reaction, the following elementary reactions are important[1]:

$$H_2 \Leftrightarrow H_{ad} + H_{ad} \tag{3-1}$$

$$H_{ad} \Leftrightarrow H^+ + e^- \tag{3-2}$$

The first reaction is a dissociative chemisorption step known as a Tafel reaction, and the second reaction is a charge transfer "Volmer" reaction. Equation 3-1 suggests that the hydrogen first absorbs on the electrode surface, and then dissociates into the H atoms. Equation 3-2 actually produces the proton and electron. The first reaction takes into consideration that the reacting species must first be absorbed on the electrode surface before the chemical reaction can occur. The rate of the electrode reaction may be influenced by the diverse absorbed species and the number of vacant sites. The surface coverage can be defined as the fraction of the electrode surface covered by adsorbed species[2]:

$$\theta_i = \frac{C_{i,ad}}{\sum_j (C_{i,ad})_s} \qquad (3\text{-}3)$$

where "ad" is the species that is absorbed on the electrode surface, and "s" is the concentration, $C_{i,ad}$ is at the saturation of the electrode surface. The reaction rate equations can be written more descriptively to include the metal adsorption sites (M)[3]:

$$H_2 + 2M \underset{k_{1,b}}{\overset{k_{1,f}}{\rightleftharpoons}} 2(H - M) \qquad (3\text{-}4)$$

$$H - M \underset{k_{2,b}}{\overset{k_{2,f}}{\rightleftharpoons}} H^+ + e^- + M \qquad (3\text{-}5)$$

Therefore, the expression for the rate of reaction can be written as follows[4]:

$$\omega''_{H_{ad}} = -k_{2,f}\theta_H + k_{2,b}C_{H^+}C_{e^-}(1 - \theta_H) \qquad (3\text{-}6)$$

where $(1 - \theta_H)$ is the metal surface not covered by adsorbed hydrogen. The equations presented in this section begin to illustrate some of the factors involved with the electrochemical reactions occurring at the electrode/electrolyte interface.

3.3 Charge Transfer

The quickness of the electrochemical reaction to proceed is dependent upon the rate that electrons are created or consumed. Therefore, the current is a direct measure of the electrochemical reaction rate. From Faraday's law, the rate of charge transfer is:

$$i = \frac{dQ}{dt} \qquad (3\text{-}7)$$

where Q is the charge and t is the time. If each electrochemical reaction results in the transfer of n electrons per unit of time, then

$$\frac{dN}{dt} = \frac{i}{nF} \qquad (3\text{-}8)$$

where dN/dt is the rate of the electrochemical reaction (mol/s) and F is Faraday's constant (96,400 C/mol). Integrating this equation gives

$$\int_0^t idt = Q = nFN \qquad (3-9)$$

Equation 3-9 states that the total amount of electricity produced is proportional to the number of moles of material times the number of electrons times Faraday's constant.

EXAMPLE 3-1: Hydrogen Consumed and Current Produced

A hydrogen–oxygen fuel cell with the reaction

$$H_2 + O_2 \rightarrow H_2O$$

is operating at 60°C and 3 atm. The fuel cell runs for 120 hours. How much current does the fuel cell produce at a flow rate of 5 sccm? How many moles of H_2 are consumed?

H_2 can be treated as an ideal gas; the molar flow rate is related to the volumetric flow rate via the ideal gas law:

$$\frac{dN}{dt} = \frac{P(dV/dt)}{RT}$$

$$\frac{dN}{dt} = \frac{3\,atm \times (0.005 \text{ L/min})}{(0.082 \text{ L/atm/(molK)}) \times (333.15 \text{ K})} = 5.49 \times 10^{-4} \frac{molH_2}{min}$$

Since 2 moles of electrons are transferred for every mole of H_2 gas reacted, n = 2.

$$i = nF\frac{dN}{dt} = 2 \times (96,400C) * \left(5.49 \times 10^{-4} \frac{molH_2}{min}\right) * 1\frac{min}{60s} = 1.77A$$

The total amount of electricity produced is calculated by integrating the current load over the operation time.

$$Q_{tot} = i_1 t_1 = (1.77A) * \left(120 hours * \frac{3600\,sec}{1hr}\right) = 762,900C$$

The total number of moles of H_2 processed by the fuel cell is

$$N_{H2} = \frac{Q_{tot}}{nF} = \frac{762,900C}{2 \times 96,400C/mol} = 3.95mol\_H_2$$

Using MATLAB to solve:

```
%%%%%%%%%%%%%%%%%%%%%%%%%%%%%%%%%%%%%%%%%
% EXAMPLE 3-1: Calculating the Enthalpy of Water
% UnitSystem SI
%%%%%%%%%%%%%%%%%%%%%%%%%%%%%%%%%%%%%%%%%
% Inputs

R = 0.082;          % Ideal gas constant (L/atm/molK)
T = 333.15;         % Temperature
dV_dt = 0.005;      % Volumetric flow rate (L/min)
P = 3;              % atm
n = 2;              % mol e- per mole fuel
F = 96487;          % Faraday's constant
T = 120*3600;       % Convert to seconds

%%%%%%%%%%%%%%%%%%%%%%%%%%%%%%%%%%%%%%%%%
% Convert volumetric flow rate to molar flow rate

dN_dt = P*dV_dt/(R*T);

% Calculate the total current

i = n*F*dN_dt*(1/60);

% Total amount of electricity produced

Q = i*t

% The total number of moles of hydrogen

N_H2 = Q/(n*F)
```

3.4 Activation Polarization for Charge Transfer Reactions

An electrochemical reaction occurring at the electrode takes the following form:

$$Ox + e \xleftrightarrow{k} Re \qquad (3\text{-}10)$$

where Ox is the oxidized form of the chemical species, and Re is the reduced form of the chemical species. If the potential of the electrode is made more negative than the equilibrium potential, the reaction will form more Re. If the potential of the electrode is more positive than the equilibrium potential, it will create more Ox[5]. The forward and backward reactions take place simultaneously. The reactant consumption is proportional to the surface concentration[6]. For the forward reaction, the flux is:

$$j_f = k_f C_{Ox} \qquad (3\text{-}11)$$

where k_f is the forward reaction rate coefficient, and C_{Ox} is the surface concentration of the reactant species.

The backward reaction of the flux is described by:

$$j_B = k_b C_{Rd} \qquad (3\text{-}12)$$

where k_b is the backward reaction rate coefficient, and C_{Rd} is the surface concentration of the reactant species.

These reactions either consume or release electrons. The net current generated is the difference between the electrons released and consumed[7]:

$$i = nF(k_f C_{Ox} - k_b C_{Rd}) \qquad (3\text{-}13)$$

The net current should equal zero at equilibrium because the reaction will proceed in both directions simultaneously at the same rate[8]. This reaction rate at equilibrium is called the exchange current density, which can be expressed as:

$$K = \frac{k_f}{k_b} = \frac{C_B}{C_A} \qquad (3\text{-}14)$$

3.5 Electrode Kinetics

Most rate constants in electrolyte reactions vary with temperature where ln k is 1/T. The first person to recognize this relationship was Arrhenius, who said that it can expressed in the following form:

$$k = A \exp\left(\frac{-E_A}{RT}\right) \qquad (3\text{-}15)$$

where E_A is the activation energy and represents an energy barrier of height E_A. The exponential expresses the probability of overcoming the energy barrier, and A is related to the number of attempts to overcome the barrier. According to the Transition State Theory, an energy barrier needs to be

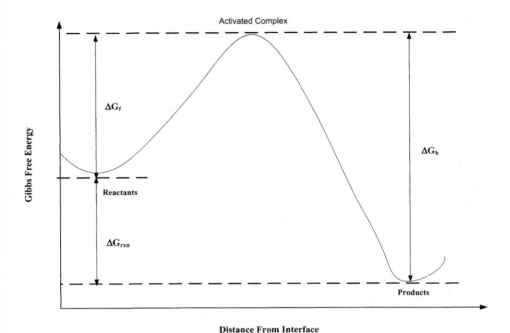

FIGURE 3-2. Gibbs free energy change compared with distance from the interface[9].

overcome for the reaction to proceed[10]. An illustration of the Gibbs free energy function with the distance from the interface is shown in Figure 3-2. The magnitude of the energy barrier to be overcome is equal to the Gibbs free energy change between the reactant and product[11]. The height of the maximum is identified as the activation energy for the forward ($E_{A,f}$) or backward ($E_{B,f}$) reaction, respectively.

E_A is also known as the standard internal energy when the transition between the minima and maxima occurs—which is during the transition state. If E_A is designated as the standard internal energy of activation ΔE, then the standard enthalpy of activation, ΔH, can be expressed as $\Delta E = \Delta(PV)$. Since $\Delta(PV)$ is usually negligible, $\Delta H \sim \Delta E$. When the standard enthalpy of activation is introduced into the equation:

$$k = A \exp\left(\frac{-\Delta H - T\Delta S}{RT}\right) \qquad (3\text{-}16)$$

Then:

$$k = A \exp\left(\frac{-G}{RT}\right) \qquad (3\text{-}17)$$

where ∆G is the standard free energy of activation when considering the reaction, where the substances are at thermal equilibrium.

The concentration of each substance can be calculated from the standard energies of activation. For each substance[12]:

$$\frac{\text{Complex}}{[A]} = \frac{\dfrac{a_A}{C^0}}{\dfrac{a_{\mp}}{C^0}} K_f = \frac{a_A}{a_{\mp}} \exp\left(\frac{-\Delta G_f}{RT}\right) \tag{3-18}$$

where C^0 is the concentration of the standard-state, and a_A and $a_{\mp}$ are the dimensionless activity coefficients for A and the standard-state, respectively.

The activated complexes decay into A or B according to a rate constant, k, and can be grouped into four categories[13]:

1. Those created from A, and converting back into A
2. Those converting into B
3. Those converting into A
4. Those created from B, converting back to A

The rate of transforming A into B can be expressed as:

$$k_f[A] = f_{AB}k'[\text{Complex}] \tag{3-19}$$

where k' is the combined rate constant. This can apply also for B converting to A. Since $k_f[A] = k_b[B]$ at equilibrium, f_{AB} and f_{BA} should ideally be the same; therefore, they can be estimated to each be $^1/_2$. If Equation 3-18 is substituted into Equation 3-19, then[13]:

$$k_f = \frac{\kappa k'}{2} \exp\left(\frac{-\Delta G_f}{RT}\right) \tag{3-20}$$

where κ is the transmission coefficient and represents the fractions f_{AB} and f_{BA}. The quantity $\dfrac{\kappa k}{2}$ depends upon the shape of the energy surface in the region of the complex, but for simple cases, this can be estimated using Boltzmann's and Plank's constants, and expressed in the following form:

$$k = \frac{k_B T}{h} \exp\left(\frac{-\Delta G}{RT}\right) \tag{3-21}$$

where k_B is the Boltzmann's constant (1.38049×10^{-23} J/K), h is Plank's constant (6.621×10^{-34} Js), and ∆G is the Gibbs energy of activation (kJ/mol).

Equation 3-21 is valid for a system at a certain temperature and pressure and is not dependent upon reactant and product concentrations. The change in Gibbs free energy due to the reaction equilibrium is illustrated in Figure 3-3.

3.5.1 Butler-Volmer Model of Electrode Kinetics

The potential of an electrode affects the kinetics of reactions occurring at its surface. The way that k_f and k_b depend upon potential can be used to control reactivity. In the previous section, it was illustrated that reactions can be visualized in terms of reaction progress on an energy surface. When electrodes are considered, the shape of the surface is a function of electrode potential.

If the potential is changed to a new value, ΔE, the relative energy of the electron on the electrode changes by $-F\Delta E = -F(E - E_0)$; therefore, the curve moves up or down that amount. Figure 3-4 shows the effect for a positive E. The Gibbs free energy can be considered to consist of both chemical and electrical terms because it occurs in the presence of an electrical field[13,14]. For a reduction reaction:

$$\Delta G = \Delta G_{AC} + \alpha_{Rd}FE \tag{3-22}$$

For an oxidation reaction:

$$\Delta G = \Delta G_{AC} - \alpha_{Ox}FE \tag{3-23}$$

where ΔG_{AC} is the activated complex of the Gibbs free energy, α is the transfer coefficient, F is Faraday's constant, and E is the potential.

The transfer coefficient is a measure of symmetry of the energy barrier. This can be illustrated by considering the geometry of the curves at the intersection region as shown in Figure 3-5. The angles can be defined by[15]:

$$\tan \theta = \alpha FE/x \tag{3-24}$$

$$\tan \phi = (1 - \alpha)FE/x \tag{3-25}$$

Therefore,

$$\alpha = \frac{\tan \theta}{\tan \phi + \tan \theta} \tag{3-26}$$

If the intersection is symmetrical or at equilibrium, $\phi = \theta$ and $\alpha = \frac{1}{2}$. If the reaction is not at equilibrium, $0 \leq \alpha < \frac{1}{2}$ or $\frac{1}{2} < \alpha \leq 1$.

There is some confusion in the literature between the transfer coefficient (α) and the symmetry factor (β). The symmetry factor is typically

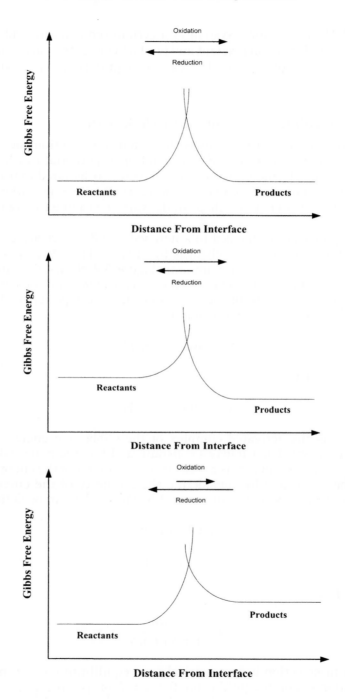

FIGURE 3-3. Gibbs free energy change due to the reaction equilibrium[16].

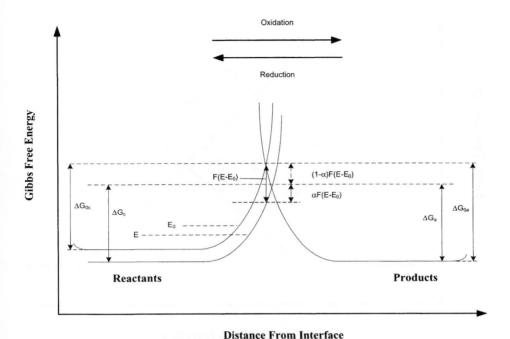

FIGURE 3-4. Effect of the potential change on the standard free energies of activation[17].

used for single-step reactions, but since the typical process is multistep, an experimental parameter called the transfer coefficient[18] is used.

The value of α is typically between 0 and 1, and more specifically, between 0.3 to 0.7 for most electrochemical reactions depending upon the activation barrier[19]. The transfer coefficient is usually approximated by 0.5 in the absence of actual measurements. The relationship between α_{Rd} and α_{ox} is

$$\alpha_{Rd} - \alpha_{Ox} = \frac{n}{v_{times}} \qquad (3\text{-}27)$$

where n is the number of electrons transferred, and v_{times} is the number of times the stoichiometric step must take place for the reaction to occur[20].

The forward and backward oxidation reaction rate coefficients are

$$k_f = k_{0,f} \exp\left[\frac{-\alpha_{Rd}FE}{RT}\right] \qquad (3\text{-}28)$$

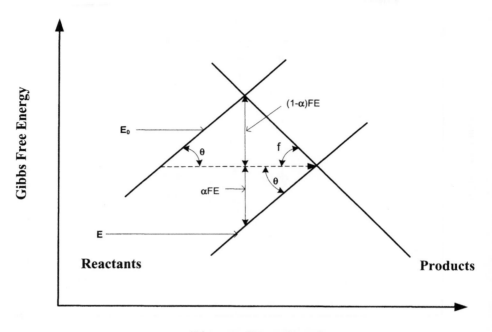

Distance From Interface

FIGURE 3-5. Relationship between the angle of intersection and the transfer coefficient[21].

$$k_b = k_{0,b} \exp\left[-\frac{\alpha_{Ox}FE}{RT}\right] \qquad (3\text{-}29)$$

If these equations are introduced into Equation 3-7, the net current is

$$i = nF\left\{k_{0,f}C_{Ox} \exp\left[\frac{-\alpha_{Rd}FE}{RT}\right] - k_{0,b}C_{Rd} \exp\left[\frac{\alpha_{Ox}FE}{RT}\right]\right\} \qquad (3\text{-}30)$$

Since the reaction proceeds in both directions simultaneously, the net current at equilibrium is equal to zero[22]. The exchange current density is the rate at which these reactions proceed at equilibrium:

$$i_0 = nFk_{0,f}C_{Ox} \exp\left[\frac{-\alpha_{Rd}FE_r}{RT}\right] = nFk_{0,b}C_{Rd} \exp\left[\frac{-\alpha_{Ox}FE_r}{RT}\right] \qquad (3\text{-}31)$$

The exchange current density measures the readiness of the electrode to proceed with the chemical reaction. It is a rate constant for electrochemical reactions, and is a function of temperature, catalyst loading, and catalyst-specific surface area. The higher the exchange current density, the lower the barrier is for the electrons to overcome, and the more active the surface of the electrode. The exchange current density can usually be determined experimentally by extrapolating plots of log i versus v_{act} to $v_{act} = 0$. The higher the exchange current density, the better the fuel cell performance. The effective exchange current density at any temperature and pressure is given by the following equation:

$$i_0 = i_0^{ref} a_c L_c \left(\frac{P_r}{P_r^{ref}} \right)^{\gamma} \exp\left[-\frac{E_c}{RT}\left(1 - \frac{T}{T_{ref}} \right) \right] \tag{3-32}$$

where i_0^{ref} is the reference exchange current density per unit catalyst surface area (A/cm^2), a_c is the catalyst-specific area, L_c is the catalyst loading, P_r is the reactant partial pressure (kPa), P_r^{ref} is the reference pressure (kPa), γ is the pressure coefficient (0.5 to 1.0), E_c is the activation energy (66 kJ/mol for O_2 reduction on Pt), R is the gas constant [8.314 J/(mol∗K)], T is the temperature, K, and T_{ref} is the reference temperature (298.15 K).

Manipulating these equations slightly gives the current-overpotential equation:

$$i = i_0 \frac{C_R^{\star}}{C_R^{0\star}} \exp(\alpha n F v_{act}/(RT)) - \frac{c_P^{\star}}{c_P^{0\star}} \exp(-(1-\alpha)n F v_{act}/(RT)) \tag{3-33}$$

where $c_R^{\star}$ and $c_P^{\star}$ are arbitrary concentrations, and i_0 is measured as the reference reactant and product concentration values $c_R^{0\star}$ and $c_P^{0\star}$. The first term describes the cathodic compartment at any potential, and the next term gives the anodic contribution. Sometimes this equation is called the Butler-Volmer equation, although the most common form of the Butler-Volmer equation is shown by Equation 3-33.

If the currents are kept low so that the surface concentrations do not differ much from the bulk values, then the Butler-Volmer equation becomes:

$$i = i_0 \left\{ \exp\left[\frac{-\alpha_{Rd}F(E - E_r)}{RT} \right] - \exp\left[\frac{\alpha_{Ox}F(E - E_x)}{RT} \right] \right\} \tag{3-34}$$

The Butler-Volmer equation is valid for both anode and cathode reaction in a fuel cell. It states that the current produced by an electrochemical reaction increases exponentially with activation overpotential[23]. This equation also says that if more current is required from a fuel cell, voltage will be lost. The Butler-Volmer equation applies to all single-step reactions, and some modifications to the equation must be made in order to use it for multistep approximations.

Activation Losses as a Function of Exchange Current Density

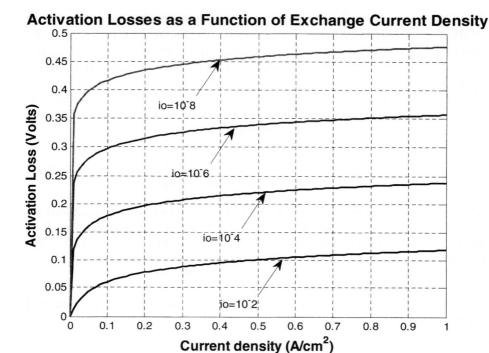

FIGURE 3-6. Effect of the exchange current density on the activation losses.

If the exchange current density is low, the kinetics become sluggish, and the activation overpotential will be larger for any particular net current as shown in Figure 3-6. If the exchange current is very large, the system will supply large currents with insignificant activation overpotential. If a system has an extremely small exchange current density, no significant current will flow unless a large activation overpotential is applied. The exchange current can be viewed as an "idle" current for charge exchange across the interface. If only a small net current is drawn from the fuel cell, only a tiny overpotential will be required to obtain it. If a net current is required that exceeds the exchange current, the system has to be driven to deliver the charge at the required rate, and this can only be achieved by applying a significant overpotential. When this occurs, this is a measure of the system's ability to deliver a net current with significant energy loss.

The Butler-Volmer equation can also be written as:

$$i = i_0 \exp\left[\frac{\alpha nF\upsilon_{act}}{RT}\right] - i_0 \exp\left[\frac{\alpha nF\upsilon_{act}}{RT}\right]$$

(3-35)

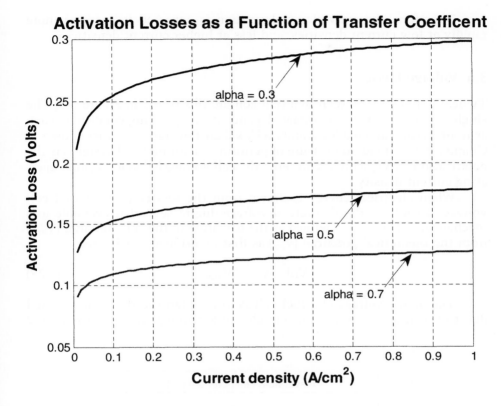

FIGURE 3-7. Effect of the transfer coefficient on the activation losses.

where i is the current density per unit catalyst surface area (A/cm^2), i_0 is the exchange current density per unit catalyst surface area (A/cm^2), v_{act} is the activation polarization (V), n is the number electrons transferred per reaction (–), R is the gas constant [8.314 J/(mol * K)], and T is the temperature (K). The transfer coefficient is the change in polarization that leads to a change in reaction rate for fuel cells and is typically assumed to be 0.5. Figure 3-7 illustrates the effect of the transfer coefficient on the activation losses.

When $\alpha = 0.5$, the equation can be rearranged to give:

$$i = 2i_0 \sinh\left[\frac{\alpha nF v_{act}}{2RT}\right] \qquad (3\text{-}36)$$

In order to obtain the activation polarization based upon the Butler-Volmer equation; the equation needs to be rearranged to give v_{act}:

$$v_{act} = \frac{2RT}{nF} \sinh^{-1}\left[\frac{i}{2i_0}\right] \qquad (3\text{-}37)$$

Therefore, the activation polarization is expected to increase more rapidly at low current densities, and less at higher current densities.

3.6 Voltage Losses

Typical voltage losses seen in a fuel cell are illustrated in Figure 3-8. The single fuel cell provides a voltage dependent on operating conditions such as temperature, applied load, and fuel/oxidant flow rates. As first shown in Chapter 2, the standard measure of performance for fuel cell systems is the polarization curve, which represents the cell voltage behavior against operating current density.

When electrical energy is drawn from the fuel cell, the actual cell voltage drops from the theoretical voltage due to several irreversible loss mechanisms. The loss is defined as the deviation of the cell potential (V_{irrev}) from the theoretical potential (V_{rev}) as first mentioned in Chapter 2.

$$V(i) = V_{rev} - V_{irrev} \tag{3-38}$$

The actual voltage of a fuel cell is lower than the theoretical model due to species crossover from one electrode through the electrolyte and

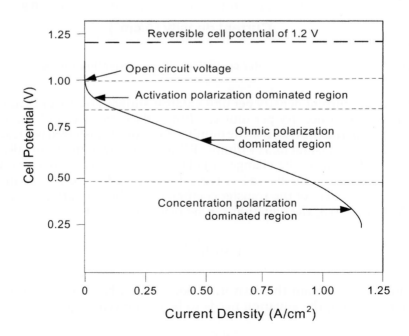

FIGURE 3-8. Generalized polarization curve for a fuel cell.

internal currents. The three major classifications of losses that result in the drop from open-circuit voltage is (1) activation polarization, (2) ohmic polarization, and (3) concentration polarization[24]. Therefore, the operating voltage of the cell can be represented as the departure from ideal voltage caused by these polarizations:

$$V(i) = V_{rev} - v_{act\_anode} - v_{act\_cath} - v_{ohmic} - v_{conc\_anode} - v_{conc\_cath} \quad (3\text{-}39)$$

where v_{act}, v_{ohmic}, v_{conc} represent activation, ohmic (resistive), and mass concentration polarization. As seen in Equation 3-39, activation and concentration polarization occur at both the anode and cathode, while the resistive polarization represents ohmic losses throughout the fuel cell.

The equation for the fuel cell polarization curve is the relationship between the fuel cell potential and current density, as illustrated in Figure 3-9, and can be written as:

$$E = E_r - \frac{RT}{\alpha_c F} \ln\left(\frac{i}{i_{0,c}}\right) - \frac{RT}{\alpha_a F} \ln\left(\frac{i}{i_{0,a}}\right) - \frac{RT}{nF} \ln\left(\frac{i_{L,c}}{i_{L,c} - i}\right) - \frac{RT}{nF} \ln\left(\frac{i_{L,a}}{i_{L,a} - i}\right) - iR_i \quad (3\text{-}40)$$

The shorter version of the equation is

$$E = E_r - \frac{RT}{\alpha F} \ln\left(\frac{i + i_{loss}}{i_0}\right) - \frac{RT}{nF} \ln\left(\frac{i_L}{i_L - i}\right) - iR_i \quad (3\text{-}41)$$

The voltage overpotential required to overcome the energy barrier for the electrochemical reaction to occur is activation polarization. As described previously this type of polarization dominates losses at low current density and measures the catalyst effectiveness at a given temperature. This type of voltage loss is complex because it involves the gaseous fuel, the solid metal catalyst, and the electrolyte. The catalyst reduces the height of the activation barrier, but a loss in voltage remains due to the slow oxygen reaction. The total activation polarization overpotential often ranges from 0.1 to 0.2 V, which reduces the maximum potential to less than 1.0 V even under open-circuit conditions[25]. Activation overpotential expressions can be derived from the Butler-Volmer equation. The activation overpotential increases with current density and can be expressed as:

$$\Delta V_{act} = E_r - E = \frac{RT}{\alpha F} \ln\left(\frac{i}{i_0}\right) \quad (3\text{-}42)$$

where i is the current density, and i_0, is the reaction exchange current density.

The activation losses can also be expressed simply as the Tafel equation:

$$\Delta V_{act} = a + b \ln(i) \tag{3-43}$$

where

$$a = -\frac{RT}{\alpha F} \ln(i_0) \text{ and } b = -\frac{RT}{\alpha F}$$

The equation for the anode and cathode activation overpotential can be represented by:

$$v_{act\_anode} + v_{act\_cath} = \frac{RT}{nF\alpha} \ln\left(\frac{i}{i_0}\right)\bigg|_{anode} + \frac{RT}{nF\alpha} \ln\left(\frac{i}{i_0}\right)\bigg|_{cath} \tag{3-44}$$

where n is the number of exchange protons per mole of reactant, F is Faraday's constant, and α is the charge transfer coefficient used to describe the amount of electrical energy applied to change the rate of the electrochemical reaction[26]. The exchange current density, i_0, is the electrode activity for a particular reaction at equilibrium. In PEM fuel cells, the anode i_0 for hydrogen oxidation is very high compared to the cathode i_0 for oxygen reduction; therefore, the cathode contribution to this polarization is often neglected. Intuitively, it seems like the activation polarization should increase linearly with temperature based upon Equation 3-44, but the purpose of increasing temperature is to decrease activation polarization. In Figure 3-9, increasing the temperature would cause a voltage drop within the activation polarization region.

The ohmic and concentration polarization regions are discussed in detail in Chapter 4 (Fuel Cell Charge Transport) and Chapter 5 (Fuel Cell Mass Transport), respectively.

EXAMPLE 3-2: Using the Activation Overpotential Equation

Use the activation overpotential equation derived from the Butler-Volmer equation to calculate and plot the activation losses for a fuel cell operating at a current density of 0.7 A/cm², $\alpha = 0.5$, and an exchange current density of $10^{-6.912}$ at (1) a temperature range from 300 to 400 K and (2) a current density from 0 to 1 A at increments of 0.01 and a temperature of 300 K.

The activation overpotential equation derived from the Butler-Volmer equation is:

$$\Delta V_{act} = \frac{RT}{\alpha F} \ln\left(\frac{i}{i_0}\right)$$

Using MATLAB to solve:

```
%%%%%%%%%%%%%%%%%%%%%%%%%%%%%%%%%%%%%%%%%%
% EXAMPLE 3-2: Using the Activation Overpotential Equation
% UnitSystem SI
%%%%%%%%%%%%%%%%%%%%%%%%%%%%%%%%%%%%%%%%%%
% Inputs
R = 8.314;          % Ideal gas constant (J/molK)
F = 96487;          % Faraday's constant
Alpha = 0.5;        % Transfer coefficient
io = 10^-6.912;     % Exchange current density
%%%%%%%%%%%%%%%%%%%%%%%%%%%%%%%%%%%%%%%%%%
% Part a: Constant Current Density of 0.7 A with a temperature
range from
i = 0.7;            % Current
T = 300:400;        % Temperature
% Activation Losses
B = R.*T./(2.*Alpha.*F);
V_act = b.*log10(i./io);    % Tafel equation
figure1 = figure('Color',[1 1 1]);
hdlp = plot(T,V_act);
title('Activation Losses as a Function of Temperature','FontSize',14,'FontWeight',
   'Bold')
xlabel('Temperature (K)','FontSize',12,'FontWeight','Bold');
ylabel('Activation Loss (Volts)','FontSize',12,'FontWeight','Bold');
set(hdlp,'LineWidth',1.5);
grid on;
% Part b: Constant temperature of 300 K with a current density
range from
% 0-1 A
i2 = 0:0.01:1;      % range of current
T2 = 300;           % Temperature
% Activation Losses
b2 = R.*T2./(2.*Alpha.*F);
V_act2 = b2.*log10(i2 ./ io);    % Tafel equation
figure2 = figure('Color',[1 1 1]);
hdlp = plot(i2,V_act2);
title('Activation Losses as a Function of Current Density','FontSize',14,'FontWeight',
   'Bold')
xlabel('Current density (A/cm^2)','FontSize',12,'FontWeight','Bold');
ylabel('Activation Loss (Volts)','FontSize',12,'FontWeight','Bold');
set(hdlp,'LineWidth',1.5);
grid on;
```

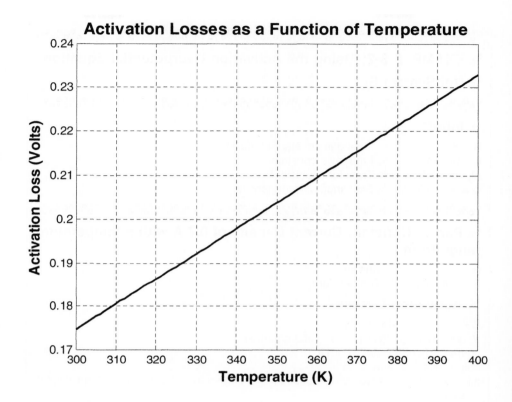

FIGURE 3-9. Activation losses as a function of temperature.

Figures 3-9 and 3-10 generated from the code in Example 3-2.

EXAMPLE 3-3: Calculating the Voltage Losses for a Polarization Curve

A 25-cm² active area hydrogen–air fuel cell stack has 20 cells, and operates at a temperature of 60 °C. Both the hydrogen and air are fed to the fuel cell at a pressure of 3 atm. Create the polarization and fuel cell power curve for this fuel cell stack.

Some useful parameters for creating the polarization curve are: the transfer coefficient, α, is 0.5, the exchange current density, i_0, is $10^{-6.912}$ A/cm², the limiting current density, i_L, is 1.4 A/cm², the amplification constant (α_1) is 0.085, the Gibbs function in liquid form, $G_{f,liq}$, is $-228{,}170$ J/mol, the constant for mass transport, k, is 1.1, and the internal resistance, R, is 0.19 Ωcm². The theoretical voltage, activation losses, ohmic losses, and concentration losses will need to be calculated for this example. The basic calculations for ohmic and concentration losses will be introduced in this example, but will be discussed in further detail in Chapters 4 and 5.

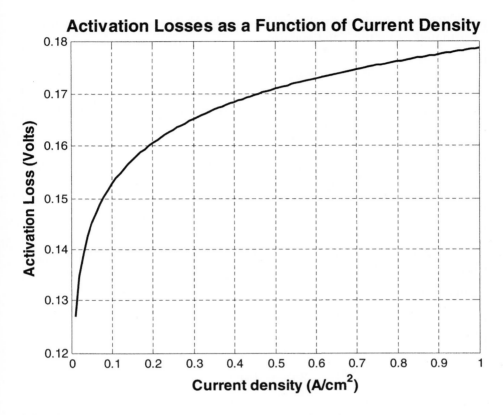

Activation Losses as a Function of Current Density

FIGURE 3-10. Activation losses as a function of current density.

(a) The first step in creating the polarization curve is to calculate the Nernst voltage and voltage losses. To calculate the Nernst voltage for this example, the partial pressures of water, hydrogen, and oxygen will be used. First calculate the saturation pressure of water:

$$\log P_{H_2O} = -2.1794 + 0.02953 * T_c - 9.1837 \times 10^{-5} * T_c^2 + 1.4454 \times 10^{-7} * T_c^3$$

$$\log P_{H_2O} = \begin{aligned} &-2.1794 + 0.02953 * 60 - 9.1837 \times 10^{-5} * 60^2 + \\ &1.4454 \times 10^{-7} * 60^3 = 0.467 \end{aligned}$$

Calculate the partial pressure of hydrogen:

$$p_{H_2} = 0.5 * (P_{H_2}/\exp(1.653 * i/(T_K^{1.334}))) - P_{H_2O} = 1.265$$

Calculate the partial pressure of oxygen:

$$p_{O_2} = (P_{air}/\exp(4.192 * i/(T_K^{1.334}))) - P_{H_2O} = 2.527$$

The voltage losses will now be calculated. The activation losses are estimated using the Tafel equation:

$$V_{act} = -b * \log\left(\frac{i}{i_o}\right) \quad \text{where} \quad b = \frac{R * T}{2 * \alpha * F}$$

The ohmic losses (see Chapter 4) are estimated using Ohm's law:

$$V_{ohmic} = -(i * r)$$

The mass transport (or concentration losses—see Chapter 5) can be calculated using the following equation:

$$V_{conc} = alphal * i^k * \ln\left(1 - \frac{i}{i_L}\right)$$

To insure that there are no negative values calculated for V_{conc} for the MATLAB program, the mass transport losses will only be calculated if $1 - \left(\frac{i}{i_L}\right) > 0$, else $V_{conc} = 0$.

The Nernst voltage can be calculated using the following equation:

$$E_{Nernst} = -\frac{G_{f,liq}}{2 * F} - \frac{R * T_k}{2 * F} * \ln\left(\frac{P_{H_2O}}{P_{H_2} * p_{O_2}^{1/2}}\right)$$

Since all of the voltage losses had a (−) in front of each equation, the actual voltage is the addition of the Nernst voltage plus the voltage losses:

$$V = E_{Nernst} + V_{act} + V_{ohmic} + V_{conc}$$

Using MATLAB to solve:

```
%%%%%%%%%%%%%%%%%%%%%%%%%%%%%%%%%%%%%%%%%%%
% EXAMPLE 3-3: Calculating the Voltage Losses for a
Polarization Curve

% UnitSystem SI
%%%%%%%%%%%%%%%%%%%%%%%%%%%%%%%%%%%%%%%%%%%
% Inputs

R = 8.314;          % Ideal gas constant (J/molK)
F = 96487;          % Faraday's constant (Coulombs)
```

```
Tc = 80;              % Temperature in degrees C
P_H2 = 3;             % Hydrogen pressure in atm
P_air = 3;            % Air pressure in atm
A_cell=100;           % Area of cell
N_cells=90;           % Number of Cells
r = 0.19;             % Internal Resistance (Ohm-cm^2)
Alpha = 0.5;          % Transfer coefficient
Alpha1 = 0.085;       % Amplification constant
io = 10^-6.912;       % Exchange Current Density (A/cm^2)
il = 1.4;             % Limiting current density (A/cm2)
Gf_liq = -228170;     % Gibbs function in liquid form (J/mol)
k = 1.1;              % Constant k used in mass transport
```

%%

% Convert degrees C to K

```
Tk = Tc + 273.15;
```

% Create loop for current

```
loop = 1;
i = 0;
for N = 0:150
i = i + 0.01;
```

% Calculation of Partial Pressures

% Calculation of saturation pressure of water

```
x  =  -2.1794  +  0.02953.*Tc-9.1837.*(10.^-5).*(Tc.^2)  +  1.4454.*(10.^-
7).*(Tc.^3);
P_H2O = (10.^x)
```

% Calculation of partial pressure of hydrogen

```
pp_H2 = 0.5.*((P_H2)./(exp(1.653.*i./(Tk.^1.334)))-P_H2O)
```

% Calculation of partial pressure of oxygen

```
pp_O2 = (P_air./exp(4.192.*i./(Tk.^1.334)))-P_H2O
```

% Activation Losses

```
b = R.*Tk./(2.*Alpha.*F);
V_act = -b.*log10(i./io); % Tafel equation
```

% Ohmic Losses

```
V_ohmic = -(i.*r);
```

% Mass Transport Losses

```
term = (1-(i./il));
if term > 0
V_conc = Alpha1 .* (i.^k) .* log(1-(i./il));
else
V_conc = 0;
end
```

% Calculation of Nernst voltage

```
E_nernst = -Gf_liq./(2.*F) - ((R.*Tk).*log(P_H2O./(pp_H2.*(pp_O2.^0.5))))./
   (2.*F)
```

% Calculation of output voltage

```
V_out = E_nernst + V_ohmic + V_act + V_conc;
if term < 0
V_conc = 0;
break
end
if V_out < 0
V_out = 0;
break
end
figure(1)
title('Fuel cell polarization curve')
xlabel('Current density (A/cm^2)');
ylabel('Output voltage (Volts)');
plot(i,V_out,'*')
grid on
hold on
disp(V_out)
```

% Calculation of power

```
P_out = N_cells.*V_out.*i.*A_cell;
figure(2)
title('Fuel cell power')
xlabel('Current density (A/cm^2)');
ylabel('Power(Watts)');
plot(i,P_out,'*');
grid on
hold on
disp(P_out);
end
```

Figures 3-11 and 3-12 are the polarization and power curve generated from the code for Example 3-3. This polarization curve will not exactly match the actual polarization curve for this fuel cell stack—but it is a good start. The topics and code introduced in Chapters 7–10 will add to these basic concepts, and will enable one to create even more accurate polarization curves.

3.7 Internal Currents and Crossover Currents

Although it is typically assumed that the electrolyte is not electrically conductive and impermeable to gases, some hydrogen and electrons diffuse through the electrolyte. The hydrogen molecules that diffuse through the electrolyte result in a decrease in the actual electrons that travel to the load. These losses are usually very small during fuel cell operation, but can be significant when the fuel cell operates at low current densities, or when

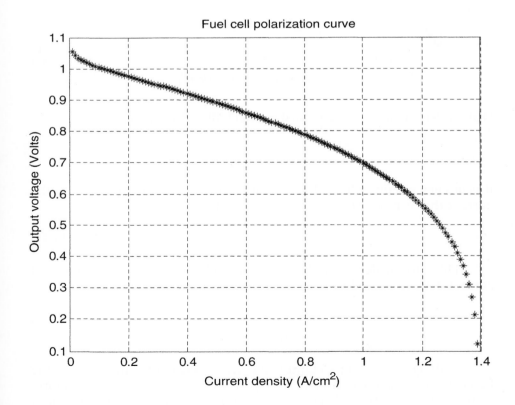

FIGURE 3-11. Polarization curve generated in MATLAB for Example 3-3.

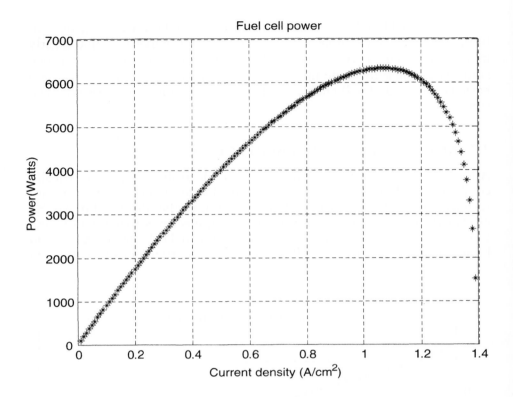

FIGURE 3-12. Power curve generated in MATLAB for Example 3-3.

it is at open circuit voltage. If the total electrical current is the sum of the current that can be used and the current that is lost, then:

$$i = i_{ext} + i_{loss} \tag{3-45}$$

The current density (A/cm^2) in the fuel cell is $i = i/A$. If the total current density is used in Equation 3-42, then:

$$E = E_r - \frac{RT}{\alpha F} \ln\left(\frac{i_{ext} + i_{loss}}{i_0}\right) \tag{3-46}$$

Hydrogen crossover and internal currents have different effects in the fuel cell. Hydrogen that diffuses through the electrolyte typically will form water and reduce the cell potential. Hydrogen crossover is a function of the electrolyte properties, such as permeability, thickness, and partial pressure[27].

Chapter Summary

Understanding the reactions at the fuel cell anode and cathode is critical when modeling fuel cells. This chapter covered the basic electrochemistry needed to predict electrode kinetics, activation losses, currents, and potentials in a fuel cell. The electrochemical reactions control the rate of power generation and are the main cause of activation voltage losses. The activation overvoltage is the voltage loss due to overcoming the catalyst activation barrier in order to convert products into reactants. The equations presented in this chapter help to predict how fast the reactants are converted into electric current, and how much energy loss occurs during the actual electrochemical reaction. In order to calculate the actual fuel cell voltage, the concepts in this chapter will be combined with the fuel cell charge and mass transport concepts in Chapters 4 and 5.

Problems

- (a) If a portable electronic device draws 2 A of current at a voltage of 12 V, what is the power requirement for the device? (b) You would like to design a device to have an operating lifetime of 72 hours. Assuming 100% fuel utilization, what is the minimum amount of H_2 fuel (in grams) required?
- A hydrogen–air fuel cell has the following polarization curve parameters: $i_0 = 0.005$, $\alpha = 0.5$, and $R_i = 0.20$ Ohm-cm^2. The fuel cell operates at 60 °C and 2 bar. (a) Calculate the cell voltage at 0.7 A/cm^2. (b) Calculate the voltage gain if the cell is going to be operated at 3 bar.
- Calculate the expected current density at 0.65 V if an MEA is prepared with a catalyst-specific area of 600 cm^2/mg and with Pt loading of 1 mg/cm^2, where the cell operates at 50 °C and 200 kPa. The cathode exchange current density is 1×10^{-10} A/cm^2 of platinum surface. What potential gain may be expected at the same current density if the Pt loading on the cathode is increased to 2 mg/cm^2?
- A hydrogen/air fuel cell operates at 25 °C and 1 atm. The exchange current density at these conditions is 0.002 mA/cm^2 of the electrode area. Pt loading is 0.5 mg/cm^2 and the charge transfer coefficient is 0.5. The electrode area is 16 cm^2. (a) Calculate the theoretical fuel cell potential at these conditions. (b) The open-circuit voltage for this fuel cell is 1.0 V. Calculate the current density loss due to hydrogen crossover or internal currents.
- What is the limiting current density of a 100-cm^2 fuel cell with a hydrogen flow rate of 1 g/s?

Endnotes

[1] Li, X. *Principles of Fuel Cells*. 2006. New York: Taylor & Francis Group.
[2] Ibid.
[3] Ibid.

[4]   Ibid.
[5]   O'Hayre, R., S.-W. Cha, W. Colella, and F.B. Prinz. *Fuel Cell Fundamentals.* 2006. New York: John Wiley & Sons.
[6]   Barbir, F. *PEM Fuel Cells: Theory and Practice.* 2005. Burlington, MA: Elsevier Academic Press.
[7]   Ibid.
[8]   Ibid.
[9]   Spiegel, C.S. Designing and Building Fuel Cells. 2007. New York: McGraw-Hill.
[10] Atkins, P.W. *Physical Chemistry*, 6th ed. 1998. Oxford: Oxford University Press.
[11] Li, *Principles of Fuel Cells.*
[12] Bard, A.J., and L.R. Faulkner. *Electrochemical Methods.* 1980. New York: John Wiley & Sons.
[13] Atkins, *Physical Chemistry*, 6th ed.
[14] O'Hayre, *Fuel Cell Fundamentals.*
[15] Bard, *Electrochemical Methods.*
[16] Ibid.
[17] Ibid.
[18] Barbir, *PEM Fuel Cells: Theory and Practice.*
[19] O'Hayre, *Fuel Cell Fundamentals.*
[20] Barbir, *PEM Fuel Cells: Theory and Practice.*
[21] Bard, *Electrochemical Methods.*
[22] Barbir, *PEM Fuel Cells: Theory and Practice.*
[23] O'Hayre, *Fuel Cell Fundamentals.*
[24] Mench, M.M., C.-Y. Wang, and S.T. Tynell. *An Introduction to Fuel Cells and Related Transport Phenomena.* Department of Mechanical and Nuclear Engineering, Pennsylvania State University. Draft. Available at: http://mtrl1 .mne.psu.edu/Document/jtpoverview.pdf Accessed March 4, 2007.
[25] Lin, B. Conceptual design and modeling of a fuel cell scooter for urban Asia. 1999. Princeton University, masters thesis.
[26] Mench, *An Introduction to Fuel Cells and Related Transport Phenomena.*
[27] Barbir, *PEM Fuel Cells: Theory and Practice.*

Bibliography

Appleby, A., and F. Foulkes. *Fuel Cell Handbook.* 1989. New York: Van Nostrand Reinhold.

Chen, E. Thermodynamics and electrochemical kinetics, in G. Hoogers (editor), *Fuel Cell Technology Handbook.* 2003. Boca Raton, FL: CRC Press.

Mench, M.M., Z.H. Wang, K. Bhatia, and C.Y. Wang. *Design of a Micro-Direct Methanol Fuel Cell.* 2001. Electrochemical Engine Center, Department of Mechanical and Nuclear Engineering, Pennsylvania State University, Pennsylvania.

Rowe, A., and X. Li. Mathematical modeling of proton exchange membrane fuel cells. *J. Power Sources.* Vol. 102, 2001, pp. 82–96.

Springer et al. Polymer electrolyte fuel cell model. *J. Electrochem. Soc.* Vol. 138, No. 8, 1991, pp. 2334–2342.

Sousa, R., Jr., and E. Gonzalez. Mathematical modeling of polymer electrolyte fuel cells. *J. Power Sources.* Vol. 147, 2005, pp. 32–45.

You, L., and H. Liu. A two-phase flow and transport model for PEM fuel cells. *J. Power Sources.* Vol. 155, 2006, pp. 219–230.

CHAPTER 4

Fuel Cell Charge Transport

4.1 Introduction

The electrochemical reactions that occur in the fuel cell catalyst layers are one of the most important concepts to understand when trying to model a fuel cell. In addition to the activation losses, there are also losses during the transport of charge through the fuel cell. Electronic charge transport describes the movement of charges from the electrode where they are produced, to the load where they are consumed. The two major types of charged particles are electrons and ions, and both electronic and ionic losses occur in the fuel cell. The electronic loss between the bipolar, cooling, and contact plates is due to the degree of contact that the plates make with each other due to the compression of the fuel cell stack. Ionic transport is far more difficult to predict and model than fuel cell electron transport. The ionic charge losses occur in the fuel cell membrane when H^+ ions travel through the electrolyte. This chapter will cover the fuel cell electronic and ionic charge transport and voltage losses due to transport resistance. The specific topics to be covered are:

- Voltage loss due to charge transport
- Electron conductivity of metals
- Ionic conductivity of polymer electrolytes

Charge transport resistance results in a voltage loss for fuel cells called ohmic loss. Common methods of reducing ohmic losses include making electrolytes as thin as possible, and employing high conductivity materials that are well connected to each other.

4.2 Voltage Loss Due to Charge Transport

Every material has an intrinsic resistance to charge flow. The material's natural resistance to charge flow causes ohmic polarization, which results in a loss in cell voltage. All fuel cell components contribute to the total electrical resistance in the fuel cell, including the electrolyte, the catalyst layer, the gas diffusion layer, bipolar plates, interface contacts, and terminal connections. The reduction in voltage is called "ohmic loss," and includes the electronic (R_{elec}) and ionic (R_{ionic}) contributions to fuel cell resistance. This can be written as:

$$v_{ohmic} = iR_{ohmic} = i(R_{elec} + R_{ionic}) \qquad (4\text{-}1)$$

R_{ionic} dominates the reaction in Equation 4-1 because ionic transport is more difficult than electronic charge transport. R_{ionic} represents the ionic resistance of the electrolyte, and R_{elec} includes the total electrical resistance of all other conductive components, including the bipolar plates, cell interconnects, and contacts.

The material's ability to support the flow of charge through the material is its conductivity. The electrical resistance of the fuel cell components is often expressed in the literature as conductance (σ), which is the reciprocal of resistance:

$$\sigma = \frac{i}{R_{ohmic}} \qquad (4\text{-}2)$$

where the total cell resistance (R_{ohmic}) is the sum of the electronic and ionic resistance. Resistance is characteristic of the size, shape, and properties of the material, as expressed by Equation 4-3:

$$R = \frac{L_{cond}}{\sigma A_{cond}} \qquad (4\text{-}3)$$

where L_{cond} is the length (cm) of the conductor, A_{cond} is the cross-sectional area (cm^2) of the conductor, and σ is the electrical conductivity (ohm^{-1} cm^{-1}). The current density, j, (A/cm^2), can be defined as:

$$j = \frac{i}{A_{cell}} \qquad (4\text{-}4)$$

or

$$j = n_{carriers} q v_{drift} = \sigma \xi \qquad (4\text{-}5)$$

where A_{cell} is the active area of the fuel cell, $n_{carriers}$ is the number of charge carriers (carriers/cm^3), q is the charge on each carrier (1.6×10^{-19} C), v_{drift} is the average drift velocity (cm/s) where the charge carriers move, and ξ is the electric field. The general equation for conductivity is:

$$\sigma = nq\frac{v}{\xi} \qquad (4\text{-}6)$$

The term $\dfrac{v}{\xi}$ can be defined as the mobility, u_i. A more specific equation for material conductivity can be characterized by two major factors: the number of carriers available, and (2) the mobility of those carriers in the material, which can be written as:

$$\sigma_i = (|z_i| * F) * c_i * u_i \tag{4-7}$$

where c_i is the number of moles of charge carriers per unit volume, u_i is the mobility of the charge carriers within the material, z_i is the charge number (valence electrons) for the carrier, and F is Faraday's constant.

Fuel cell performance will improve if the fuel cell resistance is decreased. The fuel cell resistance changes with area. When studying ohmic losses, it is helpful to compare resistances on a per-area basis using current density. Ohmic losses can be calculated from current density using Equation 4-8:

$$v_{ohmic} = j(ASR_{ohmic}) = j(A_{cell}R_{ohmic}) \tag{4-8}$$

where ASR_{ohmic} is area-specific resistance of the fuel cell. The conduction mechanisms are different for electronic versus ionic conduction. In a metallic conductor, valence electrons associated with the atoms of the metal become detached and are free to move around in the metal. In a typical ionic conductor, the ions move from site to site, hopping to ionic charge sites in the material. The number of charge carriers in an electronic conductor is much higher than an ionic conductor. Electron and ionic transport is shown in Figures 4-1 and 4-2.

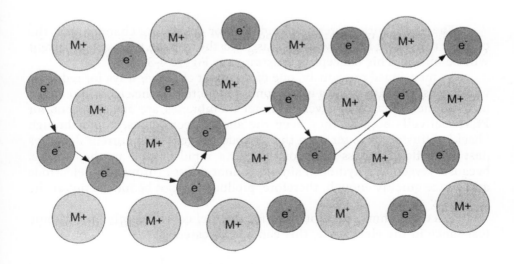

FIGURE 4-1. Electron transport in a metal.

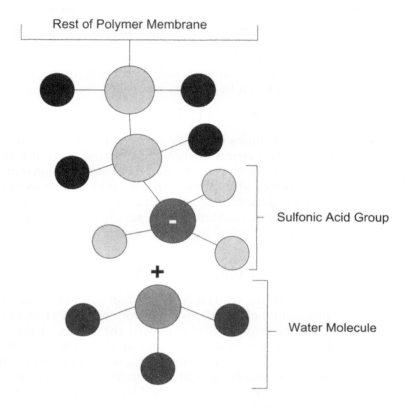

FIGURE 4-2. Ionic transport in a polymer membrane.

Therefore, with increasing land area, or decreasing channel area, the contact resistance losses will decrease and the voltage for a given current will be higher. This concept is illustrated in Figure 4-3.

As mentioned previously, one of the most effective ways for reducing ohmic loss is to either use a better ionic conductor for the electrolyte layer, or a thinner electrolyte layer. Thinner membranes are advantageous for PEM fuel cells because they keep the anode electrode saturated through "back" diffusion of water from the cathode. At very high current densities (fast fluid flows), mass transport causes a rapid dropoff in the voltage, because oxygen and hydrogen simply cannot diffuse through the electrode and ionize quickly enough, therefore, products cannot be moved out at the necessary speed[1].

Since the ohmic overpotential for the fuel cell is mainly due to ionic resistance in the electrolyte, this can be expressed as:

$$v_{ohmic} = iR_{ohmic} = jA_{cell}\left(\frac{\delta_{thick}}{\sigma A_{fuelcell}}\right) = \frac{j\delta_{thick}}{\sigma} \tag{4-9}$$

where A_{cell} is the active area of the fuel cell, δ_{thick} is the thickness of the electrolyte layer, and σ is the conductivity. As seen from Equation 4-9 and Figure 4-4, the ohmic potential can be reduced by using a thinner electrolyte layer, or using a higher ionic conductivity electrolyte.

Table 4-1 shows a summary and comparison of electronic and ionic conductors and the fuel cell components that are classified under each type.

TABLE 4-1
Comparison of Electronic and Ionic Conduction for Fuel Cell Components

Materials	Conductivity	Fuel Cell Components
Electronic Conductors		
Metals	10^3 to 10^7	Bipolar plates, gas diffusion layer, contacts, interconnects, end plates
Semiconductors	10^{-3} to 10^4	Bipolar plates, end plates
Ionic Conductors		
Solid/polymer electrolytes	10^{-1} to 10^3	PEMFC Nafion electrolyte

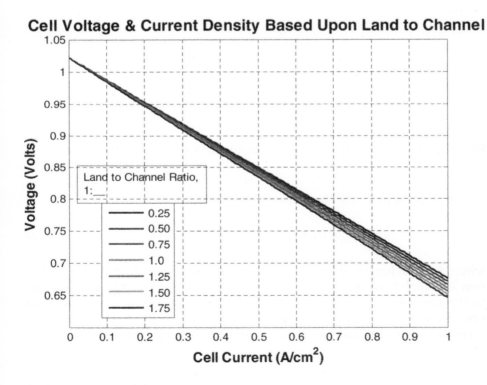

FIGURE 4-3. Cell voltage and current density based upon land to channel.

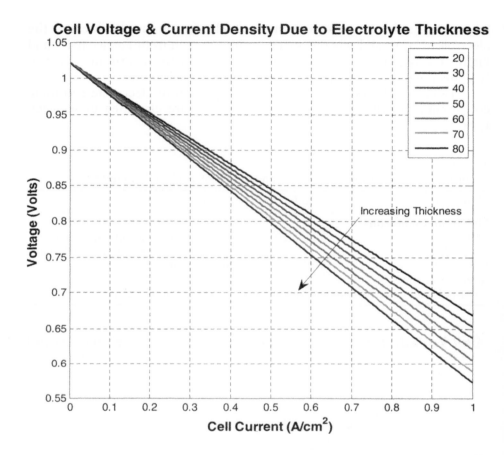

FIGURE 4-4. Cell voltage and current density due to electrolyte thickness.

The total fuel cell ohmic losses can be written as:

$$v_{\text{ohmic}} = jA\sum R = iA\left[\frac{l_a}{\sigma_a A} + \frac{l_e}{\sigma_e A} + \frac{l_c}{\sigma_c A}\right] \qquad (4\text{-}10)$$

where l is the length or thickness of the material. The first term in Equation 4-9 applies to the anode, the second to the electrolyte, and the third to the cathode. In the bipolar plates, the "land area" can vary depending upon flow channel area. As the land area is decreased, the contact resistance increases since the land area is the term in the denominator of the contact resistance:

$$R_{\text{contact}} = \frac{R_{\text{contact}}}{A_{\text{contact}}} \quad \text{and} \quad A_{\text{contact}} \approx \text{Land Area} \qquad (4\text{-}11)$$

EXAMPLE 4-1: Calculating the Ohmic Voltage Loss

Determine the ohmic voltage loss for a 100 cm^2 PEMFC that has an electrolyte membrane with a conductivity of $0.20 \ \Omega^{-1} \text{ cm}^{-1}$ and a thickness of 50 microns (μm). The current density is 0.7 A/cm^2 and R_{elec} for the fuel is $0.005 \ \Omega$. Plot the ohmic voltage losses for electrolyte thicknesses of 25, 50, 75, 100, and 150 microns (μm).

First, calculate R_{ionic} based upon electrolyte dimensions to calculate v_{ohmic}. The current of the fuel cell is

$$I = iA = 0.7 \text{A/cm}^2 \times 100 \text{cm}^2 = 70 \text{A}$$

$$R = \frac{L}{\sigma A} = \frac{0.0050 \text{cm}}{(0.10 \Omega^{-1} \text{cm}^{-1}) * (100 \text{cm}^2)} = 5 \times 10^{-4} \ \Omega$$

$$v_{ohmic} = I(R_{elec} + R_{ionic}) = 70 \text{A} * (0.005 \Omega + 5 \times 10^{-4} \Omega) = 0.385 \text{V}$$

If this equation is calculated for thinner and thicker membranes, one will notice that the ohmic loss is reduced with thinner membranes.

Using MATLAB to solve:

```
%%%%%%%%%%%%%%%%%%%%%%%%%%%%%%%%%%%%%%%
% EXAMPLE 4-1: Calculating the Ohmic Voltage Loss
% UnitSystem SI
%%%%%%%%%%%%%%%%%%%%%%%%%%%%%%%%%%%%%%%
% Inputs

i = 0.7;            % Current density (A/cm^2)
A = 100;            % Area (cm^2)
L = 0.0050;         % Electrolyte thickness (cm)
sigma = 0.1;        % Conductivity (ohms/cm)
R_elec = 0.005;     % Electrical resistance (ohms)

%%%%%%%%%%%%%%%%%%%%%%%%%%%%%%%%%%%%%%%
% Calculate the total current

I = I*A;

% Calculate the total ionic resistance

R_ohmic = L / (sigma*A);
```

% Calculate the Ohmic Voltage Loss

```
V_ohm = I. * (R_elec + R_ohmic)
i = 0:0.01:1;      % Current range
L1 = 0.0025;       % Electrolyte thickness of 25 microns
L2 = 0.0050;       % Electrolyte thickness of 50 microns
L3 = 0.01;         % Electrolyte thickness of 100 microns
L4 = 0.015;        % Electrolyte thickness of 150 microns
```

% Calculate the total current

```
I = I * A;
```

% Calculate the ohmic voltage loss

```
R_ionic1 = L1/(sigma * A); V_ohm1 = I.*(R_elec + R_ionic1);
R_ionic2 = L2/(sigma * A); V_ohm2 = I.*(R_elec + R_ionic2);
R_ionic3 = L3/(sigma * A); V_ohm3 = I.*(R_elec + R_ionic3);
R_ionic4 = L4/(sigma * A); V_ohm4 = I.*(R_elec + R_ionic4);
```

% Plot the ohmic loss as a function of electrolyte thickness

```
figure1 = figure('Color',[1 1 1]);
hdlp = plot(i,V_ohm1,i,V_ohm2,i,V_ohm,i,V_ohm3,i,V_ohm4);
title('Ohmic Loss as a Function of Electrolyte Thickness','FontSize',14,'FontWeight',
    'Bold')
xlabel('Current Density (A/cm^2)','FontSize',12,'FontWeight','Bold');
ylabel('Ohmic Loss (V)','FontSize',12,'FontWeight','Bold');
legend('L = 0.0025','L = 0.0050','L = 0.0075','L = 0.001','L = 0.015')
set(hdlp,'LineWidth',1.5);
grid on;
```

Figure 4-5 illustrates the ohmic loss as a function of electrolyte thickness when the current density is 0.8 A/cm² and the active area is 25 cm².

EXAMPLE 4-2: Calculating the Ohmic Voltage Loss

Calculate the ohmic voltage losses for two fuel cell sizes at a current density of 0.7 A/cm²: (a) $A_1 = 16$ cm², $R_1 = 0.05$ Ω; (b) $A_2 = 49$ cm², $R_2 = 0.02$ Ω; (c) for A = 1 to 100 cm², $R_1 = 0.05$. Plot the ohmic loss as a function of fuel cell area.

(a)

$$ASR_1 = R_1 A_1 = (0.05\Omega)(16 cm^2) = 0.8\Omega cm^2$$

The ohmic loss can be calculated as follows:

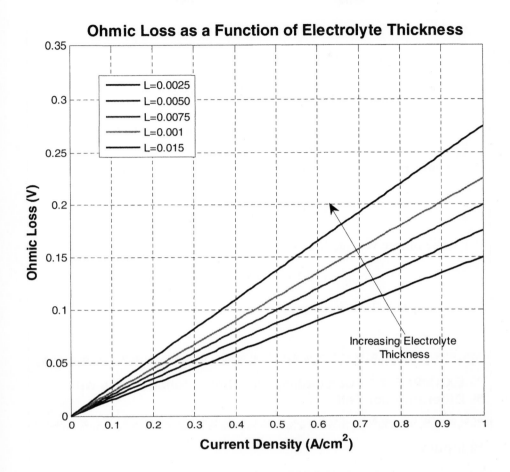

FIGURE 4-5. Ohmic loss as a function of electrolyte thickness.

$$v_{\text{ohmic}_1} = j(\text{ASR}_1) = \left(0.7\frac{S}{\text{cm}^2}\right)0.8\Omega\text{cm}^2 = 0.56\text{V}$$

Convert the current densities into fuel cells with currents:

$$i_1 = jA_1 = 0.7\frac{A}{\text{cm}^2}*16\text{cm}^2 = 11.2\text{A}$$

The ohmic voltage losses are

$$v_{\text{ohmic}_1} = i_1(R_1) = 11.2\text{A}*0.05\Omega = 0.56\text{V}$$

(b)

$$\text{ASR}_2 = R_2 A_2 = (0.02\Omega)(49\text{cm}^2) = 0.98\Omega\text{cm}^2$$

The ohmic loss can be calculated as follows:

$$\upsilon_{\text{ohmic}_2} = j(\text{ASR}_1) = \left(0.7\frac{A}{\text{cm}^2}\right)0.98\Omega\text{cm}^2 = 0.686\text{V}$$

Convert the current densities into fuel cells with currents:

$$i_1 = jA_1 = 0.7\frac{A}{\text{cm}^2} * 49\text{cm}^2 = 34.3\text{A}$$

The ohmic voltage losses are

$$\upsilon_{\text{ohmic}_2} = i_2(R_2) = 34.3\text{A} * 0.02\Omega = 0.686\text{V}$$

Using MATLAB to solve:

```
%%%%%%%%%%%%%%%%%%%%%%%%%%%%%%%%%%%%%%%%%
% EXAMPLE 4-2: Calculating the Ohmic Voltage Loss with
% Different Fuel Cell
%%%%%%%%%%%%%%%%%%%%%%%%%%%%%%%%%%%%%%%%%
% Inputs
i = 0.7;        % Current Density (A/cm^2)
A1 = 16;        % Area 1 (cm^2)
R1 = 0.05;      % Resistance (ohms)
A2 = 49;        % Area 2 (cm^2)
R2 = 0.02;      % Resistance (ohms)
%%%%%%%%%%%%%%%%%%%%%%%%%%%%%%%%%%%%%%%%%
% Part a: Ohmic voltage losses for first fuel cell size
ASR1 = R1 * A1;
% Calculate the ohmic voltage loss
V_ohm1 = i.*ASR1;
% Calculate the total current
I1 = i*A1;
```

```
% Calculate the ohmic voltage loss

V_ohm1a = I1.*R1
```

% Part b: Ohmic voltage losses for second fuel cell size

```
ASR2 = R2*A2;
```

% Calculate the ohmic voltage loss

```
V_ohm2 = i.*ASR2;
```

% Calculate the total current

```
I2 = i*A2;
```

% Calculate the ohmic voltage loss

```
V_ohm2b = I2.*R2
```

% Part c:

```
A = 1:100;
ASR = R1*A;
```

% Calculate the ohmic voltage loss

```
V_ohm = i.*ASR;
```

% Calculate the total current

```
I = i*A;
```

% Calculate the ohmic voltage loss

```
V_ohm = I.*R1
```

% Plot of the ohmic losses as a function of fuel cell area

```
figure1 = figure('Color',[1 1 1]);
hdlp = plot(A,V_ohm);
title('Ohmic Loss as a Function of Fuel Cell Area','FontSize',14,'FontWeight','Bold')
xlabel('Fuel Cell Area (cm^2)','FontSize',12,'FontWeight','Bold');
ylabel('Ohmic Loss (V)','FontSize',12,'FontWeight','Bold');
set(hdlp,'LineWidth',1.5);
grid on;
```

The plot of the ohmic losses as a function of fuel cell area for Example 4-2 is shown in Figure 4-6.

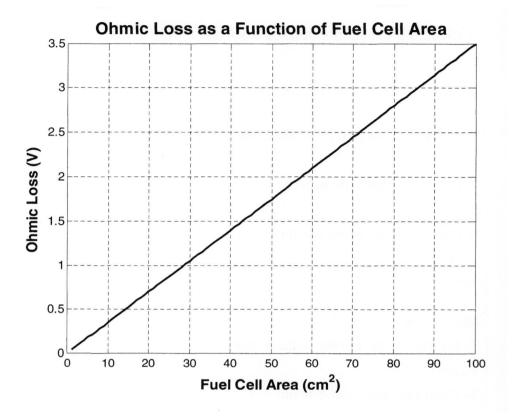

FIGURE 4-6. Ohmic loss as a function of fuel cell area.

4.3 Electron Conductivity of Metals

The electronic conductivity of the metals used in a fuel cell is important because it affects the charge transfer of electrons. Fuel cell components that are typically made of metal include the flow field plates, current collectors, and interconnects. A common expression for the mobility of free electrons in a metal conductor can be written as:

$$u = \frac{q\tau}{m_e} \qquad (4\text{-}12)$$

where τ gives the mean free time between scattering events, m_e is the mass of the electron (m = 9.11×10^{-31} kg), and q is the elementary electron charge in coulombs (q = 1.68×10^{-19} C).

Inserting Equation 4-12 into the equation for conductivity (4-7):

$$\sigma = \frac{|z_e|c_e q\tau}{m_e} \qquad (4\text{-}13)$$

Carrier concentration in a metal can be calculated from the density of free electrons[2]. Each metal atom contributes approximately one electron.

4.4 Ionic Conductivity of Polymer Electrolytes

Ionic transport in polymer electrolytes follows the exponential relationship:

$$\sigma T = \sigma_0 e^{-E_a/kT} \tag{4-14}$$

where σ_0 represents the conductivity at a reference state, and E_a is the activation energy (eV/mol). As seen in Equation 4-14, the conductivity increases exponentially with increasing temperature.

A good conductive polymer should have a fixed number of charge sites and open space. The charged sites have a negative charge, and provide a temporary resting place for the positive ion. Increasing the number of charged sites raises the ionic conductivity, but an excessive number of charged side chains may reduce the stability of the polymer. In addition, increasing the free volume in the polymer allows more space for the ions to move. In polytetrafluoroethylene (PTFE)-based polymer membranes like Nafion, ions are transported through the polymer membrane by hitching onto water molecules that move through the membrane. This type of membrane has high conductivity and is the most popular membrane used for PEM fuel cells. Nafion has a similar structure to Teflon, but includes sulfonic acid groups ($SO_3^-H^+$) that provide sites for proton transport. Figure 4-7 shows the chemical molecule of Nafion.

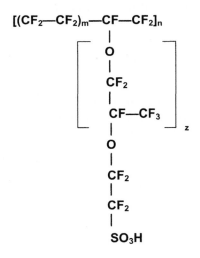

FIGURE 4-7. Chemical structure of Nafion.

The conductivity of Nafion is dependent upon the amount of hydration, and can vary with the water content. Hydration can be achieved by humidifying the gases or by relying upon the water generated at the cathode. In the presence of water, the protons form hydronium complexes (H_3O^+), which transport the protons in the aqueous phase. When the Nafion is fully hydrated, its conductivity is similar to liquid electrolytes.

The volume of Nafion can increase up to 22% when fully hydrated[3]. Since the conductivity and the hydration of the membrane are correlated, the water content can be determined through membrane conductivity. The humidity can be quantified through water vapor activity a_{water_vap}:

$$a_{water\_vap} = \frac{p_w}{p_{sat}} \qquad (4\text{-}15)$$

where p_w represents the partial pressure of water vapor in the system, and p_{sat} represents saturation water vapor pressure for the system at the temperature of operation[4].

The amount of water that the membrane can hold also depends upon the membrane pre-treatment. For example, at high temperatures, the water uptake by the Nafion membrane is much lower due to changes in the polymer at high temperatures. The relationship between water activity on the faces of the membrane and water content can be described by:

$$\lambda = 0.043 + 17.18 a_{water\_vap} - 39.85 (a_{water\_vap})^2 + 36 (a_{water\_vap})^3 \qquad (4\text{-}16)$$

Water uptake results in membrane swelling, which changes the membrane thickness along with its conductivity. Springer et al.[5] correlated the ionic conductivity (σ) (in S/cm) to water content and temperature with the following relation:

$$\sigma = (0.005139\lambda - 0.00326) \exp\left[1268 \left(\frac{1}{303} - \frac{1}{T} \right) \right] \qquad (4\text{-}17)$$

Since conductivity is proportional to resistance, the resistance of the membrane changes with water saturation and thickness. The total resistance of a membrane (R_m) is found by integrating the local resistance over the membrane thickness:

$$R_m = \int_0^{t_m} \frac{dz}{\sigma[\lambda(z)]} \qquad (4\text{-}18)$$

where t_m is the membrane thickness, λ is the water content of the membrane, and σ is the conductivity (S/cm) of the membrane. Since the protons typically have one or more water molecules associated with them, the conductivity and hydration both change simultaneously. This phenomenon of the number of water molecules that accompanies each proton is called the electroosmotic drag (n_{drag}), which is:

$$n_{drag} = n_{drag}^{Sat} \frac{\lambda}{22}$$ (4-19)

where n_{drag}^{Sat} is the electroosmotic drag (usually between 2.5 ± 0.2), and λ is the water content (which ranges from 0 to 22 water molecules per sulfonate group, and when $\lambda = 22$, Nafion is fully hydrated). The water drag flux from the anode to the cathode with a net current j is[6]:

$$J_{H_2O,drag} = 2n_{drag} \frac{j}{2F}$$ (4-20)

where $J_{H_2O,drag}$ is the molar flux of water due to the electroosmotic drag (mol/scm^2), and j is the current density of the fuel cell (A/cm^2).

The electroosmotic drag moves water in the fuel cell from the anode to the cathode. Since the reaction at the cathode produces water, it tends to build up at the cathode, and some water travels back through the membrane. This is known as "back diffusion," and it usually occurs because the amount of water at the cathode is many times greater than at the anode. The water back-diffusion flux can be determined by:

$$J_{H_2O,backdiffusion} = \frac{\rho_{dry}}{M_m} D_\lambda \frac{d\lambda}{dz}$$ (4-21)

where ρ_{dry} is the dry density (kg/m^3) of Nafion, M_n is the Nafion equivalent weight (kg/mol), D_λ is the water diffusivity, and z is the direction through the membrane thickness.

The total amount of water in the membrane is a combination of the electroosmotic drag and back diffusion, and can be calculated using Equation 4-22:

$$J_{H_2O,backdiffusion} = 2n_{drag}^{SAT} \frac{j}{2F} \frac{\lambda}{22} - \frac{\rho_{dry}}{M_m} D_\lambda(\lambda) \frac{d\lambda}{dz}$$ (4-22)

The concepts introduced by Equations 4-15 to 4-22 are illustrated by Example 4-3.

EXAMPLE 4-3: Calculating the Ohmic Voltage Loss Due to the Membrane

A hydrogen fuel cell operates at 80°C at 1 atm. It has a Nafion 112 membrane of 50 μm, and the following equation can be used for the water content across the membrane: $\lambda(z) = 5 + 2\exp(100z)$. This fuel cell has a current density of 0.8 A/cm^2, and the water activites at the anode and cathode are 0.8 and 1.0, respectively. Estimate the ohmic overvoltage loss across the membrane.

Convert the water activity on the Nafion surfaces to water contents:

$$\lambda_a = 0.043 + 17.18*0.8 - 39.85*0.8^2 + 36*0.8^3 = 7.2$$

$$\lambda_c = 0.043 + 17.18*1 - 39.85*1^2 + 36*1^3 = 14.0$$

Using these values as boundary conditions, Equation 4-23 can be arranged to create:

$$\frac{d\lambda}{dz} = \left(2n_{drag}^{SAT}\frac{\lambda}{22} - \alpha\right)\frac{jM_m}{2F\rho_{dry}D_\lambda}$$

$$D_\lambda = 10^{-6}\exp\left[2416\left(\frac{1}{303} - \frac{1}{353}\right)\right]*2.563 -$$
$$0.33*10 + 0.0264*10^2 - 0.000671*10^3$$

$$\lambda(z) = \frac{11\lambda}{2.5} + C\exp\left(\frac{\left(0.7\dfrac{A}{cm^2}1.0\dfrac{kt}{mol}*2.5\right)}{22*96500\dfrac{C}{mol} - 0.00197\dfrac{kg}{cm^3}3.81*10^{-6}\dfrac{cm^2}{s}}\right)$$

$$\lambda(z) = 4.4\alpha + 2.30*\exp(109.8z)$$

C is determined from the boundary conditions where $\lambda(0) = 7.2$ and $\lambda(0.0125) = 14.0$, and λ varies across the membrane.

The conductivity profile of the membrane is

$$\sigma(z) = 0.005193(5 + 2\exp(100z) - 0.00326)\times\exp 1268\left(\frac{1}{303} - \frac{1}{333}\right)$$

$$\sigma(z) = 0.04107 + 0.01878\exp(100z)$$

The resistance of the membrane is

$$R_m = \int_0^{t_m}\frac{dz}{\sigma(\lambda(z))} = \int_0^{0.0050}\frac{dz}{0.04107 + 0.01878\exp(100z)} = 0.15\ \Omega cm^2$$

The ohmic overvoltage due to the membrane resistance in this fuel cell is

$$V_{ohm} = j \times R_m = 0.8A/cm^2 \times 0.15\Omega cm^2 = 0.12V$$

Using MATLAB to solve:

```
%%%%%%%%%%%%%%%%%%%%%%%%%%%%%%%%%%%%%%%%
```

% EXAMPLE 4-3: Calculating the Ohmic Voltage Loss Due to the membrane

```
%%%%%%%%%%%%%%%%%%%%%%%%%%%%%%%%%%%%%%%%
```

% Inputs

```
global T; global F; global C; global alpha; global den_dry; global Sigma_a;
global z; global A; global n; global i; global Mn; global D;
Tc = 20:10:80;          % Temperature in Celsius
T = Tc + 273.15;        % Temperature in Kelvin
z = 0.005;              % Membrane Thickness (cm)
aw_a = 0.8;             % Water activity at the anode
aw_c = 1;               % Water activity at the cathode
n = 2.5;                % Electro-osmotic drag coefficient
i = 0.8;                % Current Density (A/cm^2)
Mn = 1;                 % Nafion equivalent weight(kg/mol)
F = 96487;              % Faraday's constant
den_dry = 0.00197;      % membrane dry density(kg/cm^3)
C = 2.3;                % Constant dependent upon boundary conditions
alpha = 1.12;           % Ratio of water flux to hydrogen flux
```

```
%%%%%%%%%%%%%%%%%%%%%%%%%%%%%%%%%%%%%%%%
```

% Convert the water activity on the Nafion surfaces to water contents

```
lambda_anode = 0.043 + (17.81*(aw_a)) - (39.85*(aw_a^2)) + (36* aw_a^3));
lambda_cathode = 0.043 + (17.81*(aw_c)) - (39.85*(aw_c^2)) + (36*(aw_c^3));
```

% Calculate the Water Diffusivity

```
D = (10.^-6).*exp(2416.*(1./303-1./T)).*(2.563 - (0.33.*10) + (0.0264.*10.^2) -
    (0.000671.*10.^3));
delta_lambda = ((11.*alpha)./n) + C.*exp(((i.*Mn.*n)./(22.*F.*den_dry.*D)).*z);
Sigma_a = exp(1268.*((1./303) - (1./T))).*(0.005139.*delta_lambda -
    0.00326);%S/m
Sigma_c = exp(1268.*((1./303) - (1./T))).*(0.005139.*delta_lambda -
    0.00326);%S/m
Re_a = quad('thick',0,0.00500)
V_ohm = i*Re_a
```

% Plot

```
z = 0:0.002:0.0125;
delta_lambda = ((11.*alpha)./n) + C.*exp(((i.*Mn.*n)./(22.*F.*den_dry.*D)).*z);
Sigma = exp(1268.*((1./303) - (1./T))).*(0.005139.*delta_lambda -
    0.00326);%S/m
```

```
% Plot the membrane thickness and water content
figure1 = figure('Color',[1 1 1]);
hdlp = plot(z,delta_lambda);
title('Membrane Thickness and Water Content','FontSize',14,'FontWeight','Bold')
xlabel('Membrane Thickness (cm)','FontSize',12,'FontWeight','Bold');
ylabel('Water Content(H2O/SO3)','FontSize',12,'FontWeight','Bold');
set(hdlp,'LineWidth',1.5);
grid on;

% Plot the membrane thickness and local conductivity
figure2 = figure('Color',[1 1 1]);
hdlp = plot(z,Sigma);
title('Membrane Thickness and Local Conductivity','FontSize',14,'FontWeight','Bold')
xlabel('Membrane Thickness (cm)','FontSize',12,'FontWeight','Bold');
ylabel('Local Conductivity(S/cm)','FontSize',12,'FontWeight','Bold');
set(hdlp,'LineWidth',1.5);
grid on;

%%%%%%%%%%%%%%%%%%%%%%%%%%%%%%%%%%%%%%%%%%

function y = thick(z);
T = 353;
y = 1./(exp(1268*((1/303) − (1/T)))*(0.005139*((11*1.12/2.5) + 2.3*exp(((0.7*
   2.5)/(22* . . . 96500*0.00197*3.81*10^-6))*z)) − 0.00326)); % S/m
```

The figures generated in Example 4-3 are illustrated by Figures 4-8 and 4-9.

Chapter Summary

The transport of charges through, the fuel cell layers (except the membrane) occurs through conduction. Therefore, ohmic losses occur due to the lack of proper contact by the gas diffusion layer, bipolar plates, cooling plates, contacts, and interconnect. However, the largest ohmic loss occurs during the transport of ions through the membrane. To decrease the ionic losses through the membrane, either the membrane needs to become more conductive or the membrane needs to become thinner. It is usually easier to make the membrane thinner because developing high conductivity electrolytes is very challenging. The challenge occurs in creating a material that not only is highly conductive, but also stable in a chemical environment and able to withstand the required fuel cell temperatures. The electrolyte equations presented in this chapter are applicable for Nafion, but if another type of electrolyte is employed, the equations may need to be altered to suit the chemistry.

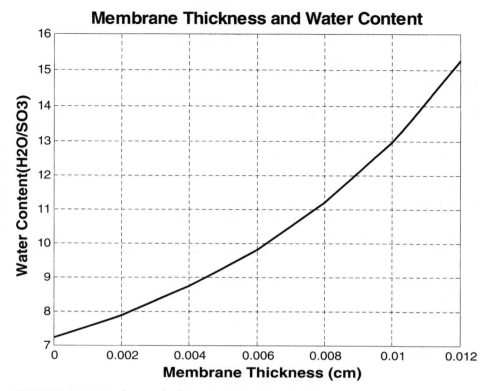

FIGURE 4-8. Membrane thickness and water content.

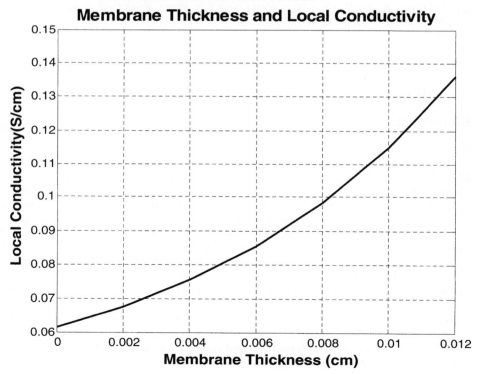

FIGURE 4-9. Membrane thickness and local conductivity.

Problems

- A 10-cm^2 fuel cell has $R_{elec} = 0.01$ Ω and $\sigma_{electrolyte} = 0.10$ Ω^{-1} cm^{-1}. If the electrolyte is 100 μm thick, predict the ohmic voltage losses for the fuel cell at j = 500 mA/cm^2.
- Estimate the ohmic overpotential for a fuel cell operating at 70 °C. The external load is 1 A/cm^2, and it uses a 50-μm-thick membrane. The humidity levels $a_{w,anode}$ and $a_{w,cathode}$ are 1.0 and 0.5, respectively.
- A fuel cell is operating at 0.8 A/cm^2 and 60 °C. Hydrogen gas at 30 °C and 50% relative humidity is provided to the fuel cell at a rate of 2 A. The fuel cell area is 10 cm^2, and the drag ratio of water molecules to hydrogen is 0.7. The hydrogen exhaust exits the fuel cell at 60 °C and p = 1 atm.
- In a PEMFC, the water activities on the anode and cathode sides of a Nafion 115 membrane are 0.7 and 0.9, respectively. The fuel cell is operating at a temperature of 60 °C and 1 atm with a current density of 0.8 A/cm^2. Estimate the ohmic overvoltage loss across the membrane.

Endnotes

[1]   Lin, B. Conceptual design and modeling of a fuel cell scooter for urban Asia. 1999. Princeton University, masters thesis.
[2]   O'Hayre, R., S.-W. Cha, W. Colella, and F.B. Prinz. *Fuel Cell Fundamentals.* 2006. New York: John Wiley & Sons.
[3]   Ibid.
[4]   Ibid.
[5]   Springer et al. Polymer electrolyte fuel cell model. *J. Electrochem. Soc.* Vol. 138, No. 8, 1991, pp. 2334–2342.
[6]   O'Hayre, *Fuel Cell Fundamentals.*

Bibliography

Barbir, F. *PEM Fuel Cells: Theory and Practice.* 2005. Burlington, MA: Elsevier Academic Press.
Li, X. *Principles of Fuel Cells.* 2006. New York: Taylor & Francis Group.
Mench, M.M., C.-Y. Wang, and S.T. Tynell. *An Introduction to Fuel Cells and Related Transport Phenomena.* Department of Mechanical and Nuclear Engineering, Pennsylvania State University. Draft. Available at: http://mtrl1.mne.psu .edu/Document/jtpoverview.pdf Accessed March 4, 2007.
Mench, M.M., Z.H. Wang, K. Bhatia, and C.Y. Wang. *Design of a Micro-Direct Methanol Fuel Cell.* 2001. Electrochemical Engine Center, Department of Mechanical and Nuclear Engineering, Pennsylvania State University.
Rowe, A., and X. Li. Mathematical modeling of proton exchange membrane fuel cells. *J. Power Sources.* Vol. 102, 2001, pp. 82–96.
Sousa, R., Jr., and E. Gonzalez. Mathematical modeling of polymer electrolyte fuel cells. *Journal of Power Sources.* Vol. 147, 2005, pp. 32–45.
You, L., and H. Liu. A two-phase flow and transport model for PEM fuel cells. *J. Power Sources.* Vol. 155, 2006, pp. 219–230.

CHAPTER 5

Fuel Cell Mass Transport

5.1 Introduction

In order to produce electricity, a fuel cell must be supplied continuously with fuel and oxidant. In addition, product water must be removed continually to insure proper fuel and oxidant at the catalyst layers to maintain high fuel cell efficiency. Voltage losses occur in the fuel cell due to activation losses (Chapter 3), ohmic losses (Chapter 4), and mass transport limitations—which is the topic of this chapter. Mass transport is the study of the flow of species, and can significantly affect fuel cell performance. Losses due to mass transport are also called "concentration losses," and can be minimized by optimizing mass transport in the flow field plates, gas diffusion layer, and catalyst layer. This chapter covers both the macro and micro aspects of mass transport. The specific topics to be covered are:

- Fuel cell mass balances
- Convective mass transport from flow channels to electrode
- Diffusive mass transport in electrodes
- Convective mass transport in flow field plates
- Mass transport equations in the literature

In conventional fuel cells, the flow field plates have channels with dimensions in millimeters or centimeters. Due to the size of these channels, mass transport is dominated by convection and the laws of fluid dynamics. Convection is the movement of fluid flow due to density gradients or hydrodynamic transport, and is characterized by laminar or turbulent flow and stagnant regions. This type of flow dominates mass transfer in the flow channels. High fuel and oxidant flow rates sometimes insure good distribution of reactants, but if the flow rate is too high, the fuel may move too fast to diffuse through the GDL and catalyst layers. In addition, delicate fuel cell components such as the membrane can rupture.

Mass transport in the fuel cell GDL and catalyst layers is dominated by diffusion due to the tiny pore sizes of these layers (4 to 10 microns). In a flow channel, the velocity of the reactants is usually slower near the walls; therefore, this aids the flow change from convective to diffusive. The mass transport theory described in this chapter will help the reader to write mass balances, predict fuel cell flow rates, and calculate the mass transfer in the flow channels, electrodes, and membrane.

5.2 Fuel Cell Mass Balances

Before convective and diffusive flows are covered, the overall mass flows through the fuel cell need to be discussed. These flow calculations, or mass balances, are critical for determining the correct flow rates for a fuel cell. In order to properly determine these mass flow rates, the mass that flows into and out of each process unit (or control volume) in the fuel cell sub-systems, stack, or fuel cell layer need to be accounted for. The procedure for formulating a mass balance can be applied to any type of system, and is as follows:

1. A flow diagram must be drawn and labeled. Enough information should be included on the flow diagram to have a summary of each stream in the process. This includes known temperatures, pressures, mole fractions, flow rates, and phases.
2. The appropriate mass balance equation(s) must be written in order to determine the flow rates of all stream components and to solve for any desired quantities.

An example flow diagram is shown in Figure 5-1. Hydrogen enters the cell at temperature, T, and pressure, P, with the mass flow rate, m_{H2}. Oxygen enters the fuel cell from the environment at a certain T, P, and m_{O2}. The hydrogen and oxygen react completely in the cell to produce

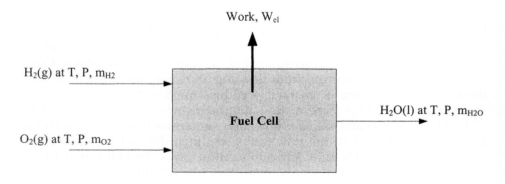

FIGURE 5-1. Detailed flowchart to obtain mass balance equation.

water, which exits at a certain T, P, and m_{H2O}. This reaction can be described by:

$$2H_2(g) + O_2(g) \rightarrow 2H_2O(l)$$

W_{el} in Figure 5-1 is the work available through chemical availability. The generic mass balance for the fuel cell in this example is:

$$m_{H2} + m_{O2} = m_{H_2O} + W_{el} \tag{5-1}$$

The formal definition for material balances in a system (or control volume) can be written as:

Input (enters through system boundaries)
+ Generation (producted within the system)
− Output (leaves through system boundaries)
− Consumption (consumed within the system)
= Accumulation (buildup within the system)

Generally, the fuel cell mass balance requires that the sum of all of the mass inputs is equal to the mass outputs, which can be expressed as:

$$\sum(m_i)_{in} = \sum(m_i)_{out} \tag{5-2}$$

where i is the mass going into and out of the cell, and can be any species, including hydrogen, oxygen, and water. The flow rates at the inlet are proportional to the current and number of cells. The cell power output is:

$$W_{el} = n_{cell}V_{cell}I \tag{5-3}$$

where n_{cell} is the number of cells, V_{cell} is the cell voltage, and I is the current. All of the flows are proportional to the power output and inversely proportional to the cell voltage:

$$I \cdot n_{cell} = \frac{W_{el}}{V_{cell}} \tag{5-4}$$

The inlet flow rates for a PEM fuel cell are as follows:

The hydrogen mass flow rate is

$$m_{H2,in} = S_{H2}\frac{M_{H2}}{2F}I \cdot n_{cell} \tag{5-5}$$

The oxygen mass flow rate (g/s) is

$$m_{O2,in} = S_{O2}\frac{M_{O2}}{4F}I \cdot n_{cell} \qquad (5\text{-}6)$$

The air mass flow rate (g/s) is

$$m_{air,in} = \frac{S_{O2}}{r_{O2}}\frac{M_{air}}{4F}I \cdot n_{cell} \qquad (5\text{-}7)$$

The nitrogen mass flow rate (g/s) is

$$m_{N2,in} = S_{O2}\frac{M_{N2}}{4F}\frac{1-r_{O2,in}}{r_{O2,in}}I \cdot n_{cell} \qquad (5\text{-}8)$$

Water vapor in the hydrogen inlet is

$$m_{H2OinH2,in} = S_{H2}\frac{M_{H2O}}{2F}\frac{\varphi_{an}P_{vs(T_{an,in})}}{P_{an}-\varphi_{an}P_{vs(T_{an,in})}}I \cdot n_{cell} \qquad (5\text{-}9)$$

Water vapor in the oxygen inlet is

$$m_{H2OinO2,in} = S_{O2}\frac{M_{H2O}}{4F}\frac{\varphi_{ca}P_{vs(T_{an,in})}}{P_{ca}-\varphi_{ca}P_{vs(T_{an,in})}}I \cdot n_{cell} \qquad (5\text{-}10)$$

Water vapor in the air inlet (g/s) is

$$m_{H2Oinairin} = \frac{S_{O2}}{r_{O2}}\frac{M_{H2O}}{4F}\frac{\varphi_{ca}P_{vs(T_{an,in})}}{P_{ca}-\varphi_{ca}P_{vs(T_{an,in})}}I \cdot n_{cell} \qquad (5\text{-}11)$$

The outlet flow rates for a PEM fuel cell are as follows:

The unused hydrogen flow rate is

$$m_{H2,out} = (S_{H2}-1)\frac{M_{H2}}{2F}I \cdot n_{cell} \qquad (5\text{-}12)$$

The oxygen flow rate at the outlet is equal to the oxygen supplied at the inlet minus the oxygen consumed in the fuel electrochemical reaction:

$$m_{O2,out} = (S_{O2}-1)\frac{M_{O2}}{4F}I \cdot n_{cell} \qquad (5\text{-}13)$$

The nitrogen flow rate at the exit is the same as the inlet because nitrogen does not participate in the fuel cell reaction:

$$m_{N2,out} = m_{N2in} = S_{O2} \frac{M_{N2}}{4F} \frac{1 - r_{O2in}}{r_{O2in}} I \cdot n_{cell} \qquad (5\text{-}14)$$

The depleted air flow rate is then simply a sum of the oxygen and nitrogen flow rates:

$$m_{air,out} = \left[(S_{O2} - 1)M_{O2} + S_{O2} \frac{1 - r_{O2in}}{r_{O2in}} M_{N2} \right] \frac{I \cdot n_{cell}}{4F} \qquad (5\text{-}15)$$

The oxygen volume fraction at the outlet is much lower than the inlet volume fraction:

$$r_{O2,out} = \frac{S_{O2} - 1}{\dfrac{S_{O2}}{r_{O2,in}} - 1} \qquad (5\text{-}16)$$

The additional outlet liquid and vapor water flow rates and balances for a PEM fuel cell are described by equations 5-17 through 5-20:

The water vapor content at the anode outlet is the smaller of the total water flux:

$$m_{H2Oin,H2out,V} = min\left[(s_{H2} - 1)\frac{M_{H2O}}{2F} \frac{P_{vs(T_{out,an})}}{P_{an} - \Delta P_{an} - P_{vs(T_{out,an})}} I \cdot n_{cell}, m_{H2Oin,H2out} \right] \qquad (5\text{-}17)$$

where ΔP_{an} is the pressure drop on the anode side. The amount of liquid water is the difference between the total water present and the water vapor:

$$M_{H2Oin,H2out,L} = m_{H2Oin,H2out} - m_{H2Oin,H2out,V} \qquad (5\text{-}18)$$

Water content in the cathode exhaust is equal to the amount of water brought into the cell, plus the water generated in the cell, along with the water transported across the membrane:

$$m_{H2OinAirout} = m_{H2OinAirin} + m_{H2Ogen} + m_{H2OED} - m_{H2OBD} \qquad (5\text{-}19)$$

The water vapor content at the cathode outlet is:

$$m_{H2Oin,Airout,V} = min\left[\left(\frac{S_{O2} - r_{O2,in}}{r_{O2,in}} \right)\frac{M_{H2O}}{4F} \frac{P_{vs(T_{out,an})}}{P_{ca} - \Delta P_{ca} - P_{vs(T_{out,an})}} I \cdot n_{cell}, m_{H2Oin,Airout} \right] \qquad (5\text{-}20)$$

EXAMPLE 5-1: Water Injection Flow Rate

A hydrogen–air PEM fuel cell generates 500 watts (W) at 0.7 V. Dry hydrogen is supplied in a dead-end mode at 20 °C. The relative humidity of the air at the fuel cell inlet is 50% at a pressure of 120 kPa. Liquid water is injected at the air inlet to help cool the fuel cell. The oxygen stoichiometric ratio is 2, and the outlet air is 100% saturated at 80 °C and atmospheric pressure. What is the water injected flow rate (g/s)?

The water mass balance is

$$m_{H2O\_air\_in} + m_{H2O\_Inject} + m_{H2O\_gen} = m_{H2O\_in\_air\_out}$$

In order to calculate the amount of water in air, the saturation pressure needs to be calculated. To calculate the saturation pressure (in Pa) for any temperature between 0 °C and 100 °C:

$$p_{vs} = e^{aT^{-1}+b+cT+dT^2+eT^3+f\ln(T)}$$

where a, b, c, d, e, and f are the coefficients.

a = –5800.2206, b = 1.3914993, c = –0.048640239, d = 0.41764768 × 10^{-4}, e = –0.14452093 × 10^{-7}, and f = 6.5459673

with T = 293.15, pvs = 2.339 kPa.

The amount of water in air can be calculated:

$$m_{H2O,airin} = \frac{S_{O2}}{r_{O2}} \frac{M_{H2O}}{nF} \frac{\phi_{ca}P_{vs(T_{ca,in})}}{P_{ca} - \phi_{ca}P_{vs(T_{ca,in})}} I \cdot n_{cell}$$

$$m_{H2O,airin} = \frac{2}{0.2095} \frac{18.015}{(4*96,485As/mol)} \frac{0.50*2.339}{120kPa - 0.50*2.339} \frac{500W}{0.7V} *1$$

$$m_{H2O,airin} = 0.00313g/s$$

Water generated is

$$m_{H2O,gen} = \frac{I}{nF} M_{H2O} = \frac{714.29}{2 \times 96,485} 18.015 = 0.0667g/s$$

Water vapor in air out is

$$m_{H2Oin,H2out,V} = \left[\left(\frac{S_{O2} - r_{O2,in}}{r_{O2,in}} \right) \frac{M_{H2O}}{4F} \frac{P_{vs(T_{out,ca})}}{P_{ca} - \Delta P_{ca} - P_{vs(T_{out,ca})}} I \cdot n_{cell} \right]$$

First calculate the saturation pressure:

$T = 80°C = 353.15K$, $Pvs = 47.67kPa$, $Pca - \Delta Pca = 101.325kPa$

$$m_{H2Oin,H2out,V} = \left[\left(\frac{2-0.2095}{0.2095}\right)\frac{18.015}{(4*96,485)}\frac{47.67}{(101.325-47.67)}\frac{500}{0.7}*1\right]$$

$$m_{H2Oin,airout} = 0.253g/s$$

The water balance is therefore:

$$m_{H2O\_air\_in} + m_{H2O\_Inject} + m_{H2O\_gen} = m_{H2O\_in\_air\_out}$$

$$0.003\,13 + m_{H2Oinject} + 0.066\,7 = 0.253$$

$$m_{H2Oinject} = 0.223\,17g/s$$

Using MATLAB to solve:

```
%%%%%%%%%%%%%%%%%%%%%%%%%%%%%%%%%%%%%%%
% EXAMPLE 5-1: Water Injection Flow Rate
%%%%%%%%%%%%%%%%%%%%%%%%%%%%%%%%%%%%%%%%%
% Inputs

T_H2_in = 20 + 273.15;      % Hydrogen inlet temperature
T_air_out = 80 + 273.15;    % Air inlet temperature
phi = 0.5;                  % Relative humidity
P = 120;                    % Pressure (kPa)
n_cell = 1;                 % No. of cells
Power = 500;                % Power (watts)
V = 0.7;                    % Voltage (V)
M_H2O = 18.015;             % Molecular weight of water
F = 96,485;                 % Faraday's law
S_O2 = 2;                   % Oxygen stoichiometric ratio
r_O2 = 0.2095;              % Mole fraction of oxygen in air
r_O2_in = r_O2;

%%%%%%%%%%%%%%%%%%%%%%%%%%%%%%%%%%%%%%%%%
% Calculate the current

I = Power/V;
n = 4;
Pvs_in = Pvs(T_H2_in);      % Calculate the saturation pressure at T_H2_in
```

```matlab
% Calculate the amount of water in air
m_h2o_air_in = S_O2*phi*M_H2O*Pvs(T_H2_in)*I*n_cell/ (r_O2*n*F*(P -
   phi*Pvs_in));
n = 2;
% Calculate the water generated
m_h2o_gen = I*M_H2O/(n*F);
DeltaPca = 18.675;
% Calculate the saturation pressure at T_air_out
Pvs_out = Pvs(T_air_out);
% Calculate the water vapor in the air outlet
m_h2o_in_air_out = (S_O2 - r_O2_in)/r_O2_in*M_H2O/(4*F)*Pvs_out/(P -
   DeltaPca - Pvs_out)*I*n_cell;
% The water injected flow rate is:
m_h2o_inject = m_h2o_in_air_out - m_h2o_air_in - m_h2o_gen;
%%%%%%%%%%%%%%%%%%%%%%%%%%%%%%%%%%%%%%%%%%
function result = Pvs(T);
T = T + 273.15;
a = -5800.2206;
b = 1.3914993;
c = -0.048640239;
d = 0.000041764768;
e = -0.000000014452093;
f = 6.5459673;
result = exp(a/T + b + c*T + d*T*T + e*T*T*T + f*log(T))/1000;
```

EXAMPLE 5-2: Calculating Mass Flow Rates

A PEM fuel cell with 100 cm^2 of active area is operating at 0.50 A/cm^2 at a voltage of 0.70. The operating temperature is 75 °C and 1 atm with air supplied at a stoichiometric ratio of 2.5. The air is humidified by injecting hot water (75 °C) before the stack inlet. The ambient air conditions are 1 atm, 22 °C, and 70% relative humidity. Calculate the air flow rate, the amount of water required for 100% humidification of air at the inlet, and the heat required for humidification.

The oxygen consumption is

$$N_{O2} = \frac{I}{4F} = \frac{0.50 \text{A/cm}^2 \times 100 \text{cm}^2}{4 \times 96,485} = 0.129 \times 10^{-3} \text{mol/s}$$

The oxygen flow rate at the cell inlet is

$$N_{O2} = SN_{O2,cons} = 2.5 \times 0.129 \times 10^{-3} = 0.324 \times 10^{-3} \text{mol/s}$$

$$N_{air} = N_{O2,act}\frac{1}{r_{O2}} = \frac{0.324 \times 10^{-3}}{0.21} = 1.54 \times 10^{-3} \text{mol/s}$$

$$m_{air} = N_{air}m_{air} = 1.54 \times 10^{-3}\text{mol/s} \times 28.85\text{g/mol} = 0.0445\text{mol/s}$$

The amount of water in air at the cell inlet where $\varphi = 1$ is

$$m_{H2O} = x_s m_{air} \quad \text{and} \quad x_s = \frac{m_{H2O}}{m_{air}}\frac{p_{vs}}{P - p_{vs}}$$

where p_{vs} is the saturation pressure at 348.15 K, and P is the total pressure (101.325 kPa).

$$p_{vs} = e^{aT^{-1}+b+cT+dT^2+eT^3+f\ln(T)} = 38.6\text{kPa}$$

$$x_s = \frac{m_{H2O}}{m_{air}}\frac{p_{vs}}{P - p_{vs}} = \frac{18}{28.85}\frac{38.6}{(101.325 - 38.6)} = 0.384\text{g}_{H2O}/\text{g}_{air}$$

$$m_{H2O} = x_s m_{air} = 0.384\text{g}_{H2O}/\text{g}_{air} \times 0.128\text{g}_{air}/\text{s} = 0.0491\text{g}_{H2O}/\text{s}$$

The amount of water in ambient air at 70% RH and 295.15 K is

$$x_s = \frac{m_{H2O}}{m_{air}} = \frac{\varphi p_{vs}}{P - \varphi p_{vs}} = \frac{18}{28.85}\frac{0.70 \times 2.645}{101.325 - 0.7 \times 2.645} = 0.011\,\text{g}_{H2O}/\text{g}_{air}$$

$$m_{H2O} = x_s m_{air} = 0.0116\text{g}_{H2O}/\text{g}_{air} \times 0.128\text{g}_{air}/\text{s} = 0.001486\text{g}_{H2O}/\text{s}$$

The amount of water needed for humidification of air is

$$m_{H2O} = 0.0491 - 0.001486 = 0.047614\text{g}_{H2O}/\text{s}$$

The heat required for humidification can be calculated from the heat balance.

$$H_{air,in} + H_{H2O,in} + Q = H_{air,out}$$

The enthalpy of wet/moist air is

$$h_{vair} = c_{p,air}t + x(c_{p,v}t + h_{fg})$$

Humidified air:

$$h_{vair} = 1.01 \times 75 + 0.384 \times (1.87 \times 75 + 2500) = 1\,089.61J/g$$

Ambient air:

$$h_{vair} = 1.01 \times 22 + 0.0116 \times (1.87 \times 22 + 2500) = 51.70J/g$$

Water:

$$h_{H2O} = 4.18 \times 75 = 313.5J/g$$

$$Q = 1089.61J/g \times 0.128g/s - 51.70J/g \times 0.128g/s - 313.5J/g \times 0.0157g/s$$
$$= 127.93W$$

Using MATLAB to solve:

```
%%%%%%%%%%%%%%%%%%%%%%%%%%%%%%%%%%%%%%%%%%%
% EXAMPLE 5-2: Calculating Mass Flow Rates
%%%%%%%%%%%%%%%%%%%%%%%%%%%%%%%%%%%%%%%%%%%
% Inputs
area = 100;                    % Fuel cell active area (cm^2)
i = 0.5;                       % Current density (A/cm^2)
stoichiometric_ratio = 2.5;    % Stoichiometric ratio
T_operating = 75 + 273.15;     % Operating temperature
T_ambient_air = 22 + 273.15;   % Temperature of ambient air
phi_ambient = 0.7;             % Relative humidity
P = 101.325;                   % Pressure
M_H2O = 18.015;                % Molecular weight of water
M_AIR = 28.85;                 % Molecular weight of air
r_O2 = 0.2095;                 % Percent of oxygen in air
F = 96485;                     % Faraday's constant
%%%%%%%%%%%%%%%%%%%%%%%%%%%%%%%%%%%%%%%%%%%
% Calculate current
I = i*area;
% Oxygen consumption
N_O2_CONS = I/(4*F);
% The oxygen flow rate at the cell inlet
N_O2_Act = stoichiometric_ratio*N_O2_CONS;
```

% The air flow rate at the cell inlet

N_Air = N_O2_Act/r_O2;

% The molar flow rate of air is

m_air = M_AIR * N_Air;

% The saturation pressure at the operating temperature

Pvs_op = Pvs(T_operating);

% Mole fraction of water in air

x_s_op = M_H2O/M_AIR * Pvs_op/(P − Pvs_op);
m_air2 = 0.128;

% Amount of water needed for the humidification of air

m_h2o_in_air = x_s_op * m_air2;

% Saturation pressure of air

Pvs_amb = Pvs(T_ambient_air);

% Mole fraction of water in ambient air

x_s_amb = (M_H2O/M_AIR) * phi_ambient * Pvs_amb/(P − phi_ambient * Pvs_amb);

% Mass of water in ambient air

m_h2o_in_amb_air = x_s_amb * m_air2;

% Amount of water needed for humidification of air

m_h2o_needed = m_h2o_in_air − m_h2o_in_amb_air;
c_p_air = 1.01;
c_p_v = 1.87;
c_p_water = 4.18;
h_fg = 2500;

% Heat required for humidified and ambient air, and water

h_humidified_air = c_p_air * T_operating + x_s_op * (c_p_v * T_operating + h_fg);
h_ambient_air = c_p_air * T_ambient_air + x_s_amb * (c_p_v * T_ambient_air + h_fg);
h_water = c_p_water * T_operating;

% Total heat required

Q = h_humidified_air * 0.128 − h_ambient_air * 0.128 − h_water * 0.0157;

```
%%%%%%%%%%%%%%%%%%%%%%%%%%%%%%%%%%%%%%%%%%%%%%%%
% Calculate the situation pressure a) the operating temp
function result = Pvs(T);
T = T + 273.15;
a = -5800.2206;
b = 1.3914993;
c = -20.048640239;
d = 0.000041764768;
e = -0.000000014452093;
f = 6.5459673;
result = exp(a/T + b + c*T + d*T*T + e*T*T*T + f*log(T))/1000;
```

5.3 Convective Mass Transport from Flow Channels to Electrode

Figure 5-2 illustrates convective flow in the reactant flow channel and diffusive flow through the gas diffusion and catalyst layers. The reactant is supplied to the flow channel at a concentration C_0, and it is transported from the flow channel to the concentration at the electrode surface, C_s, through convection. The rate of mass transfer is then:

$$\dot{m} = A_{elec}h_m(C_0 - C_s) \qquad (5\text{-}21)$$

where A_{elec} is the electrode surface area, and h_m is the mass transfer coefficient.

The value of h_m is dependent upon the wall conditions, the channel geometry, and the physical properties of species i and j. H_m can be found from the Sherwood number:

$$h_m = Sh\frac{D_{i,j}}{D_h} \qquad (5\text{-}22)$$

where Sh is the Sherwood number, D_h is the hydraulic diameter, and D_{ij} is the binary diffusion coefficient for species i and j. The Sherwood number depends upon channel geometry, and can be expressed as:

$$Sh \equiv \frac{h_H D_h}{k} \qquad (5\text{-}23)$$

where Sh = 5.39 for uniform surface mass flux ($\dot{m}$ = constant), and Sh = 4.86 for uniform surface concentration (C_s = constant).

The binary diffusion coefficient for hydrogen, oxygen, and water is given in Appendix G. If the binary diffusion coefficient needs to be calcu-

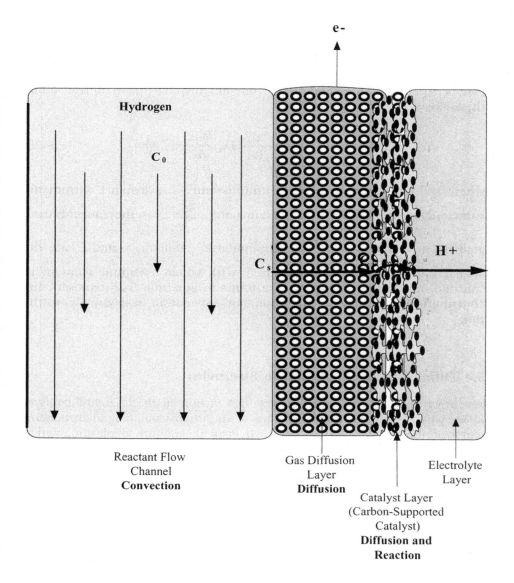

FIGURE 5-2. Fuel cell layers (flow field, gas diffusion layer, catalyst layer) that have convective and diffusive mass transport.

lated at a different temperature than what is shown in Appendix G, the following relation can be used:

$$D_{i,j}(T) = D_{i,j}(T_{ref}) * \left(\frac{T}{T_{ref}} \right)^{3/2} \qquad (5\text{-}24)$$

where T_{ref} is the temperature that the binary diffusion coefficient is given at, and T is the temperature of the fuel.

There are several equations that are commonly used in the literature that govern the mass transfer to the electrode. One of these is the Nernst-Planck equation. This equation for one-dimensional mass transfer along the x-axis is:

$$J_i(x) = -D_i\frac{\partial C_i(x)}{\partial x} - \frac{z_i F}{RT}D_i C_i\frac{\partial \phi(x)}{\partial x} + C_i v(x) \tag{5-25}$$

where $J_i(x)$ is the flux of species i $(mol/(s*cm^2))$ at a distance x from the surface, D_i is the diffusion coefficient (cm^2), $\frac{\partial C_i(x)}{\partial x}$ is the concentration gradient at distance x, $\frac{\partial \phi(x)}{\partial x}$ is the potential gradient, z_i and C_i are the charge, and v(x) is the velocity (cm/s) with which a volume element in solution moves along the axis. The terms in Equation 5-25 represent the contributions to diffusion, migration, and convection, respectively, to the flux.

5.4 Diffusive Mass Transport in Electrodes

As shown in Figure 5-2, the diffusive flow occurs at the GDL and catalyst layer, where the mass transfer occurs at the microlevel. The electrochemical reaction in the catalyst layer can lead to reactant depletion, which can affect fuel cell performance through concentration losses. In turn, the reactant depletion will also cause activation losses. The difference in the catalyst layer reactant and product concentration from the bulk values determines the extent of the concentration loss.

Using Fick's law, the rate of mass transfer by diffusion of the reactants to the catalyst layer $(\dot{m})$ can be calculated as shown in Equation 5-26:

$$\dot{m} = -D\frac{dC}{dx} \tag{5-26}$$

where D is the bulk diffusion coefficient, and C is the concentration of reactants.

The diffusional transport through the gas diffusion layer at steady-state is:

$$\dot{m} = A_{elec}D^{eff}\frac{C_s - \dot{C}_i}{\delta} \tag{5-27}$$

where C_i is the reactant concentration at the GDL/catalyst interface, δ is the gas diffusion layer thickness, and D^{eff} is the effective diffusion co-efficient for the porous GDL, which is dependent upon the bulk diffusion coefficient D, and the pore structure. Assuming uniform pore size, and that gas diffusion layer is free from flooding of water, D^{eff} can be defined as:

$$D^{eff} = D\phi^{3/2} \tag{5-28}$$

where ϕ is the electrode porosity. The total resistance to the transport of the reactant to the reaction sites can be expressed by combining Equations 5-21 and 5-27:

$$\dot{m} = \frac{C_0 - C_i}{\left(\dfrac{1}{h_m A_{elec}} + \dfrac{\delta}{D^{eff} A_{elec}} \right)} \tag{5-29}$$

where $\dfrac{1}{h_m A_{elec}}$ is the resistance to the convective mass transfer, and $\dfrac{L}{D^{eff} A_{elec}}$ is the resistance to the diffusional mass transfer through the gas diffusion layer.

When a fuel cell is started up, it begins producing electricity at a fixed current density, i. The reactant and product concentrations in the fuel cell are constant. When the current begins to be produced, the electrochemical reaction leads to the depletion of reactants at the catalyst layer. The flux of reactants and products will match the consumption/depletion rate of reactants and products at the catalyst layer, and can be described using the following equation:

$$i = \frac{nF\dot{m}}{A_{elec}} \tag{5-30}$$

where i is the fuel cell operating current density, F is the Faraday constant, n is the number of electrons transferred per mol of reactant consumed, and $\dot{m}$ is the rate of mass transfer by diffusion of reactants to the catalyst layer. Substituting Equation 5-29 into Equation 5-30 yields:

$$i = -nF\frac{C_0 - C_i}{\left(\dfrac{1}{h_m} + \dfrac{\delta}{D^{eff}} \right)} \tag{5-31}$$

The reactant concentration in the GDL/catalyst interface is less than the reactant concentration in the flow channels, which depends upon i, δ, and D^{eff}. As the current density increases, the concentration losses become greater. These concentration losses can be improved if the GDL thickness is reduced, or the porosity or effective diffusivity is increased.

The limiting current density of the fuel cell occurs when the current density becomes so large the reactant concentration falls to zero. The limiting current density (i_L) can be calculated if the minimum concentration at the GDL/catalyst layer interface is $C_i = 0$ as follows:

$$i_L = -nF\frac{C_0}{\left(\dfrac{1}{h_m} + \dfrac{\delta}{D^{eff}}\right)} \tag{5-32}$$

The limiting current density can be increased by insuring that C_0 is high through good flow field design, optimal GDL and catalyst layer porosity and thickness, and ideal operating conditions. The limiting current density is from 1 to 10 A/cm^2. The fuel cell cannot produce a higher current density than its limiting current density. However, the fuel cell voltage may fall to zero due to other types of losses before the limiting current density does.

The Nernst equation introduced in Chapter 2 shows the relationship between the thermodynamic voltage of the fuel cell, and the reactant and product concentrations at the catalyst sites:

$$E = E_r - \frac{RT}{nF}\ln\frac{\prod a_{products}^{v_i}}{\prod a_{reactants}^{v_i}} \tag{5-33}$$

In order to calculate the incremental voltage loss due to reactant depletion in the catalyst layer, the changes in Nernst potential using c_R^* values instead of c_R^0 values are represented as follows:

$$\upsilon_{conc} = E_{r,Nernst} - E_{Nernst} \tag{5-34}$$

$$\upsilon_{conc} = \left(E_r - \frac{RT}{nF}\ln\frac{1}{C_0}\right) - \left(E_r - \frac{RT}{nF}\ln\frac{1}{C_i}\right) \tag{5-35}$$

$$\upsilon_{conc} = \frac{RT}{nF}\ln\frac{C_0}{C_i} \tag{5-36}$$

where $E_{r,Nerst}$ is the Nernst voltage using C_0 values, and E_{Nernst} is the Nernst voltage using C_i values. Combining Equations 5-35 and 5-36:

$$\frac{i}{i_L} = 1 - \frac{C_i}{C_0} \tag{5-37}$$

Therefore, the ratio C_0/C_i (the concentration at the GDL/catalyst layer interface) can be written as:

$$\frac{C_0}{C_i} = \frac{i_L}{i_L - i}$$

(5-38)

Substituting Equation 5-38 into Equation 5-36 yields:

$$v_{conc} = \frac{RT}{nF} \ln\left(\frac{i_L}{i_L - i}\right)$$

(5-39)

which is the expression for concentration losses, and is only valid for $i < i_L$.

The Butler-Volmer equation from Chapter 3 describes how the reaction kinetics affect concentration and fuel cell performance. The reaction kinetics are dependent upon the reactant and product concentrations at the reaction sites:

$$i = i_0 \frac{c_R^\star}{C_R^{0\star}} \exp(\alpha nF v_{act}/(RT)) - \frac{c_P^\star}{c_P^{0\star}} \exp(-(1-\alpha)nF v_{act}/(RT))$$

(5-40)

where $c_R^\star$ and $c_P^\star$ are arbitrary concentrations, and i_0 is measured as the reference reactant and product concentration values $c_R^{0\star}$ and $c_P^{0\star}$. In the high current-density region, the second term in the Butler-Volmer equation drops out, and the expression then becomes:

$$i = i_0 \frac{c_R^\star}{c_R^{0\star}} \exp(\alpha nF v_{act}/(RT))$$

(5-41)

In terms of activation overvoltage using $c_R^\star$ instead of $c_R^{0\star}$:

$$v_{conc} = \frac{RT}{\alpha nF} \frac{c_R^{0\star}}{c_R^\star}$$

(5-42)

The ratio can be written as:

$$\frac{c_R^0}{c_R^\star} = \frac{i_L}{i_L - i}$$

(5-43)

The total concentration loss can be written as:

$$v_{conc} = \left(\frac{RT}{nF}\right)\left(1 + \frac{1}{\alpha}\right)\frac{i_L}{i_L - i}$$

(5-44)

Fuel cell mass transport losses may be expressed by the following equation:

$$v_{conc} = c \ln \frac{i_L}{i_L - i}$$

(5-45)

where c is a constant and can have the approximate form:

$$c = \frac{RT}{nF}\left(1 + \frac{1}{\alpha}\right) \qquad (5\text{-}46)$$

In actuality, the fuel cell behavior often has a larger value than what the equation predicts. Due to this, c is often obtained empirically. The concentration losses appear at high current densities, and significant concentration losses can severely limit fuel cell performance.

5.5 Convective Mass Transport in Flow Field Plates

The flow channels in fuel cell flow field plates are designed to evenly distribute reactants across a fuel cell to help keep mass transport losses to a minimum. Flow field designs are discussed in detail in Chapter 10. In the next section, the control volume method first introduced in Section 5.2 is used to calculate the mass transfer rates in the flow channels.

5.5.1 Mass Transport in Flow Channels

The mass transport in flow channels can be modeled using a control volume for reactant flow from the flow channel to the electrode layer as shown in Figure 5-3.

The rate of convective mass transfer at the electrode surface ($\dot{m}_s$) can be expressed as:

$$\dot{m}_s = h_m(C_m - C_s) \qquad (5\text{-}47)$$

where C_m is the mean concentration of the reactant in the flow channel (averaged over the channel cross-section, and decreases along the flow direction, x), and C_s is the concentration at the electrode surface.

As shown in Figure 5-3, the reactant moves at the molar flow rate, $A_c C_m v_m$ at the position x, where A_c is the channel cross-sectional area and v_m is the mean flow velocity in the flow channel. This can be expressed as:

$$\frac{d}{dx}(A_c C_m v_m) = -\dot{m}_s w_{elec} \qquad (5\text{-}48)$$

where w_{elec} is the width of the electrode surface. If the flow in the channel is assumed to be steady, and the velocity and the concentration are constant, then:

$$\frac{d}{dx}C_m = \frac{-\dot{m}_s}{v_m w_{flow}} \qquad (5\text{-}49)$$

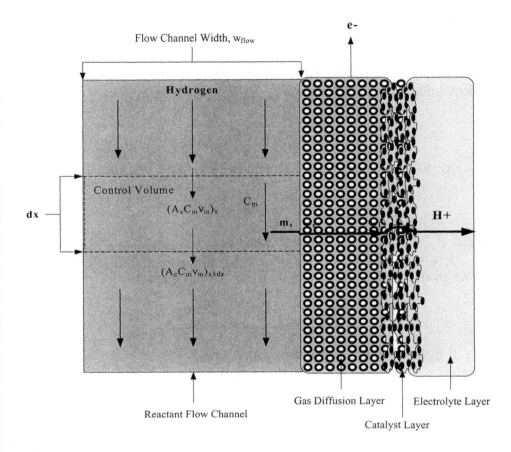

FIGURE 5-3. Control volume for reactant flow from the flow channel to the electrode layer.

When the current density is small ($i < 0.5\ i_L$), it can be assumed constant. Using Faraday's law, $\dot{m}_s = \dfrac{i}{nF}$ and integrating:

$$C_m(x) = C_{m,in}(x) - \frac{\left(\dfrac{i}{nF}\right)}{v_m w_{flow}} x \qquad (5\text{-}50)$$

where $C_{m,in}$ is the mean concentration at the flow channel inlet.

If the current density is large ($i > 0.5\ i_L$), the condition at the electrode surface can be approximated by assuming the concentration at the surface (C_s) is constant. This can be written as follows:

$$\frac{d}{dx}(C_m - C_s) = \frac{-h_m}{v_m W_{flow}}(C_m - C_s) \tag{5-51}$$

After integrating from the channel inlet to location x in the flow channel, Equation 5-51 becomes:

$$\frac{C_m - C_s}{(C_m - C_s)_{in}} = \exp\frac{-h_m x}{v_m W_{flow}} \tag{5-52}$$

At the channel outlet, x = H, and Equation 5-52 becomes:

$$\frac{C_{m,out} - C_s}{C_{m,in} - C_s} = \exp\frac{-h_m H}{v_m W_{flow}} \tag{5-53}$$

where $C_{m,out}$ is the mean concentration at the flow channel outlet.

A simple expression can be derived if the entire flow channel is assumed to be the control volume, as shown in Figure 5-4:

$$\dot{m}_s = v_m W_{flow} W_{elec}(C_{in} - C_{out})$$
$$\dot{m}_s = v_m W_{flow} W_{elec}(\Delta C_{in} - \Delta C_{out}) \tag{5-54}$$

If C_s is constant, substituting for $W_{flow} W_{elec}$:

$$\dot{m}_s = A h_m \Delta C_{lm} \tag{5-55}$$

where

$$\Delta C_{lm} = \frac{\Delta C_{in} - \Delta C_{out}}{\ln\left(\dfrac{\Delta C_{in}}{\Delta C_{out}}\right)} \tag{5-56}$$

The local current density corresponding to the rate of mass transfer is:

$$i(x) = nFh_m(C_m - C_s)\exp\left(\frac{-h_m x}{v_m W_{flow}}\right) \tag{5-57}$$

The current density averaged over the electrode surface is:

$$\bar{i} = nFh_m \Delta C_{lm} \tag{5-58}$$

The limiting current density when C_s approaches 0 is:

$$i_L(x) = nFh_m C_{m,in} \exp\left(\frac{-h_m x}{v_m W_{flow}}\right) \tag{5-59}$$

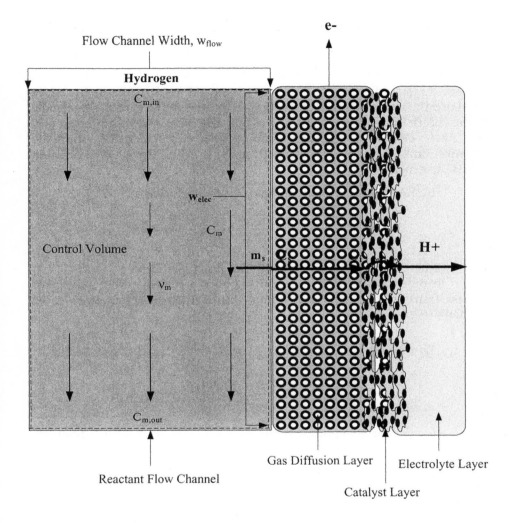

FIGURE 5-4. Entire channel as the control volume for reactant flow from the flow channel to the electrode layer.

$$\bar{i}_L = nFh_m \left[\frac{\Delta C_{in} - \Delta C_{out}}{\ln\left(\dfrac{\Delta C_{in}}{\Delta C_{out}}\right)} \right] \qquad (5\text{-}60)$$

As seen from Equations 5-57 to 5-60, both the current density and limiting current density decrease exponentially along the channel length.

EXAMPLE 5-3: Determine Current Density Distribution

A fuel cell operating at 25 °C and 1 atm uses bipolar plates with flow fields to distribute the fuel and oxidant to the electrode surface. The channels have a depth of 1.5 mm, with a distance of 1 mm apart. Air is fed parallel to the channel walls for distribution to the cathode electrode. The length of the flow channel is 18 cm, and the air travels at a velocity of 2 m/s. Determine the distribution of the current density due to the limitation of the convective mass transfer $i_L(x)$ and the average limiting current density $\bar{i}_L$.

The Reynolds number can be calculated as follows:

$$\mathrm{Re} = \frac{\rho v_m D}{\mu} = \frac{v_m D}{v} = \frac{2\mathrm{m/s}*2*1\times10^{-3}\mathrm{m}}{15.89\times10^{-6}\mathrm{m}^2/\mathrm{s}} = 251.73$$

Since 251.73 is less than 2000, the flow is laminar.

In order to calculate the limiting current density, the convective mass transfer coefficient and binary diffusivity coefficient need to be calculated:

$$D_{i,j}(T) = D_{i,j}(T_{ref})*\left(\frac{T}{T_{ref}}\right)^{3/2} = D_{O2-N2}(273)*\left(\frac{298}{273}\right)^{3/2} = 0.21\times10^{-4}\mathrm{m}^2/\mathrm{s}$$

$$h_m = \mathrm{Sh}\frac{D_{i,j}}{D_h} = 4.86*\frac{0.21\times10^{-4}\mathrm{m}^2/\mathrm{s}}{2*1\times10^{-3}\mathrm{m}} = 0.0105\mathrm{m/s}$$

The concentration of O_2 at the channel inlet, with a mole fraction of O_2 is $X_{O2} = 0.21$, is:

$$C_{O2,in} = X_{O2}*\left(\frac{P}{RT}\right) = 0.21*\frac{101,325}{8.314\mathrm{J/molK}*298\mathrm{K}} = 8.588\mathrm{mol/m}^3$$

The limiting current density based upon the rate of O_2 transfer:

$$i_L(x) = nFh_m C_{m,in}\exp\left(\frac{-h_m x}{v_m W_{flow}}\right)$$

$$i_L(x) = 4*96,487\frac{C}{\mathrm{molO}_2}*0.0105\frac{m}{s}*8.588\frac{\mathrm{molO}_2}{m^3}*$$
$$\exp\left(\frac{0.0105\mathrm{m/s}}{1*10^{-3}*2\mathrm{m/s}}\right) = 3.4802*10^4\times\exp(-5.25x)$$

The limiting current density will be 3.4802 A/cm² at the channel inlet $(x = 0)$ and 3.3022 A/cm² at the channel outlet $(x = 18$ cm$)$. In order to calculate $\bar{i}_L$, the outlet concentration of oxygen needs to be calculated:

$$C_{m,out} = C_{m,in} * \exp\left(\frac{-h_m x}{v_m W_{flow}}\right)$$

$$= 8.588\frac{mol}{m^3} * \exp\left(\frac{-0.0105m/s * 0.18m}{1*10^{-3} * 2m/s}\right)$$

$$= 3.338 mol/m^3$$

Then,

$$\bar{i}_L = nFh_m\left[\frac{\Delta C_{in} - \Delta C_{out}}{\ln\left(\frac{\Delta C_{in}}{\Delta C_{out}}\right)}\right]$$

$$= 4 * 96,487\frac{C}{molO_2} * 0.0105\frac{m}{s}\left[\frac{(8.588 - 3.338)\frac{molO_2}{m^3}}{\ln\left(\frac{8.588}{3.338}\right)}\right]$$

$$= 2.251\frac{A}{cm^2}$$

Using MATLAB to solve:

```
%%%%%%%%%%%%%%%%%%%%%%%%%%%%%%%%%%%%%%%%%
% EXAMPLE 5-3: Determining the Current Density Distribution
%%%%%%%%%%%%%%%%%%%%%%%%%%%%%%%%%%%%%%%%%
% Inputs

T = 25;                             % Fuel cell operating temperature in degrees C
gap_distance = 0.001;               % Rib distance (m)
channel_length = 0.18;              % Channel length (m)
air_velocity = 2;                   % Air velocity (m/s)
T = T + 273.15;                     % Convert temperature to K
reference_temperature = 273        % Reference temperature
meu = 15.89e-6;
D_O2_N2_at_reference = 1.84e-5;    %Diffusion coefficient
Sh = 4.86;                          % Sherwood number
X_O2 = 0.21;                        % Mole fraction of oxygen
D_k = 2/1000;
```

```
R = 8.314;                    % Ideal gas constant
P = 101.325;                  % Pressure
v_m = 2;
n = 4;
F = 96487;                    % Faraday's constant
%%%%%%%%%%%%%%%%%%%%%%%%%%%%%%%%%%%%%%%%%%%%
% Calculate Reynold's number

RE = air_velocity * v_m * gap_distance/meu

% Convective mass transfer coefficient

D_o2_n2 = D_O2_N2_at_reference * (temperature/reference_temperature)^1.5

% Binary diffusivity coefficient

h_m = Sh * D_o2_n2/D_k

% The concentration of oxygen at the channel inlet

C_O2_in = 1000 * X_O2 * P/(R * temperature)  % 1000 * is to convert from molK
   to mol
v_m = 10e-3

% The concentration of oxygen at the channel outlet

C_O2_out = C_O2_in * exp(-h_m * channel_length/(v_m * air_velocity))

% Limiting current density

limiting_current_density = n * F * h_m * (C_O2_in − C_O2_out)/log(C_O2_in/C_O2_
   out)/10000;  % /10000 to convert to cm2
```

5.6 Mass Transport Equations in the Literature

There are many equations used in the literature to determine mass flux and concentration losses. Knowing which equation to use is not as important as determining how to model the system effectively. Knowing how to solve the concentration gradients and species distributions requires knowledge of multicomponent diffusion, and can be a challenging task.

In order to precisely solve the mass balance equations (especially in the electrode layers), the mass flux must be determined. The concentration losses are incorporated into a model as the reversible potential decreases due to a decrease in the reactant's partial pressure. There are three basic approaches for determining the mass flux (N): Fick's law, the Stefan-Maxwell equation, and the Dusty Gas Model.

5.6.1 Fick's Law

The simplest diffusion model is Fick's law, which is used to describe diffusion processes involving two gas species. A form of Fick's law was introduced in Equation 5-26. The standard notation for Fick's law is the binary notation, and can be written as:

$$N_i = -cD_{i,j}\nabla X_i \qquad (5\text{-}61)$$

A multicomponent version of Fick's law is shown in Equation 5-62:

$$N_i = -cD_{i,m}\nabla X_i + X_i \sum_{j=1}^{n} N_j \qquad (5\text{-}62)$$

where c is the total molar concentration. If three or more gas species are present, such as N_2, O_2, and H_2O, a multicomponent diffusion model such as the Maxwell-Stefan equation must be used, or the binary diffusion coefficients must be expanded to tertiary diffusion coefficients.

5.6.2 The Stefan-Maxwell Equation

The Stefan-Maxwell equation is the only diffusion equation that separates diffusion from convection in a simple way. The flux equation is replaced by the difference in species velocities. The Stefan-Maxwell model is more rigorous, is commonly used in multicomponent species systems, and is employed quite extensively in the literature. The main disadvantage is that it is difficult to solve mathematically. It may be used to define the gradient in the mole fraction of components:

$$\nabla y_i = RT\sum \frac{y_i N_j - y_j N_i}{pD_{ij}^{eff}} \qquad (5\text{-}63)$$

where y_i is the gas phase mol fraction of species i, and N_i is the superficial gas phase flux of species i averaged over a differential volume element, which is small with respect to the overall dimensions of the system, but large with respect to the pore size. D_{ij}^{eff} is the binary diffusion coefficient, and can be defined by:

$$D_{ij}^{eff} = \frac{a}{p}\left(\frac{T}{\sqrt{T_{c,i}T_{c,j}}}\right)^b (p_{c,i}p_{c,j})^{1/3}(T_{c,i}T_{c,j})^{5/12}\left(\frac{1}{M} + \frac{1}{M_j}\right)^{1/2}\varepsilon^{1.5} \qquad (5\text{-}64)$$

where T_c and p_c are the critical temperature and pressure of species i and j, M is the molecular weight of species, A = 0.0002745 for diatomic gases, H_2, O_2, and N_2, and a = 0.000364 for water vapor, and B = 1.832 for diatomic gases, H_2, O_2, and N_2, and b = 2.334 for water vapor. The Stefa-Maxwell Equation is discussed in more detail in Chapter 8.

5.6.3 The Dusty Gas Model

The Dusty Gas Model is also commonly used in the literature, and looks similar to the Stefan-Maxwell equation except that it also takes into account Knudsen diffusion. Knudsen diffusion occurs when a particle's mean-free-path is similar to, or larger than in size, the average pore diameter (and is discussed in greater detail in Chapter 8):

$$-\nabla X_i = \frac{N_i}{D_{i,k}} + \sum_{j=1,j\neq i}^{n} \frac{X_j N_i - X_i N_j}{cD_{i,j}} \tag{5-65}$$

where $D_{i,j}$ is the Knudsen diffusion coefficient for species i. The molecular diffusivity depends upon the temperature, pressure, and concentration. The effective diffusivity depends also upon the microstructural parameters such as porosity, pore size, particle size, and tortuosity. The molecular gas diffusivity must be corrected for the porous media. A large portion of the corrections are made using the ratio of porosity to tortuosity (E/T), although in some cases, the Bruggman model is used due to the lack of information for gas transport in porous media:

$$D_{i,j}^{eff} = \left(\frac{\varepsilon}{\tau}\right)D_{i,j} \quad D_{i,j}^{eff} = \varepsilon^{1.5}D_{i,j} \tag{5-66}$$

The Dusty Gas diffusion model requires Knudsen diffusivity to be solved, while Fick's law and the Stefan-Maxwell equation require more work to incorporate Knudsen diffusion.

The Knudsen diffusion coefficient for gas species i can be calculated using Equation 5-67:

$$D_{i,k} = \frac{2\bar{r}}{3}\sqrt{\frac{8RT}{\pi M_i}} \tag{5-67}$$

where M is the molecular mass of species i, and r is the average pore radius.

EXAMPLE 5-4: Calculating the Diffusive Mass Flux

Hydrogen gas is maintained at 2 bars and 1 bar on opposite sides of a Nafion membrane that is 50 microns (μm) thick. The temperature is 20°C, and the binary diffusion coefficient of hydrogen in Nafion is 8.7×10^{-8} m^2/s. The solubility of hydrogen in the membrane is 1.5×10^{-3} kmol/m^3 bar. What is the mass diffusive flux of hydrogen through the membrane? What are the molar concentrations of hydrogen in the gas phase?

First, calculate the surface molar concentrations of hydrogen:

$$C_{A,s1} = S*p_A = 1.5 \times 10^{-3}\frac{kmol}{m^3} \times 3 bars = 4.5 \times 10^{-3}\frac{kmol}{m^3}$$

$$C_{A,s1} = S*p_A = 1.5 \times 10^{-3}\frac{kmol}{m^3} \times 1 bars = 1.5 \times 10^{-3}\frac{kmol}{m^3}$$

Calculate the molar diffusive flux:

$$N_A = \frac{D_{AB}}{L}(C_{A,s1} - C_{A,s2})$$

$$N_A = \frac{8.7 \times 10^{-8}}{0.3 \times 10^{-3}}(4.5 \times 10^{-3} - 1.5 \times 10^{-3}) = 8.7 \times 10^{-7}\frac{kmol}{s}m^2$$

Convert to a mass basis:

$$n_A = N_A*M_A = 8.7 \times 10^{-7}\frac{kmol}{s}*2\frac{kg}{kmol} = 1.74 \times 10^{-6}\frac{kg}{g}m^2$$

Calculate the molar concentrations of hydrogen in the gas phase:

$$C_A = \frac{p_A}{RT} = \frac{3}{8.314 \times 10^{-2}*293.15} = 0.121\frac{kmol}{m^3}$$

$$C_C = \frac{p_A}{RT} = \frac{1}{8.314 \times 10^{-2}*293.15} = 0.040\frac{kmol}{m^3}$$

Using MATLAB to solve:

%%%%%%%%%%%%%%%%%%%%%%%%%%%%%%%%%%%%%%

% EXAMPLE 5-4: Calculating the Mass Diffusive Flux through % the Membrane

%%%%%%%%%%%%%%%%%%%%%%%%%%%%%%%%%%%%%%

% Inputs

```
Pa = 3;              % Pressure on the anode side (bar)
Pc = 1;              % Pressure on the cathode side (bar)
D = 8.7*10^-8;       % Binary diffusion coefficent (m^2/s)
L = 0.3*10^-3;       % Thickness of the membrane (m)
```

```
M = 2;                  % (kg/kmol)
R = 8.314*10^-2;        % Ideal gas constant (m^3*bar/kmolK)
S = 1.5*10^-3;          % Solubility of hydrogen (kmol/m^3-bar)
T = 20 + 275.15;        % Temperature (K)
```

%%%

% The surface molar concentrations of hydrogen

```
C_as1 = S*Pa;
C_cs2 = S*Pc;
```

% Calculate the molar diffusive flux

```
N = (D/L)*(C_as1 – C_cs2);
```

% Convert to mass basis

```
n = N*M
```

% Calculate the molar concentrations of hydrogen in the gas phase

```
Ca = Pa/(R*T)
Cc = Pc/(R*T)
```

Chapter Summary

The study of mass transport involves the supply of reactants and products in a fuel cell. Inadequate mass transport can result in poor fuel cell performance. In order to calculate the mass flows through the fuel cell, mass balances can be written to calculate the ideal flow rates and mole fractions of any unknown species. There are two main mass transport effects encountered in fuel cells: convection in the flow structures, and diffusion in the electrodes. Convective flow occurs in the flow channels due to hydrodynamic transport, and the relatively large-size channels (~1 mm to 1 cm). Diffusive transport occurs in the electrodes because of the tiny pore sizes. Mass transport losses in the fuel cell result in the depletion of reactants at the electrode, which affects the Nernstian cell voltage and the reaction rate. Commonly used mass transport equations in the literature include Fick's law, the Stefan-Maxwell equation and the Dusty Gas Model. This chapter provided the necessary background to create mass balances on any fuel-cell component with convective or diffusive transport.

Problems

- A fuel cell is operating at 50 °C and 1 atm. Humidified air is supplied with the mole fraction of water vapor equal to 0.2 in the cathode. If the channels are rectangular with a diameter of 1.2 mm, find the maximum velocity of air.

- For the fuel cell in the above problem, calculate the maximum velocity of air if the channels are circular.
- A fuel cell is operating at 50°C and 1 atm. The cathode is using pure oxygen, and there is no water vapor present. The diffusion layer is 400 microns with a porosity of 30%. Calculate the limiting current density.
- Calculate the limiting current density for a fuel cell operating at 80°C and 1 atm. The cathode is of the same construction as in the third problem.
- Under the conditions from the third problem, estimate the fuel cell area that can be operated at 0.7 A/cm². Assume a stoichiometric number of 2.5, and that the fuel cell is made of a single straight channel with a width of 1 mm and the rib width is 0.5 mm.

Bibliography

Barbir, F. *PEM Fuel Cells: Theory and Practice*. 2005. Burlington, MA: Elsevier Academic Press.

Beale, S.B. Calculation procedure for mass transfer in fuel cells. *J. Power Sources*. Vol. 128, 2004, pp. 185–192.

Lin, B. 1999. Conceptual design and modeling of a fuel cell scooter for urban Asia. Princeton University, masters thesis.

Li, X. *Principles of Fuel Cells*. 2006. New York: Taylor & Francis Group.

Mench, M.M., C.-Y. Wang, and S.T. Tynell. *An Introduction to Fuel Cells and Related Transport Phenomena*. Department of Mechanical and Nuclear Engineering, Pennsylvania State University. Draft. Available at: http://mtrl1.mne.psu .edu/Document/jtpoverview.pdf Accessed March 4, 2007.

Mench, M.M., Z.H. Wang, K. Bhatia, and C.Y. Wang. 2001. *Design of a Micro-Direct Methanol Fuel Cell*. Electrochemical Engine Center, Department of Mechanical and Nuclear Engineering, Pennsylvania State University.

Mennola T., et al. Mass transport in the cathode of a free-breathing polymer electrolyte membrane fuel cell. *J. Appl. Electrochem*. Vol. 33, 2003, pp. 979–987.

O'Hayre, R., S.-W. Cha, W. Colella, and F.B. Prinz. 2006. *Fuel Cell Fundamentals*. New York: John Wiley & Sons.

Rowe, A., and X. Li. Mathematical modeling of proton exchange membrane fuel cells. *J. Power Sources*. Vol. 102, 2001, pp. 82–96.

Sousa, R., Jr., and E. Gonzalez. Mathematical modeling of polymer electrolyte fuel cells. *J. Power Sources*. Vol. 147, 2005, pp. 32–45.

Springer et al. Polymer electrolyte fuel cell model. *J. Electrochem. Soc*. Vol. 138, No. 8, 1991, pp. 2334–2342.

You, L., and H. Liu. A two-phase flow and transport model for PEM fuel cells. *J. Power Sources*. Vol. 155, 2006, pp. 219–230.

CHAPTER 6

Heat Transfer

6.1 Introduction

Temperature in a fuel cell is not always uniform, even when there is a constant mass flow rate in the channels. Uneven fuel cell stack temperatures are due to a result of water phase change, coolant temperature, air convection, the trapping of water, and heat produced by the catalyst layer. In order to precisely predict temperature-dependent parameters and rates of reaction and species transport, the heat distribution throughout the stack needs to be determined accurately. The calculations presented in this chapter give the heat transfer basics for fuel cells. The specific topics to be covered are as follows:

- Basics of heat transfer
- Fuel cell energy balances
- Fuel cell heat management

The first step in determining the heat distribution in a fuel cell stack is to perform energy balances on the system. The total energy balance around the fuel cell is based upon the power produced, the fuel cell reactions, and the heat loss that occurs in a fuel cell. Convective heat transfer occurs between the solid surface and the gas streams, and conductive heat transfer occurs in the solid and/or porous structures. The reactants, products, and electricity generated are the basic components to consider in modeling basic heat transfer in a fuel cell, as shown in Figure 6-1.

The general energy balance states that the enthalpy of the reactants entering the cell equals the enthalpy of the products leaving the cell plus the sum of the heat generated by the power output, and the rate of heat loss to the surroundings. The basic heat transfer calculations will aid in predicting the temperatures and heat in overall fuel cell stack and stack components.

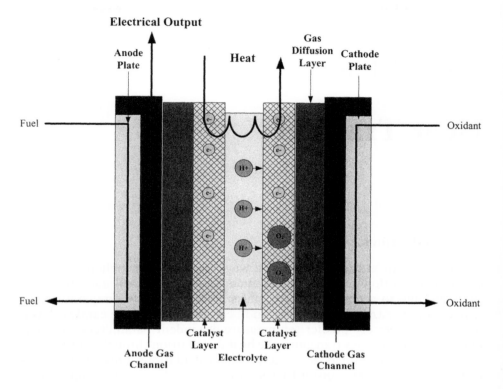

FIGURE 6-1. Stack illustration for heat flow study[1].

6.2 Basics of Heat Transfer

Conduction can be defined as the transfer of energy from more energetic particles to less energetic particles due to the interaction between the particles. A temperature gradient within a homogeneous substance results in an energy transfer through the medium, which results in a transfer of energy from more energetic to less energetic molecules. This heat transfer process can be quantified in terms of rate equations. The rate of heat transfer in the x-direction through a finite cross-sectional area, A, is known as Fourier's law, and can be expressed as:

$$q_x = -kA\frac{dT}{dx} \tag{6-1}$$

where k is the thermal conductivity, $W/(m*k)$. When the heat transfer is linear under steady-state conditions, the temperature gradient may be expressed as follows:

TABLE 6-1
Thermal Conductivity of Some Fuel Cell Materials

Material	Thermal Conductivity (W/mK) @ 300 K
Aluminum	237
Nickel	90.5
Platinum	71.5
Titanium	22
Stainless steel 316	13
Graphite	98
Carbon cloth	1.7
Teflon	0.4

$$q_x = k \frac{T_1 - T_2}{L} \qquad (6\text{-}2)$$

The thermal conductivity is a transport property that provides an indication of the rate at which energy is transferred by the diffusion process. This property is dependent upon the atomic and molecular structure of the substance. The thermal conductivity of some fuel cell materials is shown in Table 6-1.

For one-dimensional, steady-state heat conduction with no heat generation, the heat flux is constant, and independent of x. This can be expressed as:

$$\frac{d^2 T}{dx^2} = 0 \qquad (6\text{-}3)$$

When analyzing one-dimensional heat transfer with no internal energy generation and constant properties, there is an analogy between the diffusion of heat and electrical charge. A thermal resistance can be associated with the conduction of heat, and can be expressed in a plane wall as:

$$R_{cond} = \frac{T_1 - T_2}{q_x} = \frac{L}{kA} \qquad (6\text{-}4)$$

When heat is conducted through two adjacent materials with different thermal conductivities, the third boundary condition comes from a requirement that the temperature at the interface is the same for both materials[2]:

$$q = h_{tc} A \Delta T \qquad (6\text{-}5)$$

where h_{tc} is the convective heat transfer coefficient, BTU/hft^2F or W/m^2K, A is the area normal to the direction of the heat flux, ft^2 or m^2, and ΔT is the temperature difference between the solid surface and the fluid F or K.

A thermal resistance may also be associated with the heat transfer by convection at a surface. This can be expressed as:

$$R_{conv} = \frac{T_1 - T_2}{q_x} = \frac{1}{hA} \qquad (6\text{-}6)$$

Since the conduction and convection resistances are in series, they can be summed as follows:

$$R_{tot} = \frac{1}{h_1 A} + \frac{L}{kA} + \frac{1}{h_2 A} \qquad (6\text{-}7)$$

The fuel cell layers can be thought of as a "composite wall" with series thermal resistances due to layers of different materials. With a "composite" system, it is convenient to work with an overall heat transfer coefficient, U, which can be written as:

$$q = UA\Delta T \qquad (6\text{-}8)$$

Like charge transport (Chapter 4), in composite systems, the temperature drop between materials can be significant, and is called thermal contact resistance. This is defined as:

$$R_{tc} = \frac{T_A - T_B}{q_x} \qquad (6\text{-}9)$$

The thermal resistance (R_{th}) can also be expressed as:

$$R_{th} = \frac{1}{\dfrac{1}{R_c} + \dfrac{1}{R_R}} \qquad (6\text{-}10)$$

where R_c is the convective thermal resistance:

$$R_c = \frac{1}{hA_s} \qquad (6\text{-}11)$$

where h is $h = \dfrac{k}{L} Nu_L$. Figure 6-2 illustrates the temperature drop due to thermal contact resistance.

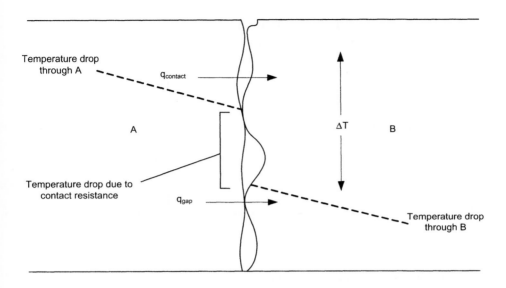

FIGURE 6-2. Temperature drop due to thermal contact resistance.

The heat lost by the stack through radiation to the surroundings is the radioactive thermal resistance (R_R) defined as:

$$R_R = \frac{1}{\sigma F A_s (T_s + T_0)(T_s^2 + T_0^2)} \qquad (6\text{-}12)$$

where σ is the Stefan-Boltzmann constant $[5.67 \times 10^{-8} \text{ W}/(\text{m}^2\text{K}^4)]$, F is the shape factor, and A_s is the stack exposed surface area, m^2.

Sometimes using the assumption that the temperature gradient is only significant for one direction is an oversimplification of the problem. For two-dimensional, steady-state conditions with no heat generation (see Figure 6-3), and constant thermal conductivity, the heat flux can be expressed as:

$$\frac{d^2T}{dx^2} + \frac{d^2T}{dy^2} = 0 \qquad (6\text{-}13)$$

Internal fuel cell heat generation can be described by the Poisson equation:

$$\frac{d^2T}{dx^2} + \frac{q_{int}}{k} = 0 \qquad (6\text{-}14)$$

where q_{int} is the rate of heat generation per unit volume.

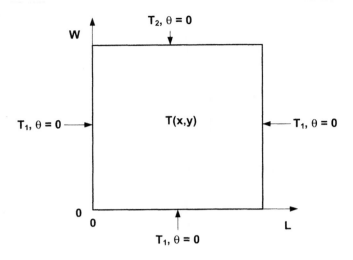

FIGURE 6-3. Two-dimensional conduction.

6.3 Fuel Cell Energy Balances

6.3.1 General Energy Balance Procedure

In order to accurately model a fuel cell system, the energy that flows into and out of each process unit in the fuel cell subsystem, and in the fuel cell itself, needs to be accounted for in order to determine the overall energy requirement(s) for the process. A typical energy balance calculation determines the cell exit temperature knowing the reactant composition, the temperatures, H_2 and O_2 utilization, the expected power produced, and the percent of heat loss. The procedure for formulating an energy balance is as follows:

1. A flowchart must be drawn and labeled. Enough information should be included on the flowchart to determine the specific enthalpy of each stream component. This includes known temperatures, pressures, mole fractions, mass flow rates, and phases.
2. Mass balance equations may need to be written in order to determine the flow rates of all stream components.
3. The specific enthalpies need to be determined for each stream component. These can be obtained from thermodynamic tables, or can be calculated if these data are not available.
4. The final step is to write the appropriate form of the energy balance equation, and solve for the desired quantity.

An example flowchart is shown in Figure 6-4. The fuel enters the cell at temperature, T, and pressure, P. Oxygen enters the fuel cell at a certain

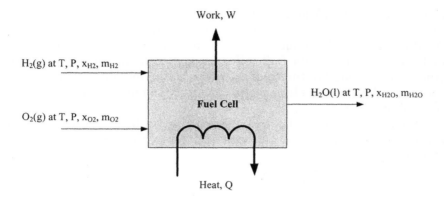

FIGURE 6-4. Detailed flowchart to obtain energy balance equation[3].

T, P, x_{O2} (mole fraction), and m_{O2} (mass flow rate). The hydrogen and oxygen react completely in the cell to produce water, which exits at a certain T, P, x (mole fraction), and m (mass flow rate). This reaction can be described by:

$$H_2 + \frac{1}{2}O_2 \rightarrow H_2O$$

Q is the heat leaving the fuel cell, and W is the work available through chemical availability. Only the energy balance equation for this example has been written (the mass balance equations have not). The specific enthalpies can be obtained through thermodynamic tables or calculations. The generic energy balance for the fuel cell in this example is:

$$\frac{W}{m_{H_2}} + \frac{Q}{m_{H_2}} = h_{H_2} + \frac{1}{2}h_{O_2} - h_{H_2O} \qquad (6\text{-}15)$$

6.3.2 Energy Balance of Fuel Cell Stack

The energy balance on the fuel cell is the sum of the energy inputs equals the sum of the energy outputs. The generic heat balance on any fuel cell stack can be written as follows[4]:

$$\sum Q_{in} - \sum Q_{out} = W_{el} + Q_{dis} + Q_c \qquad (6\text{-}16)$$

where Q_{in} is the enthalpy (heat) of the reactant gases in, Q_{out} is the enthalpy (heat) of the unused reactant gases and heat produced by the product, W_{el} is the electricity generated, Q_{dis} is the heat dissipated to the surroundings, and Q_c is the heat taken away from the stack by active cooling. Heat is carried away by reactant gases, product water, and that lost to the surroundings; the remaining heat needs to be taken from the stack through cooling. The heat generation in the fuel cell is associated with voltage losses. Most of the heat is created in the catalyst layers, in the membrane due to ohmic losses, and then in the electrically conductive solid parts of the fuel cell due to ohmic losses[5].

A good estimate for the fuel cell stack energy balance can be obtained by equating the energy of the fuel reacted to the heat and electricity generated:

$$\frac{I}{2F}H_{HHV}n_{cell} = Q_{gen} + IV_{cell}n_{cell} \qquad (6\text{-}17)$$

When all of the product water leaves the stack as liquid at room temperature, the heat generated in a fuel cell stack is

$$Q_{gen} = (1.482 - V_{cell})In_{cell} \qquad (6\text{-}18)$$

where Q_{gen} is the heat generated from the stack in watts, n_{cell} is the number of cells, and V_{cell} is the cell voltage. If all of the product water leaves the stack as vapor, the following equation can be used instead:

$$Q_{gen} = (1.254 - V_{cell})In_{cell} \qquad (6\text{-}19)$$

Equations 6-17 through 6-19 are approximations, and do not take into account the heat or enthalpy brought to, or removed from, the stack.

6.3.3 General Energy Balance for Fuel Cell

Another way of uniting the fuel cell energy balance is the sum of all the energy inputs and the sum of all of the energy outputs:

$$\sum (h_i)_{in} = W_{el} + \sum (h_i)_{out} + Q \qquad (6\text{-}20)$$

The inputs are the enthalpies of the fuel, the oxidant, and the water vapor present. The outputs are the electric power produced, enthalpies of the flows out of the fuel cell, and the heat leaving the fuel cell through coolant, convection, or radiation.

The enthalpy (J/s) for each dry gas or mixture of dry gases is

$$h = \dot{m}c_pT \qquad (6\text{-}21)$$

where $\dot{m}$ is the mass flow rate of the gas or mixture (g/s), Cp is the specific heat (J/[g*K]), and T is the temperature in °C.

If the gas has a high heating value (combustible), its enthalpy is then:

$$h = \dot{m}(c_p T + h_{HHV}^0) \tag{6-22}$$

where h_{HHV}^0 is the higher heating value of that gas (J/g) at 0 °C. The heating values are usually reported at 25 °C, therefore, the higher heating value may need to be calculated at the chosen temperature. The enthalpy of water vapor is

$$h = \dot{m}_{H2O(g)} c_{p,H2O(g)} T + h_{fg}^0 \tag{6-23}$$

The enthalpy of liquid water is

$$h = \dot{m}_{H2O(l)} c_{p,H2O(l)} T \tag{6-24}$$

The inputs and outputs of the energy balance can quickly become complicated when the heat balance is performed for each individual fuel cell layer and/or stack heating and cooling is involved.

EXAMPLE 6-1: Energy Balances

A PEM hydrogen/air fuel cell generates 1 kW at 0.8 V. Air is supplied to the fuel cell stack at 20 °C. The air outlet temperature is 70 °C. The mass flow rate for hydrogen going into the cell is 0.02 g/s and air going into the cell is 1.5 g/s. The mass flow rate of the water in air going into the fuel cell is 0.01 g/s. The mass flow rate of N_2 and O_2 coming out of the fuel cell is 0.5 and 1.5 g/s, respectively. What is the mass flow rate of the air entering the cell? What is the mass flow rate of the water in air leaving the cell? Assume that the heat generated by the fuel cell is negligible. The HHV of hydrogen is 141,600 J/g, and $h_{fg}^0 = 2500$.

The energy balance is

$$H_{H2,in} + H_{Air,in} + H_{H2O\_Air,in} = H_{Air,out} + H_{H2O\_Air,out} + W_{el}$$

The energy flows are:
Hydrogen in:

$$H_{H2,in} = m_{H2,in} \left(c_{p,H2} T_{in} + h_{HHV}^0 \right)$$

$$h_{HHV}^0 = h_{HHV}^{25} - \left(c_{p,H2} + \frac{1}{2} \frac{M_{O2}}{M_{H2}} c_{p,O2} - \frac{M_{H2O}}{M_{H2}} c_{p,H2O(l)} \right) \cdot 25$$

$$h^0_{HHV} = 141,900 - \left(14.2 + \frac{1}{2} * \frac{31.9988}{2.0158} 0.913 - \frac{18.0152}{2.0158} 4.18\right) \cdot 25 = 142,298\,J/gK$$

$$H_{H2,in} = 0.02(14.2 \times 20 + 142,298) = 2929W$$

Air in:

$$H_{Air,in} = m_{Air,in} c_{p,Air} T_{in} = m_{Air,in} \times 1.01 \times 20$$

Water vapor in air in:

$$H_{H2O\_Air,in} = m_{H2O\_Air,in} \times (c_{p,H2O} T_{in} + h^0_{fg})$$

$$H_{H2O\_Air,in} = 0.01 \times (1.85 \times 20 + 2500) = 30.42$$

Air out:

$$H_{Air,out} = m_{O2,out} c_{p,O2} T_{out} + m_{N2,out} c_{p,N2} T_{out}$$

$$H_{Air,out} = (0.5 * 0.913 * 75) + (1.5 * 1.04 * 75) = 691.96$$

The amount of water generated is:

$$I = P/V = 1000W/0.7V = 1250$$

$$m_{H2O,gen} = \frac{I}{2F} M_{H2O} = \frac{1428.6}{2 * 96,485} * 18.015 = 0.1167\,g/s$$

The water mass balance can be written as:

$$m_{H2O\_Air,in} + m_{H2O,gen} = m_{H2O\_Air,out}$$

$$m_{H2O\_Air,out} = 0.01g/s + 0.133g/s = 0.1267g/s$$

Water vapor in air out:

$$H_{H2O\_Air,out} = m_{H2O\_Air,out} \times (c_{p,H2O} T_{out} + h^0_{fg})$$

$$H_{H2O\_Air,out} = 0.143 \times (1.85 \times 75 + 2500) = 397.17$$

Energy balance is

$$H_{H2,in} + H_{Air,in} + H_{H2O\_Air,in} = H_{Air,out} + H_{H2O\_Air,out} + W_{el}$$

$$2124.5 + H_{Air,in} + 25.37 = 151.24 + 377.34 + 1000$$

$$H_{Air,in} = -870.50W$$

Substituting this back into the enthalpy calculation for $H_{Air,in}$:

$$H_{Air,in} = m_{Air,in}c_{p,Air}T_{in} = m_{Air,in} \times 1.01 \times 20$$

$$m_{Air,in} = 2.94 g/s$$

Using MATLAB to solve:

```
%%%%%%%%%%%%%%%%%%%%%%%%%%%%%%%%%%%%%%%%%%%
% EXAMPLE 6-1: Energy Balances
%%%%%%%%%%%%%%%%%%%%%%%%%%%%%%%%%%%%%%%%%%%
% Inputs
V = 0.8;                    % Cell voltage (V)
P = 1000;                   % Power (Watt)
T = 20 + 273.15;            % Air Temperature (K)
T_out = 70 + 273.15;        % Temperature out (K)
m_H2_in = 0.02;             % Hydrogen flow rate (g/s)
m_H2O_in = 0.01;            % Water flow rate (g/s)
m_O2_out = 0.5;             % Oxygen flow rate (g/s)
m_N2_out = 1.5;             % Nitrogen flow rate (g/s)
HHV_25 = 141900;            % HHV of hydrogen
cp_H2 = 14.2;               % Specific heat of H2
cp_O2 = 0.913;              % Specific heat of O2
cp_N2 = 1.04;               % Specific heat of N2
cp_AIR = 1.01;              % Specific heat of air
cp_H2O_1 = 4.18;            % Specific heat of liquid water
cp_H2O = 1.85;              % Specific heat of water
F = 96485;                  % Faraday's constant
M_O2 = 31.9988;             % Molecular weight of O2
M_H2 = 2.0158;              % Molecular weight of H2
M_H2O = 18.0152;            % Molecular weight of H2O
H_fg = 2500;

%%%%%%%%%%%%%%%%%%%%%%%%%%%%%%%%%%%%%%%%%%%
% Calculate the higher heating value of hydrogen
h_HHV_0 = HHV_25 - (cp_H2 + 0.5*cp_O2*M_O2 / M_H2 - M_H2O*cp_H2O_1
   / M_H2 )*25

% Energy flow for hydrogen in
h_H2_in = m_H2_in*(cp_H2*T + h_HHV_0)
```

% Energy flow for water in air in

h_H2O_in = m_H2O_in * (cp_H2O * T + H_fg)

% Energy flow for air out

h_AIR_out = (m_O2_out * cp_O2 + m_N2_out * cp_N2) * T_out

%The amount of water generated

I = P / V; % Current density
m_H2O_gen = I / (2 * F) * M_H2O

% Water vapor in air out

m_H2O_out = m_H2O_gen + m_H2O_in % Mass balance

%Water vapor in air out

h_H2O_out = m_H2O_out * (cp_H2O * T_out + H_fg)

% From the energy balance, energy flow for air in

h_AIR_in = h_AIR_out + h_H2O_out + P – h_H2_in – h_H2O_in

% Mass flow rate for air in

m_AIR_in = abs(h_AIR_in / (cp_AIR * T))

6.3.4 The Nodal Network

To help obtain more accurate heat transfer solutions, numerical techniques such as finite-difference, finite-element, or boundary-element methods can be used, and readily extended to up to three-dimensional problems. The nodal network solution allows the determination of the temperature at discrete points. This is accomplished by subdividing the medium of interest into a smaller number of regions, and assigning a reference point at its center. The reference point is termed a nodal point, and a nodal network is a grid or mesh. An energy balance is typically solved for the node. Figure 6-5 shows an example of the x and y locations for a two-dimensional system. The next section, 6.3.5, illustrates an example of transient conduction in a plate using a one-dimensional nodal network.

6.3.5 Transient Conduction in a Plate[6]

The process of obtaining a numerical solution to a one-dimensional, transient conduction problem will be illustrated in the context of a layer in a fuel cell stack, as shown in Figure 6-6.

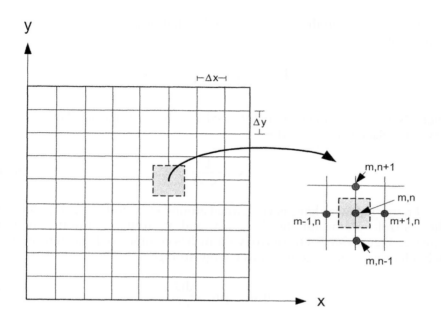

FIGURE 6-5. Two-dimensional conduction with nodal network.

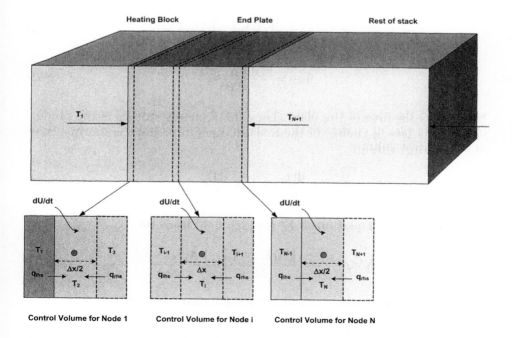

FIGURE 6-6. Nodes distributed uniformly throughout computational domain.

For the uniform distribution of nodes that is shown in Figure 6-6, the location of each node (x_i) is:

$$x_i = \frac{(i-1)}{(N-1)}L \quad \text{for } i = 1 \ldots N \tag{6-25}$$

where N is the number of nodes used for the simulation. The distance between adjacent nodes (Δx) is:

$$\Delta x = \frac{L}{N-1} \tag{6-26}$$

Energy balances have been defined around each node (control volume). The control volume for the first, last, and an arbitrary, internal node is shown in Figure 6-6. Each control volume has conductive heat transfer with each adjacent node in addition to energy storage:

$$\dot{q}_{LHS} + \dot{q}_{RHS} = \frac{dU}{dt} \tag{6-27}$$

Each term in Equation 6-27 must be approximated. The conduction terms from the adjacent nodes are modeled as:

$$\dot{q}_{LHS} = \frac{kA(T_{i-1} - T_i)}{\Delta x} \tag{6-28}$$

$$\dot{q}_{RHS} = \frac{kA(T_{i+1} - T_i)}{\Delta x} \tag{6-29}$$

where A is the area of the plate. The rate of energy storage is the product of the time rate of change of the nodal temperature and the thermal mass of the control volume:

$$\frac{dU}{dt} = A\Delta x \rho c \frac{dT_i}{dt} \tag{6-30}$$

Substituting Equations 6-27 through 6-30 leads to:

$$A\Delta x \rho c \frac{dT_i}{dt} = \frac{kA(T_{i-1} - T_i)}{\Delta x} + \frac{kA(T_{i-1} - T_i)}{\Delta x} \tag{6-31}$$

Solving for the time rate of the temperature change:

$$\frac{dT_i}{dt} = \frac{k}{\Delta x^2 \rho c}(T_{i-1} + T_{i+1} - 2T_i) \quad \text{for} \quad i = 2 \ldots (N-1) \tag{6-32}$$

The control volumes on the edges must be treated separately because they have a smaller volume and experience different energy transfers. The control volume for the node located at the outer surfaces (node N) has the following the energy balance:

$$\frac{dU}{dt} = \dot{q}_{LHS} + \dot{q}_{conv} \tag{6-33}$$

or

$$\frac{A\Delta x \rho c}{2} \frac{dT_N}{dt} = \frac{kA(T_{N-1} - T_N)}{\Delta x} + hA(T_f - T_N) \tag{6-34}$$

Solving for the time rate of temperature change for node N:

$$\frac{dT_N}{dt} = \frac{2k}{\rho c \Delta x^2}(T_{N-1} - T_N) + \frac{2h}{\Delta x \rho c}(T_f - T_N) \tag{6-35}$$

Note that the equations provide the time rate of change for the temperature of every node given the temperatures of the nodes. The energy balance for each control volume provides an equation for the time rate of change of the temperature in terms of the temperature. Therefore, the energy balance written for each control volume has a set of equations for the time rate of change.

The temperature of each node is a function both of position (x) and time (t). The index that specifies the node's position is i where i = 1 corresponds to the adiabatic plate and i = N corresponds to the surface of the plate. A second index, j, is added to each nodal temperature in order to indicate the time ($T_{i,j}$); j = 1 corresponds to the beginning of the simulation and j = M corresponds to the end of the simulation. The total simulation time is divided into M time steps; most of equal duration, Δt:

$$\Delta t = \frac{\tau_{sim}}{(M-1)} \tag{6-36}$$

The time associated with any time step is:

$$t_j = (j-1)\Delta t \quad \text{for} \quad j = 1 \ldots M \tag{6-37}$$

The initial conditions for this problem are that all of the temperatures at t = 0 are equal to T_{in}.

$$T_{i,1} = T_{in} \quad \text{for} \quad i = 1 \ldots N \tag{6-38}$$

Note that the variable T is a two-dimensional array (i.e., a matrix).

EXAMPLE 6-2: Transient Conduction through Plate

Plot the one dimensional heat transfer through a fuel cell end plate with a thickness of 0.01 and an initial temperature of 343.15 K. The plate is insulated on its left side, and is exposed to air at 298 K on the right side. The plate has the following material properties: (a) a polymer end plate with a conductivity of 0.2 W/mK, a density of 1740, and specific heat capacity of 1464 J/KgK, and (b) an aluminum end plate with a conductivity of 220 W/mK, a density of 2700 kg/m^3, and specific heat capacity of 900 J/KgK. Air has a heat transfer coefficient of 17 W/m^{2}K. Set up a grid with 10 nodes (slices) in the x-direction, and plot the temperature at each node after 10 seconds and 2 minutes.

Using MATLAB to solve:

```
%%%%%%%%%%%%%%%%%%%%%%%%%%%%%%%%%%%%%%%%%%%%
```

% EXAMPLE 6-2: Transient Conduction through Plate

```
%%%%%%%%%%%%%%%%%%%%%%%%%%%%%%%%%%%%%%%%%%%%
```

% Inputs

```
L = 0.01;          % Plate thickness (m)
k = 0.2;           % Conductivity (W/m-K)
rho = 2000;        % Density (kg/m^3)
cp = 200;          % Specific heat capacity (J/kg-K)
T_in = 343.15;     % Initial temperature (K)
T_f = 298;         % Gas temperature (K)
h = 17;            % Heat transfer coefficient (W/m^2-K)
```

```
%%%%%%%%%%%%%%%%%%%%%%%%%%%%%%%%%%%%%%%%%%%%
```

% Setup Grid

```
N = 10;                 % Number of nodes (-)
for i = 1:N
x(i)=(i-1)*L/(N-1);     % Position of each node (m)
end
DELTAx = L/(N-1);       % Distance between adjacent nodes (m)
tau_sim = 10;           % Simulation time (s)
OPTIONS = odeset('RelTol',1e-6);
[time,T] = ode45(@(time,T) dTdt_functionv(time,T,L,k,rho,cp,T_f,h),[0,tau_sim], ...
    T_in*ones(N,1),OPTIONS);
```

% Plot figure of transient heat conduction through fuel cell plates

```
surf(T);
xlabel('Number of Nodes');
zlabel('Temperature (K)');
```

```
%%%%%%%%%%%%%%%%%%%%%%%%%%%%%%%%%%%%%%%%%%%
function[dTdt]=dTdt_functionv(time,T,L,k,rho,c,T_f,h)
[N,g]=size(T);          % Determine the size of T
DELTAx=L/(N-1);    % Calculate the distance between adjacent nodes
dTdt=zeros(N,1);    % Create dTdt vector
dTdt(1)=2*k*(T(2)-T(1))/(rho*c*DELTAx^2);
for i=2:(N-1)
dTdt(i)=k*(T(i-1)+T(i+1)-2*T(i))/(rho*c*DELTAx^2);
end
dTdt(N)=2*k*(T(N-1)-T(N))/(rho*c*DELTAx^2)+2*h*(T_f-T(N))/(rho*c*DELTAx);
end
```

Figures 6-7 and 6-8 both shows graphs of the temperature at each node after 10 seconds and 2 minutes for the polymer and aluminum end plate, respectively. In Figure 6-2, one can see that only a little heat has transferred to the plate after 10 seconds for the polymer, but the temperature distribution through the aluminum end plate is exactly the same at 10 and 120 seconds. These figures show that the heat rapidly diffuses through the aluminum end plate in comparison with the polymer end plate.

6.3.6 Energy Balance for Fuel Cell Layers

Energy balances can be defined around each of the fuel cell layers to enable the study of the diffusion of heat through a particular layer as a function of time or position. Figure 6-9 shows an example of a fuel cell layer as a control volume. This section describes the mode of heat diffusion in each layer, and calculates the energy balances for the end plate, gasket, contact, flow field, GDL, catalyst, and membrane layers[7].

End Plates, Contacts, and Gasket Materials
The end plate is typically made of a metal or polymer material, and is used to uniformly transmit the compressive forces to the fuel cell stack. The end plate must be mechanically sturdy enough to support the fuel cell stack, and be able to uniformly distribute the compression forces to all of the major surfaces of each layer within the fuel cell stack. Depending upon the stack design, there also may be contact and gasket layers in the fuel cell stack. The gasket layers help to prevent gas leaks and improve stack compression. The contact layers or current collectors are used to collect electrons from the bipolar plate and gas diffusion layer (GDL).

Since there is typically no gas or liquid flows in the end plates, gaskets, or contact layers, conduction is the only mode of heat transfer. One side of each of these layers is exposed to an insulating material (or the ambient environment), and the other side is exposed to a conductive current

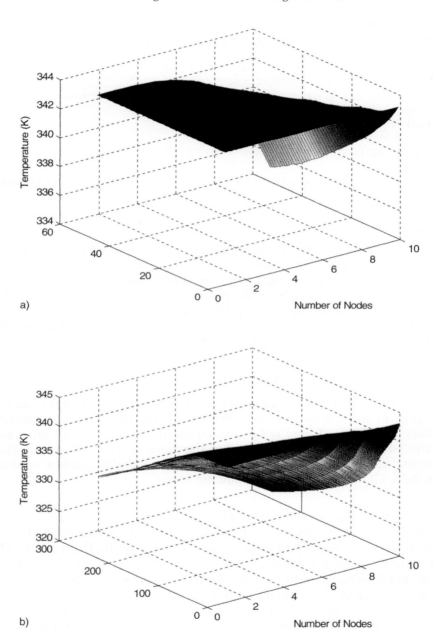

FIGURE 6-7. Transient heat conduction through polymer fuel cell plate at a) 10 and b) 120 sec.

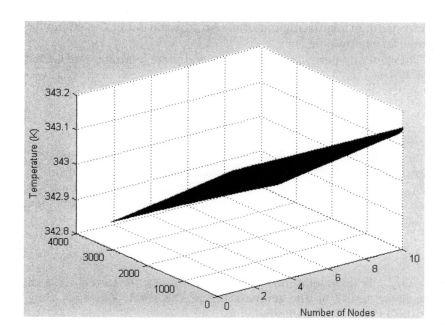

FIGURE 6-8. Transient heat conduction through aluminum fuel cell plate at 10 and 120 sec (same graph).

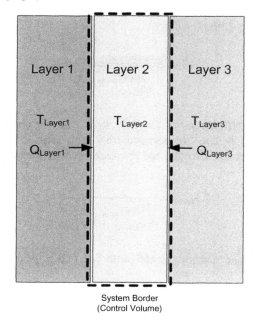

FIGURE 6-9. Energy balance around layer.

collector plate or insulating material. An illustration of the energy balance is shown in Figure 6-9.

The general energy balance for the end plate, contact, and GDL layers can be written as:

$$\left(\rho_{Layer2} A_{Layer2} t_{Layer2} cp_{Layer2}\right) \frac{dT_{Layer2}}{dt} = Q_{Layer1} + Q_{Layer3} \qquad (6\text{-}39)$$

where ρ_{Layer2} is the density of Layer2, A_{Layer2} is the area of Layer2, cp_{Layer2} is the specific heat capacity of Layer2, Q_{Layer1} is the heat flow from Layer1, and Q_{Layer3} is the heat flow from Layer3. The derivative on the left side is the rate of change of control volume temperature (dT_{Layer2}/dt). The heat flow from Layer1 to Layer2 is:

$$Q_{Layer1} = U_{Layer1} A \left(T_{Layer1} - T_{Layer2}\right) \qquad (6\text{-}40)$$

where U_{Layer1} is the overall heat transfer coefficient for Layer1, A is the area of the layer, and T is the temperature of the layer. The heat flow from Layer3 to Layer2 can be expressed as:

$$Q_{Layer3} = U_{Layer3} A \left(T_{Layer3} - T_{Layer2}\right) \qquad (6\text{-}41)$$

If the heat is coming from the surroundings, the overall heat transfer coefficient can be calculated by:

$$U_{surr} = \frac{1}{\dfrac{t_{Layer2}}{*k_{Layer2}} + \dfrac{1}{h_{surr}}} \qquad (6\text{-}42)$$

where t_{Layer2} is the thickness of Layer2, k_{Layer2} is the thermal conductivity of Layer2, and h_{surr} is the convective loss from the stack to the air. The overall heat transfer coefficient for the heat coming from Layer1 is:

$$U_{Layer1} = \frac{1}{\dfrac{t_{Layer2}}{*k_{Layer2}} + \dfrac{t_{Layer1}}{*k_{Layer1}}} \qquad (6\text{-}43)$$

The overall heat transfer coefficient for the heat coming from Layer3 is:

$$U_{Layer3} = \frac{1}{\dfrac{t_{Layer3}}{*k_{Layer3}} + \dfrac{t_{Layer2}}{*k_{Layer2}}} \qquad (6\text{-}44)$$

If the layer conducts electricity (such as the contact layer), then there is an additional heat generation in Layer2 (Q_{res_Layer2}) due to electrical resistance, which can be calculated as:

$$Q_{res\_Layer2} = (iA)^2 \frac{\rho_{res\_Layer2} t_{Layer2}}{A} \qquad (6\text{-}45)$$

where i is the current density, A is the area of the layer, ρ_{res_Layer2} is the specific resistance of the material, and t_{Layer2} is the thickness of the layer. Typically, there is no heat generated in the end plate, contact, or gasket layers. However, in some fuel cell stack designs, the end plate may be heated, therefore, an additional heat generation term would need to be added to the model formulation.

Bipolar Plate
In the fuel cell stack, the bipolar plates separate the reactant gases of adjacent cells, connect the cells electrically, and act as a support structure. The bipolar plates have reactant flow channels on both sides, forming the anode and cathode compartments of the unit cells on the opposing sides of the bipolar plate. Flow channel geometry affects the reactant flow velocities, mass transfer, and fuel cell performance. Bipolar plate materials must have high conductivity and be impermeable to gases. The material should also be corrosion-resistant and chemically inert due to the presence of reactant gases and catalysts. An illustration of the energy balance is shown in Figure 6-10.

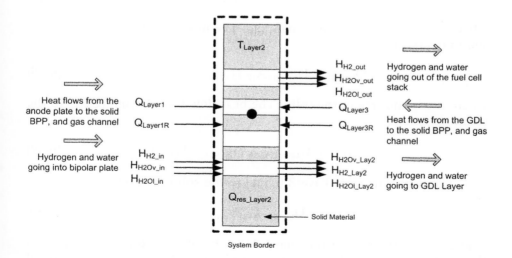

FIGURE 6-10. Flow field plate energy balance.

The bipolar plate has both conductive and convective heat transfer due to the gas channels in the plate. The percentage of the bipolar plate that has channels affects the heat transfer of the overall plate, therefore, this is accounted for by calculating the effective cross-sectional area for conduction heat transfer, A_{1R}, which represent the area of the solid material in contact with Layer1 and Layer3. The governing equation for heat transfer in an anode bipolar plate can be written as:

$$(cp_{avg}(n_{gases} + n_{liq}) + \rho_{Layer2}(A_{Layer2}t_{Layer2})cp_{Layer2})\frac{dT_{Layer2}}{dt} = Q_{Layer1} + Q_{Layer1R} +$$

$$Q_{Layer3} + Q_{Layer3R} + Q_{res\_Layer2}H_{H2\_in} + H_{H2Ov\_in} + H_{H2Ol\_in} - H_{H2\_Lay3} -$$

$$H_{H2Ov\_Lay3} - H_{H2Ol\_Lay3} - H_{H2\_out} - H_{H2Ov\_out} - H_{H2Ol\_out}$$

$$(6\text{-}46)$$

where ρ_{Layer1} is the density of Layer2, A is the area of Layer2, cp_{Layer2} is the specific heat capacity of Layer2, Q_{Layer1} is the heat flow from Layer1 to the channels, $Q_{Layer1R}$ is the heat flow from Layer1 to the solid material, Q_{Layer3} is the heat flow from Layer3 to the channels, $Q_{Layer3R}$ is the heat flow from Layer3 to the solid material, Q_{res_Layer2} is the heat generation in the layer due to electrical resistance, and H_A is the enthalpy of component A coming into or out of the Layer2. The derivative on the left side is the rate of change of control volume temperature (dT_{Layer2}/dt). The heat flows coming from the right and left layer will transfer a different amount of heat from the layer to the solid and gas flow in the channels.

The heat flow from Layer1 to the channels is:

$$Q_{Layer1} = U_{Layer1}A_{void}(T_{Layer1} - T_{Layer2}) \qquad (6\text{-}47)$$

where A_{void} is the area of the channels. The heat flow from Layer1 to the solid material is:

$$Q_{Layer1R} = U_{Layer1R}A_{1R}(T_{Layer1} - T_{Layer2}) \qquad (6\text{-}48)$$

where A_{1R} is the area of the solid. The heat flow from Layer3 (GDL) to the channels is:

$$Q_{Layer3} = U_{Layer3}A_{void}(T_{Layer3} - T_{Layer2}) \qquad (6\text{-}49)$$

The heat flow from Layer3 (GDL) to the solid material is:

$$Q_{Layer3R} = U_{Layer3R}A_{1R}(T_{Layer3} - T_{Layer2}) \qquad (6\text{-}50)$$

where A_{void} is the area of the channels in the plate, and A_{1R} is the area of the solid material. The heat generation in Layer2 (Q_{res_Layer2}) due to electrical resistance is:

$$Q_{res\_Layer2} = (iA)^2 \frac{\rho_{res\_Layer2} t_{Layer2}}{A_{1R}} \qquad (6\text{-}51)$$

where i is the current density, A is the area of the layer, ρ_{res_Layer2} is the specific resistance of the material, and t_{Layer2} is the thickness of the layer.

The enthalpy of each gas or liquid flow into or out of the layer can be defined as:

$$H_A = n_A h_A T_{Layer2} \qquad (6\text{-}52)$$

where H_A is the enthalpy of the stream entering or leaving the layer, n_A is the molar flow rate of A, and h_A is the enthalpy of A at the temperature of the layer (T_{Layer}).

The overall heat transfer coefficient terms can be calculated as:

$$U_{Layer1R} = \frac{1}{\dfrac{t_{Layer1}}{*k_{Layer1}} + \dfrac{t_{Layer2}}{*k_{Layer2}}} \qquad (6\text{-}53)$$

$$U_{Layer1} = \frac{1}{\dfrac{t_{Layer1}}{*k_{Layer1}} + \dfrac{1}{h_1}} \qquad (6\text{-}54)$$

$$U_{Layer3R} = \frac{1}{\dfrac{t_{Layer3}}{*k_{Layer3}} + \dfrac{t_{Layer2}}{*k_{Layer2}}} \qquad (6\text{-}55)$$

$$U_{Layer3} = \frac{1}{\dfrac{t_{Layer3}}{*k_{Layer3}} + \dfrac{1}{h_1}} \qquad (6\text{-}56)$$

The calculation of the thermal mass of the gas/liquid mixture is as follows[8]:

$$thermal\_mass = cp_{avg} (n_{gases} + n_{liq}) \qquad (6\text{-}57)$$

where cp_{avg} is the average specific heat of the gases (hydrogen and water) at the temperature of Layer2. The molar flow rate of the gases at the temperature of the Layer2 can be calculated by:

$$n_{gases} = \frac{PV_{gases}}{RT_{Layer2}} \qquad (6\text{-}58)$$

where $V_{gases} = \varepsilon V_{void}$ is the volume of the gases in the channel, ε is the void fraction, and V_{void} is the volume of the channel space, and is defined by:

$$V_{void} = A_{void}t_{Layer2} \tag{6-59}$$

The molar flow rate of the liquid in the channels in Layer2 is given by:

$$n_{liq} = \frac{V_{liq}\rho_{liq}}{MW_{H2O}} \tag{6-60}$$

where the volume of gases can be calculated by:

$$V_{liq} = V_{void} - V_{gases} \tag{6-61}$$

Anode/Cathode Diffusion Media
The gas diffusion layer (GDL) is located between the flow field plate and the catalyst layer. This layer allows the gases and liquids to diffuse through it in order to reach the catalyst layer. The GDL has a much lower thermal conductivity than the bipolar plates and other metal components in the fuel cell, therefore, it partially insulates the heat-generating catalyst layers. When modeling the heat transfer through this layer, the solid portion has conductive heat transfer, and the gas/liquid flow has advective heat transfer. An illustration of the energy balance is shown in Figure 6-11.

Heat is generated in the GDL due to ohmic heating. Since the GDL has high ionic conductivity, ohmic losses are negligible compared with the catalyst and membrane layers. The overall energy balance equation for the anode GDL layer can be written as:

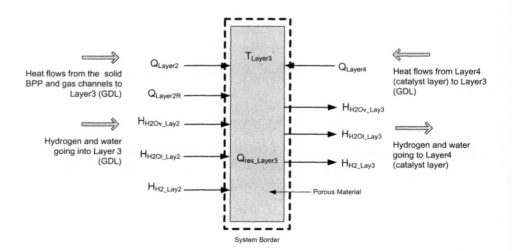

FIGURE 6-11. GDL energy balance.

$$\left(cp_{avg}\left(n_{gases} + n_{liq}\right) + \rho_{Layer3}\left(A_{Layer3}t_{Layer3}\right)cp_{Layer3}\right)\frac{dT_{Layer3}}{dt} = Q_{Layer4} + Q_{Layer2} +$$

$$Q_{Layer2R} + Q_{res\_Layer3} + H_{H2\_Lay2} + H_{H2Ov\_Lay2} + H_{H2Ol\_Lay2} - H_{H2\_Lay3} -$$

$$H_{H2Ov\_Lay3} - H_{H2Ol\_Lay3}$$

$$(6\text{-}62)$$

Anode/Cathode Catalyst Layer

The anode and cathode catalyst layer is a porous layer made of platinum and carbon. It is located on either side of the membrane layer. When modeling the heat transfer through this layer, the solid portion has conductive heat transfer, and the gas/liquid flow has advective heat transfer. Figure 6-12 shows the energy balance of the catalyst layer.

The overall energy balance equation can be written as:

$$\left(cp_{avg}\left(n_{gases} + n_{liq}\right) + \left(\rho_{a\_cat}A_{a\_cat}t_{a\_cat}\right)cp_{a\_cat}\right)\frac{dT_{a\_cat}}{dt} = Q_{a\_gdl} + Q_{mem} + Q_{a\_cat\_int} +$$

$$Q_{res\_a\_cat} + H_{H2\_3} + H_{H2Ov\_3} + H_{H2Ol\_3} - H_{H2\_4} - H_{H2Ov\_4} - H_{H2Ol\_4}$$

$$(6\text{-}63)$$

The heat generation in the catalyst layer is due to the electrochemical reaction and voltage overpotential. The heat generation term in the catalyst layer can be written as:

$$Q_{int\_Layer4} = \frac{i}{t_{Layer4}}\frac{T_{Layer4}\Delta S}{nF} + \eta \qquad (6\text{-}64)$$

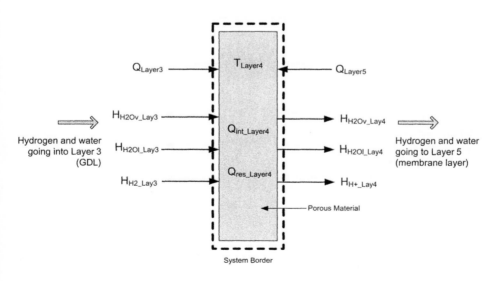

FIGURE 6-12. Catalyst energy balance.

where T_{Layer4} is the local catalyst temperature, i is the current density, t_{Layer4} is the layer thickness, n is the number of electrons, F is Faraday's constant, ΔS is the change in entropy, and η is the activation over-potential. The entropy change at standard-state with platinum catalyst is taken as ΔS = 0.104 Jmol^{-1} K^{-1} for the anode, and ΔS = −326.36 Jmol^{-1} K^{-1} for the cathode. The activation over-potential (v_{act}) was calculated based on typical Tafel kinetics for a Pt-electrode.

Membrane

The PEM fuel cell membrane layer is a persulfonic acid layer that conducts protons, and separates the anode and cathode compartments of a fuel cell. The most commonly used type is DuPont's Nafion® membranes. The dominant mode of heat transfer in the membrane is conduction. An illustration of the energy balance is shown in Figure 6-13.

The overall energy balance equation can be written as:

$$(cp_{avg}(n_{gases} + n_{lip}) + (\rho_{mem}A_{mem}t_{mem})cp_{mem})\frac{dT_{mem}}{dt} = Q_{a\_cat} + Q_{c\_cat} + Q_{res\_mem} +$$

$$H_{H+\_4} + H_{H2Ov\_4} + H_{H2Ol\_4} - H_{H2\_5} - H_{H2Ov\_5} - H_{H2Ol\_5}$$

$$(6\text{-}65)$$

The heat-generation term in the membrane consists of Joule heating only.

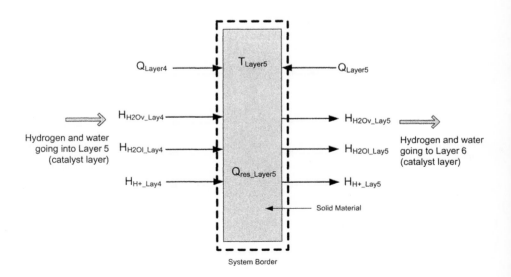

FIGURE 6-13. Membrane energy balance.

TABLE 6-2
Material Properties Used for Heat Transfer Calculations for Example 6-3

Fuel Cell Layer	Material	Thickness (m)	Area (m²)	Density (kg/m²)	Thermal Conductivity (W/m-K)	Specific Heat Capacity (J/kg-K)
End plate	Polycarbonate	0.01	0.0064	1300	0.2	1200
Gasket	Conductive rubber	0.001	0.001704	1400	1.26	1000
Flow field plate	Stainless steel	0.0005	0.003385	8000	65	500
MEA	Carbon cloth/ Pt/C/Nafion	0.001	0.0016	1300	26	864
Flow field plate	Stainless steel	0.0005	0.003385	8000	65	500
Gasket	Conductive rubber	0.001	0.001704	1400	1.26	1000
End plate	Polycarbonate	0.01	0.0064	1300	0.2	1200

EXAMPLE 6-3: Energy Balances

A PEM hydrogen/air single cell fuel cell stack operates at an initial temperature of 298.15 K. There are seven layers in the stack: two polycarbonate end plates, two rubber gaskets, stainless steel flow field plates, and an MEA. The necessary material properties for heat transfer calculations for each layer are shown in Table 6-2. Using the equations in Section 6.3.6, plot the conductive heat transfer through the stack at 60 sec, 300 sec, and 1000 sec with one and 10 nodes per layer. Neglect heat generation and losses from thermal resistance, electrochemical reactions, and mass flows through the fuel cell.

Using MATLAB to solve:

```
function [t,T] = fuelcellheat
% FUELCELLHEAT Fuel cell stack heat transfer model
% Best viewed with a monospaced font with 4 char tabs.
%%%%%%%%%%%%%%%%%%%%%%%%%%%%%%%%%%%%%%%%%
% Constants
const.tfinal = 60;      % Simulation time (s)
const.N = 7;            % Number of layers
const.T_0 = 350;        % End plate temperature
const.T_end = 350;      % End plate temperature
const.T_in = 298.15;    % Initial temperature (K)
const.h_surr = 17;      % Convective loss from stack
```

% Layers

```
% 1 – Left end plate
% 2 – Rubber gasket
% 3 – Anode Flow Field (Stainless steel)
% 4 – MEA
% 5 – Cathode Flow Field (Stainless steel)
% 6 – Rubber gasket
% 7 – Right end plate
```

% Layer parameters

```
%    1 2 3 4 5 6 7
```

% Number of temperature slices within layer (can be changed to as many slices as necessary)

```
param.M      = [10, 4, 2, 2, 2, 4, 10];
```

% Density (kg/m^3)

```
param.den    = [1300, 1400, 8000, 1300, 8000, 1400, 1300];
```

% Area (m^2)

```
param.A      = [0.0064, 0.001704, 0.003385, 0.0016, 0.003385, 0.001704, 0.0064];
```

% Thickness (m)

```
param.thick  = [0.011, 0.001, 0.0005, 0.001, 0.0005, 0.001, 0.011];
```

% Thermal Conductivity (W/m-K)

```
param.k      = [0.2, 1.26, 65, 26, 65, 1.26, 0.2];
```

% Specific heat capacity (J/Kg-k)

```
param.Cp     = [1200, 1000, 500, 864, 500, 1000, 1200];
```

%%%%%%%%%%%%%%%%%%%%%%%%%%%%%%%%%%%%%%%

% Create 1D temperature slices (M temps per layer).

% The temperatures are at the center of each slice.

% x is at the edge of each slice (like a stair plot).

```
x = 0;
layer = [];
for i=1:const.N,
x = [x, x(end) + (1:param.M(i)) * param.thick(i)/param.M(i)]; % Boundary Points
layer = [layer, i * ones(1,param.M(i))];
```

```
end
x = [-x(1), x]; % Add left hand heating block position (same width as first slice)
```

% Last point x(end) is the position of the right hand heating block

% Slice thicknesses

```
dx = diff(x); % Gives approximate derivatives between x's
```

% Heat transfer parameter (W/m^2-K)

```
left = 2:length(dx)-1;
center = 3:length(dx);
right = 3:length(dx);
layer = [0, layer];
param.U_left = [0, 1 ./ (dx(center) ./ (*param.k(layer(center))) + dx(left) ./ (*param.
    k(layer(left))))];
param.U_right = [ 1 ./ (dx(center-1)./ (*param.k(layer(center-1))) + dx(right) ./
    (*param.k(layer(right)))), 0];
param.U_left(1) = 1 ./ (dx(1) ./ (*param.k(1)) + 1/const.h_surr);
param.U_right(end)= 1 ./ (dx(end) ./ (*param.k(end)) + 1/const.h_surr);
layer = layer(2:end);
```

% Define temperature matrix

```
T = const.T_in*ones(size(x));      % Preallocate output
T(1) = const.T_0;                  % Left hand end plate temperature
T(end) = const.T_end;              % Right hand end plate temperature
options = odeset('OutputFcn',@(t,y,opt) heatplot(t,y,opt,x));
[t,T] = ode45(@(t,T) heat(t,T,x,layer',param,const), [linspace(0,const.tfinal,100)], T,
    options);
end % of function
```

```
%---------------------------------------------------------------------------
```

```
function dTdt = heat(t,T,x,layer,param,const)
```

% Heat transfer equations for fuel cell

% Make a convenient place to set a breakpoint

```
if (t > 30)
s = 1;
end

dT = diff(T);      % Gives approximate derivatives between T's
dx = diff(x);      % Gives approximate derivatives between x's

dT_left = -dT(1:end-1);
dT_right = dT(2:end);
dx = dx(2:end);
```

```
% Common energy balance terms

Q_right = param.U_right.*param.A(layer).*dT_right';
Q_left = param.U_left.*param.A(layer).*dT_left';
mass = param.den(layer).*param.A(layer).*param.Cp(layer).*dx;

%================================================
% Combine into rate of change of T
%================================================

dTdt = (Q_left + Q_right) ./ (mass);
dTdt = [0;dTdt(:);0];
end % of function

%------------------------------------------------

function status = heatplot(t,y,opt,x)
if isempty(opt)
stairs(x,y), title(['t = ',num2str(t)])
status = 0;
drawnow
end

end % of function
```

Figures 6-14 and 6-15 show the temperature plots obtained at 60, 300, and 1000 seconds using one and ten slices per fuel cell layer. As illustrated in the figures, the temperature distribution through the stack begins to become more accurate as the number of slices in the layer increases.

6.4 Fuel Cell Heat Management

The heat distribution through a fuel cell stack needs precise temperature control in order for the system to run efficiently. When a PEM fuel cell is run at higher temperatures, the kinetics will be faster, which enables a voltage gain that typically exceeds the activation voltage losses. Higher operating temperature also means more of the product water is vaporized; thus, more waste heat goes into the latent heat of vaporization, and less liquid water is left to be pushed out of the fuel cell.

Most fuel cell stacks require some type of cooling system to maintain temperature homogeneity throughout the fuel cell stack. Small and micro-fuel cells may not require a cooling system, and often can be designed to be self-cooled. Cooling can be achieved through a number of means. One of the simplest solutions for cooling a fuel cell stack is through free convection. This method does not require any complicated designs or coolant and

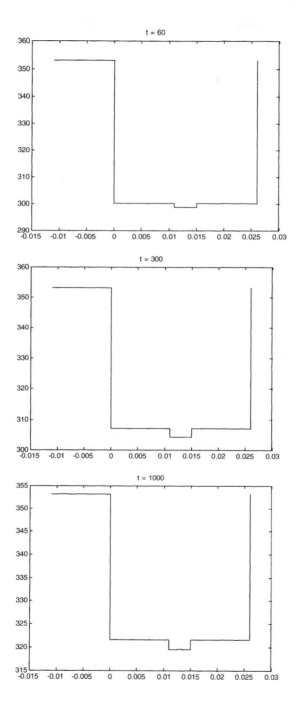

FIGURE 6-14. Temperature plots for t = 60, 300, and 1000 sec using 1 slice per layer.

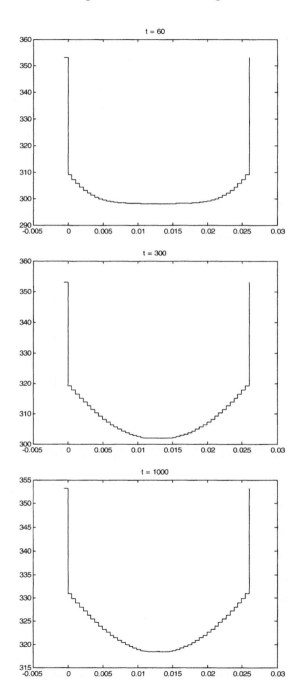

FIGURE 6-15. Temperature plots for t = 60, 300, and 1000 sec using 10 slices per layer.

can be suitable for small or low-power fuel cell stacks. Heat dissipation can be achieved through manufacturing fins or through an open cathode flow field design. Condenser cooling allows the stack to operate at higher temperatures than other cooling types. In this system, the water can be condensed from the exhaust and then reintroduced into the stack. Heat spreaders can help transfer heat to the outside of the stack through conduction, which then dissipates into the surroundings using natural or forced convection. A common method for cooling fuel cell stacks is using cooling plates. Thin cooling plates can be manufactured and inserted into the fuel cell, or additional channels can be machined in the bipolar plates to allow air, water, or coolant to flow through the channels to remove heat from the stack.

6.4.1 Air Cooling

Many PEM fuel cells use air for cooling the fuel cell. There are several methods of accomplishing this, and popular methods include putting cooling channels into the bipolar plates, or using separate fuel cell cooling plates. The air flow rate can be found from a simple heat balance. The heat transferred into the air is[9]:

$$Q = m_{coolant} c_p \left(T_{coolant,out} - T_{coolant,in} \right) \tag{6-66}$$

In order to estimate the maximum wall temperature of the channel, a heat transfer coefficient will be used. The Nusselt number is:

$$Nu = \frac{h D_h}{k} \tag{6-67}$$

where Nu is the Nusselt number, D_h is the hydraulic diameter, h is the convection heat transfer coefficient, and k is the fluid heat conductivity (W/mK). For channels that have a constant heat flux at the boundary, with a high aspect ratio and laminar flow, Incropera and DeWitt[10] report Nu = 8.23.

The hydraulic diameter (D_h) can be defined as:

$$D_h = \frac{4 A_c}{P_{cs}} \tag{6-68}$$

where A_c and P_{cs} are the cross-sectional area and the perimeter of the circular cooling channel, respectively. Some Nusselt numbers for Reynolds numbers <2300 are shown in Table 6-3.

Recall that for a circular channel, the Reynolds number must be <2300 to ensure laminar flow through the channel. The Reynolds number, Re, can be computed using Equation 6-59:

TABLE 6-3
Nusselt Numbers for Channel Aspect Ratios for Reynold's Numbers <2300[11]

Channel Aspect Ratio	Nusselt Number
1	3.61
2	4.12
4	5.33
8	6.49

$$Re = \frac{\rho v_m D_{ch}}{\mu} = \frac{v_m D_{ch}}{v} \tag{6-69}$$

where v_m is the characteristic velocity of the flow (m/s), D_{ch} is the flow channel diameter or characteristic length (m), ρ is the fluid density (kg/m³), μ is the fluid viscosity (Ns/m²), and v is the kinematic viscosity (m²/s).

This equation is altered slightly for the coolant:

$$Re = \frac{4m_{coolant}}{\mu_{gas}P_{cs}} \tag{6-70}$$

where μ is the gas or fluid viscosity (Ns/m²).

An empirical correlation from Incropera and DeWitt[12] allows both a Nusselt number and a heat transfer coefficient for air to be determined.

$$Nu = 0.664 Re1/2 * Pr1/3 (Pr > 0.6) \tag{6-71}$$

The coolant heats up as it travels along the channel, therefore, there is a temperature difference between the inlet and outlet of the flow channel. Assuming a uniform heat flux, the temperature difference between the solid and gas is:

$$Q = L_{plate}P_{cs}h(T_{surface} - T_{gas}) \tag{6-72}$$

where L_{plate} is the length of the bipolar plate, $T_{surface}$ is the temperature of the bipolar plate surface, and T_{gas} is the temperature of the gas.

The relationship between the surface temperature and cell edge can be obtained using an energy balance within the bipolar plate, cathode, and anode:

$$Q = L_{plate}P_{cs}k_{solid}\frac{(T_{edge} - T_{surface})}{t_{bc}} \tag{6-73}$$

where t_{bc} is the thickness of bipolar plate, cathode, anode, and electrolyte; k_{solid} is the solid heat conductivity (W/mK); and T_{edge} is the temperature of the cell edge. These equations assume that the temperature difference is constant along the entire channel due to the assumption of a constant heat flux.

EXAMPLE 6-4: Coolant Mass Flow Rate

A 100-W PEMFC needs to maintain a consistent temperature in order to provide adequate power to the load. The maximum operating temperature that this fuel cell is designed for is 80 °C. The fuel cell stack is cooled using natural convection with air at 22 °C. The length of the bipolar plate is 10 cm. The cooling channel and cell plus bipolar plate thickness are 10 and 0.4 cm, respectively. The width of the channel is 0.1 cm. The values for the thermal conductivity of the solid and gas are 20 W/mK and 0.0263 W/mK, respectively. The viscosity of the gas is 1.84×10^{-4} g/cms, and the specific heat is 1.0 J/gK. The heat generated per cell is 2 W. What is the air mass flow rate required?

In order to find the temperature of the surface, the channel perimeter needs to be calculated:

$$2d + 2w = 2*0.10 + 2*0.001 = 0.202 \text{m}$$

The solid surface temperature at the cooling channel exit is:

$$Q = L_{plate} P_{cs} k_{solid} \frac{(T_{edge} - T_{surface})}{t_{bc}}$$

$$T_{surface} = 80C - \frac{2W}{0.10} * \frac{1}{0.202} * \frac{0.004}{20 \frac{W}{mK}} * \frac{1C}{1K} = 79.98C$$

The hydraulic diameter is given by:

$$D_h = \frac{4A_{cross\text{-}section}}{P_{channel}} = \frac{4dw}{2(d+w)}$$

$$D_h = \frac{4}{2} \left(\frac{0.10\text{m}}{0.10\text{m} + 0.001\text{m}} \right) 0.001\text{m} = 0.002\text{m}$$

Which is close to 2w as d >> w. The heat transfer coefficient is then calculated:

$$h = 8.23 \frac{k_{gas}}{D_h} = 8.23 * \frac{0.0263 \text{W/mK}}{0.002\text{m}} = 109.31 \frac{W}{m^2 K}$$

The heat transfer coefficient is used to determine the gas exit temperature:

$$T_{gas} = T_{surface} - \frac{Q_{cell}}{hLP_{channel}}$$

$$T_{gas} = 80C - \frac{2W}{109.31W/m^2K}\left(\frac{1}{0.10m}\right)\left(\frac{1}{0.202m}\right)\frac{C}{K} = 79.07C$$

The air mass flow rate can be determined by setting $T_{coolant,out} = T_{gas}$.

$$\dot{m}_{coolant} = \frac{Q_{cell}}{C_p T_{coolant,out} - T_{coolant,in}} = \frac{2W}{1.0\frac{J}{gK}(79.07 - 22C)} = 0.035\,g/s$$

The total mass flow rate of coolant is equal to the value for one cell multiplied by the number of cells.

Using MATLAB to solve:

```
%%%%%%%%%%%%%%%%%%%%%%%%%%%%%%%%%%%%%%%%%
% EXAMPLE 6-4: Coolant Mass Flow Rate
%%%%%%%%%%%%%%%%%%%%%%%%%%%%%%%%%%%%%%%%%
w_chan = 0.001;          % Channel width (m)
t_chan = 0.10;           % Channel thickness (m)
L_bpp = 0.10;            % Bipolar plate length (m)
t_bpp_cell = 0.004;      % Bipolar plate and cell thickness (m)
T_max = 80 + 273.15;     % Maximum operating temperature (K)
T_cool = 22 + 273.15;    % Cooling temperature (K)
Q_cell = 2;              % Heat generated per cell (W)
cp = 1.0;                % Specific heat (J/gK)
k_gas = 0.0263;          % Thermal Conductivity of gas (W/mK)
k_solid = 20;            % Thermal Conductivity of solid (W/mK)
```

% Calculate the channel perimeter

```
Peri = 2*t_chan + 2*w_chan
```

% The solid surface temperature at the cooling channel exit

```
T_surface = T_max - Q_cell / Peri*t_bpp_cell / L_bpp / k_solid
```

% Calculate the hydraulic diameter

```
D_h = 4*t_chan*w_chan / Peri
```

% **Heat transfer coefficent**

h = 8.23 * k_gas / D_h

% **Gas exit temperature**

T_gas = T_surface − Q_cell / (h * L_bpp * Peri)

% **Air mass flow rate of the coolant**

m_coolant = Q_cell / (cp * (T_gas − T_cool))

6.4.2 Edge Cooling

Another commonly used method for cooling the fuel cell is to remove heat from the sides of the cell instead of between the cells. For this case, the one-dimensional heat transfer that can be in a flat plate with heat generation is:

$$\frac{d^2 T}{dx^2} + \frac{Q}{kAd_{BP}^{eff}} = 0 \tag{6-74}$$

where Q is the heat generated in the cell, W, K is the bipolar plate in-plane thermal conductivity, W/(m * K), A is the cell active area, m², and D_{BP}^{eff} is the average thickness of the bipolar plate in the active area, m.

The solution of the last equation for symmetrical cooling on both sides with $T(0) = T(L) = T_0$ is

$$T - T_0 = \frac{Q}{kAd_{BP}^{eff}} \frac{L^2}{2} \left[\frac{x}{L} - \left(\frac{x}{L}\right)^2 \right] \tag{6-75}$$

where T_0 is the temperature at the edges of the active area, and L is the width of the active area. The maximum temperature difference between the edge and the center is[13]:

$$\Delta T_{max} = \frac{Q}{kAd_{BP}^{eff}} \frac{L^2}{8} \tag{6-76}$$

The thickness of the plate at the border is d_{BP}. According to Fourier's law, the temperature will be

$$T_0 - T_b = \frac{Q}{2kA} \frac{L}{d_b} b \tag{6-77}$$

where t_b is the temperature at the edge of the bipolar plate. The total temperature difference between the center of the plate and the edge of the plate is

$$\Delta T_{max} = \frac{Q}{kA} L \left(\frac{L}{8 d_{BP}^{eff}} + \frac{b}{2 d_{BP}} \right) \qquad (6\text{-}78)$$

Chapter Summary

The calculation of heat transfer through the fuel cell stack is very important because it affects reaction kinetics, water loss, and membrane hydration and heat loss. All of these characteristics ultimately determine the fuel cell power output. The general energy balance states that the enthalpy of the reactants entering the cell equals the enthalpy of the products leaving the cell plus the sum of heat generated, the power output, and the rate of heat loss to the surroundings. Conducting energy balances on fuel cell systems is especially important because design of the subsystems and fuel cell stack vary.

Problems

- A fuel cell with a 25-cm² active area generates 0.8 A/cm² at 0.70 V. The air at the inlet is completely saturated at 80 °C and 1 atm. The oxygen stoichiometric ratio is 2.0. Calculate the heat generated, assuming that hydrogen is supplied in a dead-ended mode.
- Calculate the temperature at the center of a flow field (3 cm) of a fuel cell operating at 0.65 V and 0.60 A/cm². The bipolar plate is made of graphite with k = 22 W/mK; it is 2.2 mm thick in the active area, and 3 mm thick at the border. The border around the active area is 4 mm wide.
- Calculate the heat generated for a fuel cell with a 100-cm² active area that generates 1 A/cm² at 0.60 V. The fuel cell operates at 70 °C and 3 atm. The oxygen stoichiometric ratio is 2.5.
- A fuel cell operates at 0.7 V and 0.8 A/cm². Calculate the temperature distribution through the gas diffusion layer-bipolar plate on the cathode side. The ionic resistance through the membrane is 0.12 Ohm-cm². The heat is removed from the plate by a cooling fluid at 25 °C, with a heat transfer coefficient, h = 1600 W/m²k. The electrical resistivity of the gas diffusion layer and the bipolar plate is 0.07 Ohm-cm and 0.06 Ohm-cm, respectively. The contact resistance between the gas diffusion layer and the bipolar plate is 0.006 Ohm-cm. The effective thermal conductivity of the GDL and the bipolar plate is 16 W/mK and 19 W/mK, respectively. The thickness of the GDL and the bipolar plate is 0.30 and 2.5 mm, respectively.

Endnotes

[1] Spiegel, C.S. Designing and Building Fuel Cells. 2007. New York: McGraw-Hill.
[2] Barbir, F. *PEM Fuel Cells: Theory and Practice.* 2005. Burlington, MA: Elsevier Academic Press.
[3] Spiegel, *Designing and Building Fuel Cells.*
[4] Barbir, *PEM Fuel Cells: Theory and Practice.*
[5] Ibid.
[6] Nellis, G. ME 564: Elementary Heat Transfer. Course Notes. College of Engineering. University of Wisconsin-Madison. http://courses.engr.wisc.edu/ecow/get/me/564/nellis/ Last updated 02/27/08. Last Accessed 03/10/08.
[7] Sundaresan, Meena. A Thermal Model to Evaluate Sub-Freezing Startup for a Direct Hydrogen Hybrid Fuel Cell Vehicle Polymer Electrolyte. Fuel cell start and system. March 2004. Ph.D. Dissertation. University of California Davis, California, U.S.A.
[8] Ibid.
[9] Barbir, *PEM Fuel Cells: Theory and Practice.*
[10] Incropera, F., and D. deWitt. 1996. *Fundamentals of Heat and Mass Transfer.* 4th ed. New York: Wiley & Sons.
[11] Kutz, M. 2006. *Heat Transfer Calculations.* New York: McGraw-Hill.
[12] Incropera, *Fundamentals of Heat and Mass Transfer.* 4th ed.
[13] Barbir, *PEM Fuel Cells: Theory and Practice.*

Bibliography

Chase, M.W., Jr. et al. 1985. *JANAF Thermochemical Tables.* 3rd ed. American Chemical Society and the American Institute for Physics, *J. Physical and Chemical Reference Data.* Vol. 14, Supplement 1.

Chen, R., and T.S. Zhao. Mathematical modeling of a passive feed DMFC with heat transfer effect. *J. Power Sources.* Vol. 152, 2005, pp. 122–130.

Faghri, A., and Z. Guo. Challenges and opportunities of thermal management issues related to fuel cell technology and modeling. *Int. J. Heat Mass Transfer.* Vol. 48, 2005, pp. 3891–3920.

Felder, R.M., and R.W. Rousseau. 1986. *Elementary Principles of Chemical Processes.* 2nd ed. New York: John Wiley & Sons.

Fuel Cell Group, EEI, AIST. *Thermal and Physical Properties of Materials for Fuel Cells.* Available at: http://unit.aist.go.jp/energy/fuelcells/english/database/thphy1.html. Accessed March 24, 2007.

Fuel Cell Handbook. 5th ed. October 2000. EG&G Services, Parsons Incorporated, Science Applications International Corporation, U.S. Department of Energy.

Graf, C., A. Vath, and N. Nicolosos. Modeling of heat transfer in a portable PEFC system within MATLAB-Simulink. *J. Power Sources.* Vol. 155, 2006, pp. 52–59.

Hwang, J.J., and P.Y. Chen. heat/mass transfer in porous electrodes of fuel cells. *Int. J. Heat Mass Transfer.* Vol. 49, 2006, pp. 2315–2327.

Kulikovsky, A.A. Heat balance in the catalyst layer and the boundary condition for heat transport equation in a low-temperature fuel cell. *J. Power Sources.* Vol. 162, 2006, pp. 1236–1240.

Li, X. *Principles of Fuel Cells.* 2006. New York: Taylor & Francis Group.

Lin, B. 1999. Conceptual design and modeling of a fuel cell scooter for urban Asia. Princeton University, masters thesis.

Mench, M.M., C.-Y. Wang, and S.T. Tynell. *An Introduction to Fuel Cells and Related Transport Phenomena.* Department of Mechanical and Nuclear Engineering, Pennsylvania State University. PA, USA. Draft. Available at: http://mtrl1.mne.psu.edu/Document/jtpoverview.pdf. Accessed March 4, 2007.

Mench, M.M., Z.H. Wang, K. Bhatia, and C.Y. Wang. 2001. *Design of a Micro-Direct Methanol Fuel Cell.* Electrochemical Engine Center, Department of Mechanical and Nuclear Engineering, Pennsylvania State University. PA, USA.

Middleman, S. 1998. *An Introduction to Mass and Heat Transfer. Principles of Analysis and Design.* New York: John Wiley & Sons.

O'Hayre, R., S.-W. Cha, W. Colella, and F.B. Prinz. 2006. *Fuel Cell Fundamentals.* New York: John Wiley & Sons.

Pitts, D., and L. Sissom. 1998. *Heat Transfer.* 2nd ed. Schaum's Outline Series. New York: McGraw-Hill.

Rowe, A., and X. Li. Mathematical modeling of proton exchange membrane fuel cells. *J. Power Sources.* Vol. 102, 2001, pp. 82–96.

Sousa, R., Jr., and E. Gonzalez. Mathematical modeling of polymer electrolyte fuel cells. *J. Power Sources.* Vol. 147, 2005, pp. 32–45.

Springer et al. Polymer electrolyte fuel cell model. *J. Electrochem. Soc.* Vol. 138, No. 8, 1991, pp. 2334–2342.

van den Oosterkamp, P.F. Critical issues in heat transfer for fuel cell systems. *Energy Convers. Manage.* Vol. 47, 2006, pp. 3552–3561.

You, L., and H. Liu. A two-phase flow and transport model for PEM fuel cells. *J. Power Sources.* Vol. 155, 2006, pp. 219–230.

CHAPTER 7

Modeling the Proton Exchange Structure

7.1 Introduction

The electrolyte layer is essential for a fuel cell to work properly. In PEM fuel cells (PEMFCs), the fuel travels to the catalyst layer and gets broken into protons (H^+) and electrons. The electrons travel to the external circuit to power the load, and the hydrogen protons travel through the electrolyte until they reach the cathode to combine with oxygen to form water. The PEM fuel cell electrolyte must meet the following requirements in order for the fuel cell to work properly:

- High ionic conductivity
- Present an adequate barrier to the reactants
- Be chemically and mechanically stable
- Low electronic conductivity
- Ease of manufacturability/availability
- Preferably low cost

The membrane layer contains the solid polymer membrane, liquid water, and may also contain water vapor and trace amounts of H_2, O_2, or CO_2 depending upon the purity of the H_2 coming into the system. Various models use different equations which are derived from the same governing equations. Table 7-1 summarizes the most commonly used equations for creating an accurate model of the PEM layer. Specific topics covered in this chapter include:

- Mass and species conservation
- Ion transport
- Momentum conservation
- Conservation of energy
- Other required relations

TABLE 7-1
Equations Used to Model the PEM Layer

Model Characteristic	Description/Equations
No. of dimensions	1, 2, or 3
Mode of operation	Dynamic or steady-state
Phases	Gas, liquid, or a combination of gas and liquid
Mass transport	Nernst-Planck + Schlogl, Nernst-Planck + drag coefficient, or Stefan-Maxwell equation
Ion transport	Ohm's law
Membrane swelling	Empirical or thermodynamic models
Energy balance	Isothermal or full energy balance

This chapter explains the physical characteristics, properties, and current modeling theories for the polymer exchange membranes used for PEM fuel cells.

7.2 Physical Description of the Proton Exchange Membrane

The standard electrolyte material presently used in PEMFCs is a copolymer of poly(tetrafluoroethylene) and polysulfonyl fluoride vinyl ether. The polymer is stable in both oxidative and reductive environments and has high protonic conductivity (0.2 S/cm) at typical PEMFC operating temperatures. The thickness of these membranes ranges from 50 to 175 microns (μm).

The proton-conducting membrane usually consists of a PTFE-based polymer backbone, to which sulfonic acid groups are attached. The proton-conducting membrane works well for fuel cell applications because the H^+ jumps from SO_3 site to SO_3 site throughout the material. The H^+ emerges on the other side of the membrane. The membrane must remain hydrated to be proton-conductive. This limits the operating temperature of PEMFCs to under the boiling point of water and makes water management a key issue in PEMFC development. Figures 7-1 and 7-2 illustrate the SO_3 sites in the Nafion membrane.

Perfluorosulfonic acid (PFSA) membranes, such as Nafion, have a low cell resistance (0.05 Q cm^2) for a 100-μm-thick membrane with a voltage loss of only 50 mV at 1 A/cm^2. Some disadvantages of PFSA membranes include material cost, supporting structure requirements, and temperature-related limitations. The fuel cell efficiency increases at higher temperatures, but issues with the membrane, such as membrane dehydration, reduction of ionic conductivity, decreased affinity for water, loss of mechanical strength via softening of the polymer backbone, and increased parasitic losses through high fuel permeation, become worse. PFSA membranes must be kept hydrated in order to retain proton conductivity, but the operating temperature must be kept below the boiling point of water.

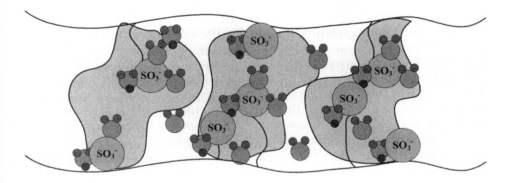

FIGURE 7-1. A pictorial illustration of Nafion.

$$[(CF_2\text{---}CF_2)_m\text{---}CF\text{---}CF_2]_n$$

$$\left[\begin{array}{c} O \\ | \\ CF_2 \\ | \\ CF\text{---}CF_3 \end{array}\right]_z$$

$$\begin{array}{c} O \\ | \\ CF_2 \\ | \\ CF_2 \\ | \\ SO_3H \end{array}$$

FIGURE 7-2. The chemical structure of Nafion.

The most popular type of electrolyte used in PEMFCs is made by DuPont and has the generic brand name Nafion. The Nafion membranes are stable against chemical attack in strong bases, strong oxidizing and reducing acids, H_2O_2, Cl_2, H_2, and O_2 at temperatures up to 125 °C. Figure 7-2 illustrates the chemical structure.

When modeling the polymer electrolyte membrane, it is typically assumed that the concentration of positive ions is fixed by electroneutrality, which means that a proton occupies every fixed SO_3^- charge site. The charge sites are assumed to be distributed homogeneously throughout the membrane, which results in a constant proton concentration in the membrane. A flux of protons, thus, results from a potential gradient and not a concentration gradient. In addition, the number of protons that can be transported is only one, which helps to simplify the governing transport equations. The permeation of reactants into the electrolyte results in mixed potentials at the electrodes, reduced performance, and possibly degradation of the catalyst. The polymer electrolyte membrane contains water and hydrogen protons; therefore, the transfer of the water and protons are important phenomena to investigate. In addition to species transfer, the primary phenomena investigated inside the membrane are energy transfer and potential conservation. For water transport, the principle driving forces are a convective force, an osmotic force (i.e., diffusion), and an electric force. The first of these results from a pressure gradient, the second from a concentration gradient, and the third from the migration of protons from anode to cathode and their effect (drag) on the dipole water molecules. Proton transport is described as a protonic current and consists of this proton-driven flux and a convective flux due to the pressure-driven flow of water in the membrane. Figure 7-3 illustrates the transport phenomena for the protons taking place within the membrane.

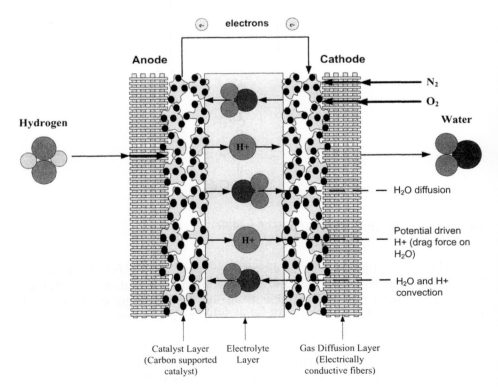

FIGURE 7-3. Membrane transport phenomena.

7.3 Types of Models

Most fuel cell models that have been developed assume that the membrane system is a single phase. The membrane system is assumed to have three main components: the membrane, protons, and water. Therefore, there are three main transport properties. This assumption neglects any other ion types that may be in the membrane and does not consider hydrogen or oxygen crossover in the membrane. The effect of hydrogen or oxygen crossover does not significantly influence water or proton transport and, therefore, can be neglected in most fuel cell models without affecting the model efficiency. Table 7-2 summarizes the types of models in the literature for fuel cell membranes.

7.3.1 Microscopic and Physical Models

There have been numerous microscopic models that have been based on statistical mechanics, molecular dynamics, and other types of macroscopic phenomena. These models are valuable because they provide a fundamental understanding of the processes such as diffusion and conduction on a microscopic level. These models also observe effects that are ignored in the macroscopic models, such as ionic and backbone moieties, and conduction through different proton–water complexes. Almost all microscopic models treat the membrane as a two-phase system. Although these models provide valuable information, they are usually too complex to be integrated into

TABLE 7-2
Types of Models

Type of Model	Description
Microscopic and physical models	Microscopic and physical models provide a fundamental understanding of many membrane processes such as diffusion and conduction in the membrane on a pore level.
Diffusive models	Diffusive models treat the membrane system as a single phase. It is assumed that the membrane is a vapor-equilibrated membrane, where the water and protons dissolve and move by diffusion. The common types are dilute and concentrated solution theory.
Hydraulic models	Hydraulic models assume that the membrane system has two phases, which are the membrane and liquid water.
Hydraulic–diffusive models	Hydraulic–diffusive models consider both diffusion and pressure-driven flow.
Combination models	Combination models include the characteristics of all of the above models and include the effects when the membrane is saturated with water and dehydrated.

an overall fuel cell model. One important membrane property that should be integrated into a macroscopic model is how the membrane structure changes as a function of water content (where λ is the moles of water per mole of sulfonic acid sites). This property is well documented in the literature and can be measured by examining the weight gain of an equilibrated membrane.

The dry membrane absorbs water in order to solvate the acid groups. The initial water content is associated strongly with the sites, and the addition of water causes the water to become less bound to the polymer and causes the water droplets to aggregate. These water clusters eventually grow and form interconnections with each other. These connections create "water channels," which are transitory, and have hydrophobicities comparable to that of the matrix. A transport pathway forms when water clusters are close together and become linked. An illustration of the water uptake of the Nafion membrane is shown in Figure 7-4. The percolation phenomenon begins around $\lambda = 2$. The next stage occurs when a complete cluster-channel network has formed. In the last stage, the channels are now filled with liquid, and the uptake of the membrane has increased without a change in the chemical potential of water. This phenomenon is known as Schroeder's paradox[1].

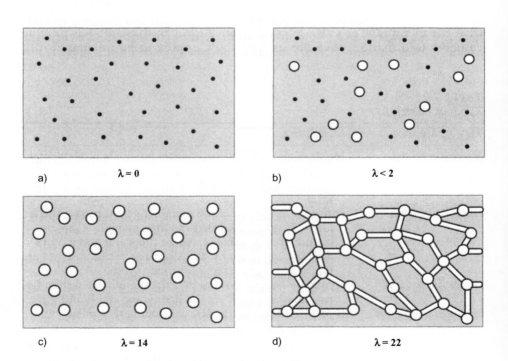

FIGURE 7-4. A pictorial illustration of the water uptake of Nafion[2].

7.3.2 Diffusive Models

Diffusive membrane models treat the membrane system as a single phase, and correspond to part c of Figure 7-4. Typically, this system has no true pores, the collapsed channels fluctuate, and the system is treated as a single, homogeneous phase where water and protons dissolve and move by diffusion. For the proton movement, Ohm's law is used:

$$i_2 = -\kappa \nabla \Phi_2 \qquad (7\text{-}1)$$

where κ is the ionic conductivity of the membrane. This can easily be integrated to yield a resistance for use in a polarization equation in a zero-dimensional model.

7.3.3 Dilute Solution Theory

The Nernst-Planck equation first introduced in Chapter 5 yields:

$$N_i = -z_i \frac{F}{RT} D_i C_i \frac{d\Phi_m}{dx} - D_i \frac{dC_i}{dx} + vC_i \qquad (7\text{-}2)$$

where N_i is the superficial flux of species i, z_i is the charge number of species i, C_i is the concentration of species i, D is the diffusion coefficient of species i, Φ_m is the electrical potential in the membrane, and v is the velocity of H_2O. The first term is a migration term, representing the motion of charged species that results from a potential gradient. The migration flux is related to the potential gradient $(-\nabla\Phi_2)$ by a charge number, z_i, concentration, c_i, and mobility, u_i. The second term relates the diffusive flux to the concentration gradient. The final term is a convective term and represents the motion of the species as the bulk motion of the solvent carries it along.

In single-phase systems, the solvent is assumed to be the membrane. Dilute solution theory only considers the interactions between each dissolved species and the solvent. The motion of each charged species is described by its transport properties, which are the mobility and the diffusion coefficient, which are related to each other by the Nernst-Einstein equation:

$$D_i = RTu_i \qquad (7\text{-}3)$$

If the solution species are very dilute, then the interactions among them can be neglected, and just the material balances can be used. If water movement in the membrane is considered, the Nernst-Planck equation will also be needed. As the protons move across the membrane, they induce a flow of water in the same direction. This electroosmotic flow is a result of the proton–water interaction and is not a dilute solution effect because the membrane is taken to be the solvent. The electroosmotic flux is proportional to the current density and can be added to the diffusive flux to get the overall flux of water:

$$N_{w,2} = \xi \frac{i_2}{F} - D_w \nabla c_{w,2} \qquad (7\text{-}4)$$

where ξ is the electroosmotic coefficient. Most single-phase models use Equation 7-4 with Ohm's law (Equation 7-1). Differences in the models arise from the functions used for the transport properties and the concentration of water in the membrane.

7.3.4 Concentrated Solution Theory

Concentrated solution theory can easily be used when an electrolyte is modeled with three species. This model can take into effect the binary interactions between all of the species. The equations for the three-species system are:

$$i_2 = -\frac{\kappa \xi}{F} \nabla \mu_{w,2} - \kappa \nabla \Phi_2 \quad \text{and} \quad N_{w,2} = \xi \frac{i_2}{F} - \alpha_w \nabla \mu_{w,2} \qquad (7\text{-}5)$$

where μ_w represents the chemical potential of water, and α_w is the transport coefficient of water. The equation for the membrane is ignored, since it is dependent on the other two equations by the Gibbs-Duhem equation. For many models in the literature, these equations were used in a Stefan-Maxwell framework.

7.3.5 Membrane Water Content

In addition to using a dilute or concentrated solution theory, functional forms for the transport parameters and the concentration of water are needed. These properties include temperature and water content. Different models determine the membrane water content in different ways. The majority of models correlate with water activity since it is easily calculated. Other models in the literature use the Flory-Huggins theory, simple mass transfer relationships, capillary arguments, or equilibrium between water and protons in the membrane.

Schroeder's paradox is an observed phenomenon that needs to be considered in any model where the membrane is not either fully hydrated or dehydrated. There are several ways in which this can be accounted for: (1) it can be ignored by assuming that the membrane is fully hydrated or only vapor filled, or (2) a relation can be made between water content and water activity. As the water content increases, the properties of the

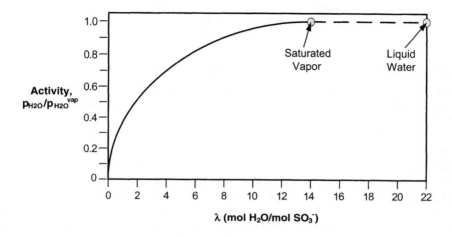

FIGURE 7-5. Water uptake isotherm at 25 °C showing the effect of Schroeder's paradox.

membrane change. Nafion exhibits a water-uptake isotherm as shown in Figure 7-5.

There also are many models that use an empirical expression for the isotherm. One of the first models to use an isotherm was that by Springer et al.[3] In that model, lambda was used to represent the amount of water flow in the membrane, and an activity coefficient was used to account for the isotherm behavior. This empirical relationship is one of the most commonly used in the literature to model the membrane, and is written in Sections 7.4.1, 7.4.6, and 7.4.7.

7.3.6 Hydraulic Models

There are also many models in the literature that assume that the membrane system is two phases. This is accomplished by assuming that the membrane has pores that are filled with liquid water; therefore, the two phases are water and membrane. The additional degree of freedom allows the inclusion of a pressure gradient in the water because of a possibly unknown stress relation between the membrane and fluid at every point in the membrane. Most of these models assume that the water is pure, and the water content of the membrane is assumed to remain constant as long as the pores are filled and the membrane has been pretreated appropriately. The first model to describe the membrane in this manner was that of Bernardi and Verbrugge, which was based on earlier work by Verbrugge and Hill. This model assumed a dilute solution approach that used the Nernst-Planck equation to describe the movement of protons. This is because there are two phases; the protons are in the water and the velocity of the water is given by Schlogl's equation:

$$v_{w,2} = -\left(\frac{k}{\mu}\right)\nabla p_L - \left(\frac{k_\Phi}{\mu}\right)z_f c_f F \nabla \Phi_2 \qquad (7\text{-}6)$$

where k and k_Φ are the effective hydraulic and electrokinetic permeability, respectively, p_L is the hydraulic or liquid pressure, μ is the water viscosity, and z_f and c_f refer to the charge and concentration of fixed ionic sites, respectively.

The movement of water can be attributed to a potential gradient and a pressure gradient. The movement of water by a pressure gradient is determined primarily by an effective permeability of water moving through the pore network. This approach is quite useful for describing fuel cell systems where the membrane is well hydrated, but it requires that the water content be uniform across the membrane, with only a pressure gradient as a driving force for water movement. Such a treatment does not necessarily lend itself to describing the flux of water resulting when there is a water-activity gradient across the membrane.

Unlike the cases of the single-phase models above, the transport properties are constant because the water content does not vary, and thus, one can expect a linear gradient in pressure. However, due to Schroeder's paradox, different functional forms might be expected for the vapor- and liquid-equilibrated membranes. The equations for the concentrated solution theory are the same for both one and two phases, except that the chemical potential is replaced by the hydraulic pressure and the transport coefficient is related to the permeability through comparison to Darcy's law. Therefore, Equation 7-5 becomes:

$$N_{w,2} = \xi\frac{i_2}{F} - \frac{k}{\mu V_w}\nabla p_L \qquad (7\text{-}7)$$

where V_w is the molar volume of water.

7.3.7 Combination Models

There is a need to be able to describe both types of behavior, diffusive and hydraulic, in a consistent manner, which also agrees with experimental data. For example, a membrane with low water content is expected to be controlled by diffusion, and an uptake isotherm needs to be used. This is due to the fact that there is not a continuous liquid pathway across the medium, and that the membrane matrix interacts significantly with the water due to binding and solvating the sulfonic acid sites. A hydraulic pressure in this system may not be defined. On the other hand, when the membrane is saturated, transport still occurs. This transport must be due to a hydraulic-pressure gradient because oversaturated activities are nonphysical. A model that combines the concepts from both the diffusive and hydraulic models would most accurately describe the membrane system. The two types of models are seen as operating fully at the limits of water concentration and must somehow be averaged between those limits. As mentioned, the hydraulic/diffusive models try to do this, but Schroeder's paradox and its effects on the transport properties are not taken into consideration.

7.4 Proton Exchange Membrane Modeling Example

In order to model the electrolyte accurately, the transport of mass, charge, and energy must be included in the model. A form of Equation 7-1 must be solved in the electrolyte for ion transport. Contact resistance between the electrode and the electrolyte can also be significant and should be incorporated into the model.

Table 7-3 lists the most common variables used for describing the membrane layer. It is important to keep in mind that the complete set of equations is seldom used in modeling; often simplifying assumptions are used. Sections 7.4.1 through 7.4.6 show an example of how these equations and variables can be used to create a model for the polymer exchange membrane.

TABLE 7-3
Fuel Cell Polymer Membrane Layer Variables and Equations

Variable	Equation	Equation No.
Overall liquid water flux (N_L)	Mass balance	7-2, 7-10, 9-4 or Chapter 5 equations
Overall membrane water flux (N_W)	Mass balance	7-2, 7-10, 9-4 or Chapter 5 equations
Gas phase component flux ($N_{G,i}$)	Mass balance	7-2, 7-10, 9-4 or Chapter 5 equations
Gas phase component partial pressure ($p_{G,i}$)	Stefan-Maxwell	5-63
Membrane water chemical potential (μ_w)	Schlogl's equation	7-6 or 7-7
Electronic phase current density (i_1)	Ohm's law	8-46
Membrane current density (i_2)	Ohm's law	7-5, 8-46, 9-2 or 9-2
Electronic phase potential (Φ_1)	Charge balance	8-46
Total gas pressure (p_G)	Darcy's law	8-50
Liquid saturation (S)	Saturation relation	8-54
Liquid pressure in membrane ($P_{L,m}$)	Darcy's law	8-52
Temperature (T)	Energy balance	Chapter 6 equations

7.4.1 Mass and Species Conservation

For both water and protons, the mass conservation equation can be represented as:

$$\frac{\partial c_i}{\partial t} = -\frac{\partial}{\partial t} N_i \qquad (7\text{-}8)$$

where i is H_2O or H^+, c_i is the molar concentration, and N_i is the molar flux due to electroosmotic driving forces and convection. In a diluted solution, N_i is given by the Nernst-Planck equation along with the Nernst-Einstein relationship:

$$N_i = J_i + c_i u^m \qquad (7\text{-}9)$$

where u^m is the mixture velocity and J_i is the diffusive flux.

In PEM fuel cells, the two important fluxes or material balances are the proton flux and the water flux. The membrane needs to stay hydrated in order to ionically conduct hydrogen; therefore, the water profile must be calculated in the electrolyte. In the Nafion membrane, two types of water flux are present: back diffusion and electroosmotic drag. From Chapter 4 and equation 7-4, both fluxes can be accounted for by the following equation:

$$J_{H2O}^M = -D_{cH2O,T}\frac{\partial c_{H2O}^m}{\partial x} + n_{drag}\frac{i_x}{F} \qquad (7\text{-}10)$$

where n_{drag} is the measured drag coefficient, i_x is the protonic current in the x direction, F is Faraday's constant, $\lambda_{H2O/SO3}$ is the water content (mol_{H2O}/mol_{SO3-}), ρ_{dry}^m is the dry membrane density (kg/m3), $D_{cH2O,T}$ is the diffusion coefficient, and M_m is the membrane molecular mass (kg/mol). The water content is not constant in this equation. The resistance of the electrolyte can be estimated using the water content, which can be described by:

$$n_{drag} = 2.5\frac{\lambda_{H2O/SO3}}{22} \qquad (7\text{-}11)$$

$$\lambda_{H2O/SO3} = \frac{c_{H2O}^m}{\dfrac{\rho_{dry}^m}{M^m} - bc_{H2O}^m} \qquad (7\text{-}12)$$

where b is the membrane extension coefficient in the x direction, which is determined experimentally, and the value b = 0.0126 is typically used.

$D_{cH2O,T}$ is the diffusion coefficient, which includes a correction for the temperature and for the water content. It is expressed in a fixed coordinate system with the dry membrane by:

$$D_{cH2O,I} = D'\left[\exp 2416\left(\frac{1}{303} - \frac{1}{T}\right)\right]\lambda_{H2O/SO3}\frac{1}{a}\frac{1}{17.81 - 78.9a + 108a^2} \quad (7\text{-}13)$$

where a is the activity of water, and D′ (m^2/s) is the diffusion coefficient measured at constant temperature, and in coordinates moving with the swelling of the membrane. D has been added to the above equation to ensure that water contents below 1.23 do not result in negative diffusion coefficients. D′ at 30 °C is written as:

$$D = 2.64227e(-13)\lambda_{H2O/SO3} \quad \text{for } \lambda_{H2O/SO3} \leq 1.23 \quad (7\text{-}14)$$

$$D = 7.75e(-11)\lambda_{H2O/SO3} - 9.5e(-11) \quad \text{for } 1.23 < \lambda_{H2O/SO3} \leq 6 \quad (7\text{-}15)$$

$$D = 2.5625e(-11)\lambda_{H2O/SO3} + 2.1625e(-10) \quad \text{for } 6 < \lambda_{H2O/SO3} \leq 14 \quad (7\text{-}16)$$

The total molar flux for water can be expressed as:

$$N_{H2O} = J_{H2O} + (c_{H2O}^m u^m) \quad (7\text{-}17)$$

where the mixture velocity u^m is given by the momentum equation below.

The mass conservation of water can be expressed as:

$$\frac{\partial c_{H2O}^m}{\partial t} = -\frac{\partial}{\partial x}J_{H2O} + -\frac{\partial}{\partial x}(c_{H2O}^m u^m) \quad (7\text{-}18)$$

where the mixture velocity u^m is given by the momentum equation.

Now, due to the assumption of electroneutrality and the homogeneous distribution of charge sites, the mass conservation of protons simplifies to:

$$\frac{\partial c_{H+}}{\partial x} = 0, \quad \frac{\partial c_{H+}}{\partial t} = 0 \quad (7\text{-}19)$$

Therefore, as soon as a current exists, the membrane is charged, and the concentration of protons remain constant. The charge of the protons equals that of the fixed charges. The diffusive molar flux for the protons (J_{H+}) can be written as:

$$J_{H+} = -\frac{F}{RT}D_{H+}c_{H+}\frac{\partial \Phi_m}{\partial x} \quad (7\text{-}20)$$

where Φ_m is the membrane proton potential, and D_H is the proton diffusivity. Combining this diffusive flux with the convective flux results in the total molar flux for the hydrogen protons, i.e.:

$$N_{H+} = J_{H+} + c_{H+}u^m \quad (7\text{-}21)$$

7.4.2 Momentum Equation

For the mixture (water and protons), the assumption is made that the momentum equation takes the form of the generalized Darcy relation:

$$u^m = -\frac{Kk_r^g}{\mu}\left[\frac{\partial p}{\partial x} - \rho g\cos\theta\right] \qquad (7\text{-}22)$$

where u^m is the mixture velocity, K is the absolute permeability of the porous medium, k_r^g is the relative permeability, g is the gravity, and θ is the angle that the x-axis (the direction of flow) makes to the direction of gravity. The mixture density and the dynamic viscosity of the mixture are written as:

$$\rho = M_{H+}c_{H+} + M_{H2O}c_{H2O} \qquad (7\text{-}23)$$

$$\mu = \frac{M_{H+}c_{H+}}{\rho}\mu_{H+} + \frac{M_{H2O}c_{H2O}}{\rho}\mu_{H2O} \qquad (7\text{-}24)$$

7.4.3 Conservation of Energy Equation

Energy is transported by conduction and convection within the three phases of the membrane (polymer, liquid/gas). The effects of ohmic losses within the membrane are taken into account by an additional source term in the energy balance equation so that energy conservation is given by:

$$\rho c_p\frac{\partial T}{\partial t} = \lambda_m\frac{\partial^2 T}{\partial x^2} - Mc_pN\frac{\partial T}{\partial x} + R_m \qquad (7\text{-}25)$$

where

$$\rho c_P = \rho_m^{dry}c_{pm} + \rho_{H2O}^m c_{p,H2O}^m + \rho_{H+}c_{pH+} \qquad (7\text{-}26)$$

$$\rho_{H+} = M_{H+}c_{H+}\rho_{H2O}^m = M_{H2O}c_{H2O}^m \qquad (7\text{-}27)$$

$$Mc_pN = M_{H2O}c_{p,H2O}^m N_{H2O} + M_{H+}c_{p,H+}N_{H+} \qquad (7\text{-}28)$$

The transient energy effect associated with mass storage within the hydrated membrane is neglected due to the fact that the dry membrane mass does not change, and is several orders of magnitude larger than that of the water that hydrates the membrane.

Substituting the expressions for N_{H2O} and N_{H+}, an expanded expression for Mc_pN can be obtained:

$$Mc_pN = M_{H2O}c^m_{p,H2O}\left(c^m_{H2O}u^m - D_{cH2O,T}\frac{\partial c^m_{p,H2O}}{\partial x} + 2.5\frac{\lambda_{H2O/SO3}}{22}\frac{i_m}{F}\right) +$$

$$M_{H+}c_{p,H+}\left(c_{H+}u^m - D_{H+}c_{H+}\frac{\partial \Phi_m}{\partial x}\right) \tag{7-29}$$

The source term, R_m, is given by

$$R_m = \frac{i^2}{\sigma_m} \tag{7-30}$$

where σ_m is the conductivity of the membrane and is written as a function of the temperature, and the water content is:

$$\sigma_m = \sigma_{m303}\exp\left[1268\left(\frac{1}{303} - \frac{1}{T}\right)\right] \tag{7-31}$$

with σ_{m303}, the conductivity of the membrane at 303 K given by:

$$\sigma_{m303} = 100*(0.005139\lambda_{H2O/SO3} - 0.00326) \quad \text{for } \lambda_{H2O/SO3} > 1 \tag{7-32}$$

7.4.4 Ion Transport

The equation for the proton potential is derived from Ohm's law, and represents the proton flux divided by the membrane conductivity. The electroneutrality assumption allows the total molar proton flux to be related directly to current density and the velocity u^m, represents the convective flux of protons. This results in the following equation:

$$\frac{\partial \Phi_m}{\partial x} = -\frac{i}{\sigma_m} + \frac{F}{\sigma_m}c_{H+}u^m \tag{7-33}$$

7.4.5 Interface Water Activity Relation

At the membrane interfaces, the water vapor activity is given by

$$a = \frac{RT}{p_{sat}(T)}c^g_{H2O} + 2s \quad a \in [0, \dots, 3] \tag{7-34}$$

where c^g_{H2O} is the water vapor concentration and s is the saturation ratio. An assumption is made that s is zero for activities less than 1, meaning that no liquid water is present in the membrane pores until the activity exceeds 1. The highest value that the first term can reach is 1; therefore, a maximum saturation ratio of 1 results in an activity of 3.

7.4.6 Membrane Water Activity Relation

The relation for the water activity within the membrane is given by the reciprocal of the sorption curve. As with the water vapor activity at the interfaces, the result from Springer et al.[4] for water vapor activity in Nafion 117 at 30°C is given by:

$$a = \frac{1}{2160}c_1 + c_2\lambda_{H2O/SO3} + (216(c_3 - c_4\lambda_{H2O/SO3} + c_5\lambda_{H2O/SO3}^2)^{1/2})^{1/3} -$$
$$134,183/2160 / c_1 + c_2\lambda_{H2O/SO3} + (216(c_3 - c_4\lambda_{H2O/SO3} + \qquad\qquad (7\text{-}35)$$
$$c_5\lambda_{H2O/SO3}^2)^{1/2})^{1/3} + \frac{797}{2160} \quad \text{for } \lambda_{H2O/SO3} \le 14$$

where c1 = −41,956e4, c2 = 139,968e3, c3 = 382,482e6, c4 = 251,739e3, and c5 = 419,904e6.

$$a = 0.7143\lambda_{H2O/SO3} - 9.0021 \quad \text{for } 14 \le \lambda_{H2O/SO3} \le 16.8 \qquad (7\text{-}36)$$

$$a = 3 \quad \text{for } 16.8 \le \lambda_{H2O/SO3} \qquad (7\text{-}37)$$

EXAMPLE 7-1: Modeling the Polymer Electrolyte Layer

Create a model for the PEM layer as a function of x (membrane position in the x direction) using Equations 7-8 to 7-37 introduced in this chapter. Set dc_H2O/dt (Equation 7-18) and dT/dt (Equation 7-25) to zero in order to solve for the steady-state distribution of the other variables. Solve the equations over 10 slices in order to get a good solution. Use the equations for the following slice-dependent state variables: the concentration of water (C_H2O), temperature (T), change in T with respect to x (dT/dx), potential (pot), and pressure (P). Plot C_H2O, T, pot, and P as a function of x.

Since there are five differential equations for each state variable, a mass matrix in MATLAB will be used to efficiently solve the differential equations simultaneously. To set up the equations, first rewrite them with all the differentials on the left side and everything else on the right side. The equations will then be arranged into an order roughly matching the order of the state variables. The terms will then be collected together, and put into a "mass" matrix m-file, which is a function of the state variables M(y). The right-hand side forms the equations you compute in the dydx function f(x,y). The resulting form of the equations is M(y) * dydx

= f(x,y) where y is the state variable and M and f are matrix functions. Use odeset to specify the mass m-file M(y) and then call one of the ode solvers.

The mass conservation of water from Equations 7-8 and 7-9 is:

$$\frac{\partial c_{H2O}^m}{\partial t} = -\frac{\partial}{\partial x}J_{H2O} + -\frac{\partial}{\partial x}(c_{H2O}^m u^m)$$

since $\frac{\partial}{\partial x}J_{H2O} = -\frac{\partial}{\partial x}(c_{H2O}^m u^m)$ when $\frac{\partial c_{H2O}^m}{\partial t} = 0$. For this example, the diffusive flux for water (Equation 7-10) will be used to calculate the water concentration:

$$J_{H_2O} = -D_{c_{H2O,T}}\frac{\partial c_{H_2O}^m}{\partial x} + 2.5\frac{\lambda_{H_2O/SO_3}}{22}\frac{i_{mx}}{F}$$

This equation can be rewritten as:

$$c_{H_2O}^{m'} = \frac{J_{H_2O}}{-D_{c_{H2O,T}}} - \frac{2.5}{-D_{c_{H2O,T}}}\frac{\lambda_{H_2O/SO_3}}{22}\frac{i_{mx}}{F}$$

The energy conservation Equation (7-25) is written as:

$$\rho c_p\frac{\partial T}{\partial t} = \lambda_m\frac{\partial^2 T}{\partial x^2} - Mc_p N\frac{\partial T}{\partial x} + R_m$$

when $\frac{\partial T}{\partial t} = 0$, then:

$$\frac{\partial^2 T}{\partial x^2} = \frac{Mc_p N}{\lambda_m}\frac{\partial T}{\partial x} - \frac{R_m}{\lambda_m}$$

rewritten as:

$$\lambda_m T'' - Mc_P NT' = -R_m$$

$$(T)' - T' = 0$$

The potential (Equation 7-38) can be expressed as:

$$\frac{\partial \Phi_m}{\partial x} = -\frac{i}{\sigma_m} + \frac{F}{\sigma_m}c_{H+}u^m$$

rewritten as:

$$\Phi'_m = -\frac{i}{\sigma_m} + \frac{F}{\sigma_m} c_{H+} u^m$$

Molar velocity of the mixture (Equation 7-22) is:

$$u^m = -\frac{K k_r^g}{\mu}\left[\frac{\partial p}{\partial x} - \rho g \cos\theta\right]$$

rewritten as:

$$p' = -\frac{u^m \mu}{K k_r^g} + \rho g \cos\theta$$

Therefore, there are five equations and five state variables. The other relations needed to calculate the parameters for the five differential equations are as follows:

Diffusive flux for hydrogen (Equation 7-20):

$$J_{H+} = -\frac{F}{RT} D_{H+} c_{H+} \frac{\partial \Phi_m}{\partial x}$$

Density for the mixture (Equation 7-23):

$$\rho_{H+} = M_{H+} c_{H+} + M_{H2O} c_{H2O}^m$$

Dynamic viscosity of the mixture (Equation 7-24):

$$\mu = \frac{M_{H+} c_{H+}}{\rho} \mu_{H+} + \frac{M_{H2O} c_{H2O}}{\rho} \mu_{H2O}$$

Diffusion coefficient (Equation 7-13):

$$D_{cH2O,I} = D'\left[\exp 2416\left(\frac{1}{303} - \frac{1}{T}\right)\right] \lambda_{H2O/SO3} \frac{1}{a} \frac{1}{17.81 - 87.9a + 108a^2}$$

D' at 30 °C is written as (Equation 7-14):

$D = 2.642276e(-13)\lambda_{H2O/SO3}$ for $\lambda_{H2O/SO3} \le 1.23$

$D = 7.75e(-11)\lambda_{H2O/SO3} - 9.5e(-11)$ for $1.23 < \lambda_{H2O/SO3} \le 6$

$D = 2.5625e(-11)\lambda_{H2O/SO3} + 2.1625e(-10)$ for $6 < \lambda_{H2O/SO3} \le 14$

Water uptake (Equation 7-12):

$$\lambda_{H2O/SO3} = \frac{c_{H2O}^m}{\dfrac{\rho_{dry}^m}{M^m} - bc_{H2O}^m}$$

Heat capacity of the phase mixture (Equations 7-26 and 7-28):

$$\rho c_p = \rho_m^{dry} c_{pm} + \rho_{H2O}^m c_{p,H2O}^m + \rho_{H+} c_{pH+}$$

$$Mc_p N = M_{H2O} c_{p,H2O}^m \left(c_{H2O}^m u^m - D_{cH2O,T} \frac{\partial c_{H2O}^m}{\partial x} + 2.5 \frac{\lambda_{H2O/SO3}}{22} \frac{i_m}{F} \right) +$$

$$M_{H+} c_{p,H+} \left(c_{H+} u^m - D_{H+} c_{H+} \frac{\partial \Phi_m}{\partial x} \right)$$

Activity of the water molecules (Equation 7-37):

$$a = \frac{1}{2160} c_1 + c_2 \lambda_{H2O/SO3} + (216(c_3 - c_4 \lambda_{H2O/SO3} +$$

$$c_5 \lambda_{H2O/SO3}^2)^{1/2})^{1/3} - 134{,}183/2160/c_1 +$$

$$c_2 \lambda_{H2O/SO3} + (216(c_3 - c_4 \lambda_{H2O/SO3} +$$

$$c_5 \lambda_{H2O/SO3}^2)^{1/2})^{1/3} + \frac{797}{2160} \quad \text{for } \lambda_{H2O/SO3} \leq 14$$

where $c_1 = -41{,}956e4$, $c_2 = 139{,}968e3$, $c_3 = 382{,}482e6$, $c_4 = 251{,}739e3$, and $c_5 = 419{,}904e6$

$$a = 0.7143 \lambda_{H2O/SO3} - 9.002\,1 \quad \text{for } 14 \leq \lambda_{H2O/SO3} \leq 16.8$$

$$a = 3 \quad \text{for } 16.8 \leq \lambda_{H2O/SO3}$$

Ohmic loss source term (Equation 7-30):

$$R_m = \frac{i^2}{\sigma_m}$$

Electrical conductivity (Equation 7-31):

$$\sigma_m = \exp\left[1268\left(\frac{1}{303} - \frac{1}{T}\right)\right]\left[0.5139 \lambda_{\frac{H2O}{SO3}} - 0.326\right], \quad \text{for } \lambda_{\frac{H2O}{SO3}} \geq 1$$

$$\sigma_m = \sigma_m\left[\lambda_{\frac{H2O}{SO3}} = 1\right], \quad \text{for } 0 < \lambda_{\frac{H2O}{SO3}} < 1$$

Using MATLAB to solve:

```matlab
%%%%%%%%%%%%%%%%%%%%%%%%%%%%%%%%%%%%%%%%%
% EXAMPLE 7-1: Modeling the Polymer Electrolyte Layer
% UnitSystem SI
%%%%%%%%%%%%%%%%%%%%%%%%%%%%%%%%%%%%%%%%%
function [x,y] = fuelcellmembrane
% FUELCELLMEMBRANE Fuel Cell membrane model

% Constants
mc = memconst;

% Set up grid. Assume each layer abuts the
% next one. The state variables are defined at the center of each slice. x is at the
% edge of each slice (like a stair plot).
x = linspace(0,mc.thick,mc.N);

% State variables: c_H2O, T, dTdx, pot, P
% Specify the constant boundary conditions at the edges of the membrane
% (scale them to match scaled state vector)

% [T dTdx pot P]
bc_anode = [343 10 0.1 202650.02] ./ [mc.scale_T mc.scale_dTdx mc.scale_pot
  mc.scale_P];

% [c_H2O]
bc_cathode = [9.5] ./ [mc.scale_c_H2O];

% Set up two-point boundary value problem
opts = bvpset('Nmax',150);
x0 = [0 mc.thick];
solinit = bvpinit(x,@meminit,[],x0,bc_anode,bc_cathode);

% Now solve two-point boundary value problem
sol = bvp4c(@memode,@memboundary,solinit,opts,bc_anode,bc_cathode);

x = sol.x;
y = sol.y' * diag(mc.scale);

end % of function
%_____
%
function yi = meminit(x,x0,bc_a,bc_b)
%Initial condition adapter function for bvpinit.
%Perform linear interpolation between boundary values.
% Initialize to constant values equal to boundary conditions
% State variables: c_H2O, T, dTdx, pot, dpotdx, sigma_m, P
yi = [bc_b';bc_a'];
end % of function
```

```
function res = memboundary(ya,yb,bc_a,bc_b)
%Boundary condition residual
% At the solution we expect y(a) = bc_a, y(b) = bc_b
% bc_a and bc_b are row vectors and ya,yb are columns, so adapt
res = [ya(2:5)-bc_a'; yb(1)-bc_b'];
end % of function

function dydx = memode(x,y,bc_a,bc_b)

%%%%%%%%%%%%%%%%%%%%%%%%%%%%%%%%%%%%%%%%%

% Polymer Electrolyte Layer

mc = memconst;

% 1 2 3 4 5
% State variables: c_H2O, T, dTdx, pot, P
c_H2O = y(1,:)*mc.scale_c_H2O;
T = y(2,:)*mc.scale_T;
dTdx = y(3,:)*mc.scale_dTdx;
pot = y(4,:)*mc.scale_pot;
P = y(5,:)*mc.scale_P;

% left hand side "Mass" matrix
lhs = zeros(5,5,size(y,2));

% Right hand side (non-differential terms)
rhs = zeros(size(y));
```

% Initial conditions

```
%%%%%%%%%%%%%%%%%%%%%%%%%%%%%%%%%%%%%%%%%

[a,lambda] = icondition(c_H2O,T,mc.R);

% Mixture density
rho = mc.M_H.*mc.c_H + mc.M_H2O.*c_H2O; %kgm^3

% Dynamic viscosity of the mixture
mu  =((mc.M_H.*mc.c_H./rho).*mc.mu_H)+((mc.M_H2O.*c_H2O./rho).*mc.mu_
    H2O); %kg/ms

%Molar velocity for the mixture:Darcy
u_m = (-mc.Kkr./mu).*(P – rho.*mc.g.*cos(mc.theta)); %m/s

%Calculate the initial conductivity at the anode
sigma_m = conductivity(T,lambda);

%%%%%%%%%%%%%%%%%%%%%%%%%%%%%%%%%%%%%%%%%

% Proton potential
% d(pot) = -i/sigma_m + F/sigma_m*c_H*u_m
dpotdx = (-mc.ii2 + mc.F.*mc.c_H.*u_m) ./ sigma_m;
```

```
lhs(3,4,:) = 1;
rhs(3,:) = dpotdx;
```

%Energy Conservation: constitutive variable: T

%%%

```
D_H2O = dcoe(a,T,lambda); % Diffusion coefficient
[rcp,McpN]  =  heatcap(mc.rho_dry,mc.cp_m,mc.M_H2O,mc.cp_H2O,mc.M_H,mc.
  c_H,...
mc.c_H2O,mc.cp_H,D_H2O,lambda,c_H2O,dpotdx,T, mc.R, mc.F,u_m,mc.ii2);
Rm = ohmic(mc.ii2,sigma_m);
```

```
% Steady state energy conservation
% lamda*d(dTdx) = M*cp*N*dTdx – Rm
lhs(1,3,:) = lambda;
rhs(1,:) = McpN.*dTdx – Rm;
```

```
% d(T) = dTdx;
lhs(2,2,:) = 1;
rhs(2,:) = dTdx;
```

```
% Diffusive flux for water: dependent variable: J_H2O
% d(c_H2O) = -(J_H2O – 2.5/22*lambda*i/F)/D_H2O
lhs(4,1,:) = 1;
rhs(4,:) = -(mc.J_H2O – 2.5*(lambda/22)*(mc.ii2/mc.F)) ./ D_H2O;
```

%%%

```
% Diffusive flux for hydrogen
% dpotdx = -(J_H*R*T)/(D_H*c_H*F);
J_H = dfluxH2(mc.D_H,mc.c_H,mc.R,T,mc.F,dpotdx);  % Fuel  Cell  membrane
  diffusive flux
```

%%%

```
% Molar velocity of the mixture
% d(P) = -u_m*mu /(K*k_rg) + rho*g*cos(theta)
lhs(5,5,:) = 1;
rhs(5,:) = -u_m*mu / (mc.Kkr) + rho*mc.g*cos(mc.theta);
```

%%%

```
% Invert the "Mass" matrix numerically
dydx = zeros(size(rhs));
for i=1:size(rhs,2)
dydx(:,i) = lhs(:,:,i) \ rhs(:,i);
end
```

```
% Scale derivative
dydx = dydx ./ mc.scale';
```

```
end % of function
```

%%%

```
function const = memconst
%MEMCONST Fuel Cell membrane model constants
% MEMCONST, by itself, returns a structure containing various fuel cell constants

const.N = 10;                     % Number points within membrane layer
const.tfinal = 60;                % Simulation time(s)

const.F = 96485.3383;             % Faraday's Constant (coulomb/mole)
const.R = 8.314472;               % Ideal gas constant (J/K-mol)
const.P_tot = 101325.01;          % Outside pressure (1 atm)
const.mw_H2O = 18.0152;           % Molecular weight of water
const.mu_H2 = 8.6e-6;             % Viscosity of wet hydrogen (Pa-s)
const.mu_air = 8.6e-6;            % Viscosity of air (Pa-s)
const.c_H = 1.2e-3;               % Mass conservation for the protons (mol/m^3)
const.b = 0.0126;
const.D_H = 4.5e-5;               % Proton Diffusivity (cm^2/s)
const.rho_dry = 2000;             % Density of membrane (kg/m^3)
const.M_mem = 1.1;                % Molecular weight of membrane (kg/mol SO3)
const.cp_m = 852.63;              % Specific Heat of membrane (J/kgK)
const.M_H2O = 18e-3;              % Molecular weight of water (kg/mol)
const.cp_H2O = 4190;              % Specific heat of water (J/kgK)
const.M_H = 1e-3;                 % Molecular weight of hydrogen (kg/mol)
const.cp_H = 20630;               % Specific Heat of Hydrogen (J/kgK)
const.mu_H = 98.8e-7;             % Viscosity of hydrogen (kg/ms)
const.rho_H2O = 972;              % Density of Hydrogen (kg/m^3)
const.g = 9.81;                   % gravitational constant (m/s^2)
const.mu_H2O = 8.91e-4;           % Viscosity of water (kg/ms)
const.theta = 90;                 % degrees
const.thick = 0.00005;            % thickness (m)
const.s = 0.02;                   % saturation ratio
const.ii2 = 0.6;                  % current density in membrane at node
const.Kkr = 1.8e-18;              % K*kr m^2
const.J_H2O = 0;                  % Water flux
const.c_H2O = 9.5;                % Initial cathode water concentration

% Scaling factor for state variables
const.scale_c_H2O = 10^4;
const.scale_T = 10^1;
const.scale_dTdx = 10^2;
const.scale_pot = 1;
const.scale_P = 10^1;
const.scale = [const.scale_c_H2O, const.scale_T, const.scale_dTdx, const.scale_
    pot, const.scale_P];

% Constants for the activity of the water molecules
const.c1 = -41956e4;
const.c2 = 139968e3;
const.c3 = 382482e6;
```

```matlab
const.c4 = 251739e3;
const.c5 = 419904e6;

%%%%%%%%%%%%%%%%%%%%%%%%%%%%%%%%%%%%%%%%%%

function [a,lambda] = icondition(c_H2O,T,R)
% ICONDITION Fuel Cell initial conditions
% ICONDITION returns the water activity and lambda

if c_H2O < 10
   s = 0;
elseif (c_H2O >= 10)
   s = 0.5;
elseif (c_H2O >= 14.5)
    s = 1;
end

a = ((R*T) / psat(T))*c_H2O + (2*s);

% Water uptake as a function of activity at the anode boundary
if ((a >=0) && (a <= 1))
   lambda = 0.043+(17.81*(a))-(39.85*(a^2))+(36*(a^3));
elseif ((a > 1) && (a <= 3))
   lambda = 14 + 1.4*(a – 1);
elseif a > 3
   lambda = 16.8;
end

%%%%%%%%%%%%%%%%%%%%%%%%%%%%%%%%%%%%%%%%%%

function sigma = conductivity(T,lambda)
% CONDUCTIVITY Fuel Cell conductivity
% CONDUCTIVITY returns the conductivity of the membrane

%Electrical conductivity
if lambda >=1
   sigma = 100*exp(1268*((1/303)-(1/T)))*(0.5139*lambda-0.326);  %Membrane
      Electrical conductivity (ohm/m)
else
   lambda = 1;
   sigma = 100*exp(1268*((1/303)-(1/T)))*(0.5139*lambda-0.326);
end

%%%%%%%%%%%%%%%%%%%%%%%%%%%%%%%%%%%%%%%%%%

function D_H2O = dcoe(a,T,lambda)
% DCOE Fuel Cell diffusion coefficient
% DCOE returns the diffusion coefficient of water in the membrane

% Diffusion coefficient at 303 K
if lambda <= 1.23 % m^2/s
   D = 2.642276e-13*lambda;
```

```
elseif (lambda > 1.23)& (lambda <= 6)
    D = 7.5e-11*lambda-9.5e-11;
elseif (lambda > 6) & (lambda <= 14)
    D = 2.5625e-11*lambda+2.1625e-10;
else % lambda > 14
    D = 1;
end

%Diffusion coefficient
D_H2O  =  D*(exp(2416*((1/303)-(1/T)))*lambda*(1/a)*(1/(17.81  -  78.9*a  +
    108*a^2))); % m/s
```

%%%%%%%%%%%%%%%%%%%%%%%%%%%%%%%%%%%%%%%

```
function [rcp,McpN] = heatcap(rho_dry,cp_m,M_H2O,cp_H2O,M_H,c_H,c_H2O_
    a,cp_H,D_H2O,lambda,dcH2O_dx,dpot_dx,T, R, F, um,ii2)
% HEATCAP Fuel Cell heat capacity of the phase mixture
% HEATCAP returns the heat capacity of the phase mixture in the membrane

% Heat capacity of the phase mixture: dependent variables: rho*cp, and M*cp*N
rcp = rho_dry*cp_m + M_H2O*c_H2O_a*cp_H2O + M_H*c_H*cp_H; % rho*cp
    J/m^3K^-1
McpN = M_H2O*cp_H2O*(c_H2O_a-D_H2O*dcH2O_dx + 2.5*(lambda/
    22)*(ii2/F))+ . . . M_H*cp_H*(c_H*um -((F/(R*T))*mc.D_H*c_H*dpot_dx));
    % M*cp*N W/m^2K
```

%%%%%%%%%%%%%%%%%%%%%%%%%%%%%%%%%%%%%%%

```
function Rm = ohmic(ii2,sigma)
% OHMIC Fuel cell ohmic losses
% DFLUX returns the fuel cell ohmic losses

% Membrane Ohmic Loss
Rm = ii2^2 / sigma; % W/m^3
```

%%%%%%%%%%%%%%%%%%%%%%%%%%%%%%%%%%%%%%%

```
function J_H2 = dfluxH2(D_H,c_H,R,T,F,dpot_dx)
%DFLUX Fuel Cell membrane diffusive flux
% DFLUX returns the diffusive flux of water in the membrane

% Diffusive flux for hydrogen
J_H2 = -(F /(R*T))*D_H*c_H*dpot_dx; % mol/m^2s
```

%%%%%%%%%%%%%%%%%%%%%%%%%%%%%%%%%%%%%%%

```
% Plot results
% Type 'load results' before running this

% x – position in membrane
% y – state variables: c_H2O, T, dTdx, pot, P
```

```
subplot(4,1,1), plot(x,y(:,1)), ylabel('c_{H_2O}')
subplot(4,1,2), plot(x,y(:,2)), ylabel('T')
subplot(4,1,3), plot(x,y(:,4)), ylabel('pot')
subplot(4,1,4), plot(x,y(:,5)), ylabel('P')
```

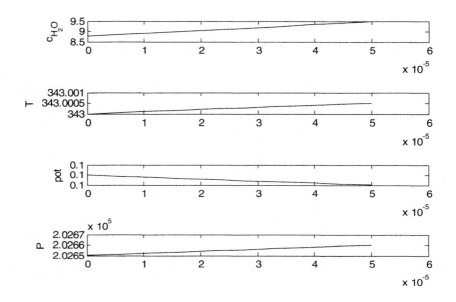

FIGURE 7-6. Plots of the state variables for Example 7-1.

Figure 7-6 shown the plots for the state variables for Example 7-1.

Chapter Summary

The electrolyte layer must be a good proton conductor, chemically stable, and able to withstand the temperatures and compression forces of the fuel cell stack. Accurately modeling the PEM layer can help improve the properties of future membrane materials. There are many types of PEM models, and choosing the right one depends upon the end goals and resources available. In order to have an accurate model, mass, energy, and charge balances must be written for the fuel cell membrane layer. In addition to these, using an empirical relationship for membrane water content may save time when creating a model. The requirements for the membrane include high ionic conductivity, adequate barrier to the reactants, and chemically and mechanically stable and low electronic conductivity. There are many choices for the PEM in the fuel cell, and the decision regarding the type chosen must depend upon many factors including, most importantly, cost and mass manufacturing capabilities.

Problems

- Calculate the resistance of the Nafion 115 membrane if the Nafion conductivity is 0.2 S/cm when 100% saturated.
- Calculate the resistance of the Nafion 115 membrane if the Nafion conductivity is 0.2 S/cm when 25% saturated.
- Calculate the permeation rate through Nafion 117 if the hydrogen pressure is 200 kPa.
- Calculate the ionic resistance and hydrogen crossover rate for Nafion 1135 if the fuel cell operates at 60°C and 1 atm.
- There are many ways in which Example 7-1 can be improved. List 3 ways that you can improve the model.

Endnotes

[1] Weber, A.Z. and J. Newman. Modeling Transport in Polymer Electrolyte Fuel cells. Chem. Rev. vol. 104, 2004, pp. 4679–4726.
[2] Ibid.
[3] Springer et al. Polymer electrolyte fuel cell model. *J. Electrochem. Soc.* Vol. 138, No. 8, 1991, pp. 2334–2342.
[4] Ibid.

Bibliography

Antoli, E. Recent developments in polymer electrolyte fuel cell electrodes. *J. Appl. Electrochem.* Vol. 34, 2004, pp. 563–576.
Barbir, F. *PEM Fuel Cells: Theory and Practice.* 2005. Burlington, MA: Elsevier Academic Press.
Chan, K., Y.J. Kim, Y.A. Kim, T. Yanagisawa, K.C. Park, and M. Endo. High performance of cup-stacked type carbon nanotubes as a Pt-Ru catalyst support for fuel cell applications. *J. Appl. Phys.* Vol. 96, No. 10, November 2004.
Chen, R., and T.S. Zhao. Mathematical modeling of a passive feed DMFC with heat transfer effect. *J. Power Sources.* Vol. 152, 2005, pp. 122–130.
Coutanceau, C., L. Demarconnay, C. Lamy, and J.M. Leger. Development of electrocatalysts for solid alkaline fuel cell (SAFC). *J. Power Sources.* Vol. 156, 2006, pp. 14–19.
EG&G Technical Services. *The Fuel Cell Handbook.* November 2004. 7th ed. Washington, DC: U.S. Department of Energy.
Fuel Cell Scientific. Available at: http://www.fuelcellscientific.com/. Provider of fuel cell materials and components.
Girishkumar, G., K. Vinodgopal, and P.V. Kamat. Carbon nanostructures in portable fuel cells: Single-walled carbon nanotube electrodes for methanol oxidation and oxygen reduction. *J. Phys. Chem. B.* Vol. 108, 2004, pp. 19960–19966.

Hahn, R., S. Wagner, A. Schmitz, and H. Reichl. Development of a planar micro fuel cell with thin film and micro patterning technologies. *J. Power Sources.* Vol. 131, 2004, pp. 73–78.

Haile, S.M. Fuel cell materials and components. *Acta Materialia.* Vol. 51, 2003, pp. 5981–6000.

Kharton, V.V., F.M.B. Marques, and A. Atkinson. Transport properties of solid electrolyte ceramics: A brief review. *Solid State Ionics.* Vol. 174, 2004, pp. 135–149.

Larminie, J., and A. Dicks. 2003. *Fuel Cell Systems Explained.* 2nd ed. West Sussex, England: John Wiley & Sons.

Li, W., C. Liang, W. Zhou, J. Qui, Z. Zhou, G. Sun, and Q. Xin. Preparation and characterization of multiwalled carbon nanotube-supported platinum for cathode catalysts of direct methanol fuel cells. *J. Phys. Chem. B.* Vol. 107, 2003, pp. 6292–6299.

Li, Xianguo. *Principles of Fuel Cells.* 2006. New York: Taylor & Francis Group.

Lin, B. Conceptual design and modeling of a fuel cell scooter for urban Asia. 1999. Princeton University, masters thesis.

Lister, S., and G. McLean. PEM fuel cell electrode: A review. *J. Power Sources.* Vol. 130, 2004, pp. 61–76.

Liu, J.G., T.S. Zhao, Z.X. Liang, and R. Chen. Effect of membrane thickness on the performance and efficiency of passive direct methanol fuel cells. *J. Power Sources.* Vol. 153, 2006, pp. 61–67.

Matsumoto, T., T. Komatsu, K. Arai, T. Yamazaki, M. Kijima, H. Shimizu, Y. Takasawa, and J. Nakamura. Reduction of Pt usage in fuel cell electrocatalysts with carbon nanotube electrodes. *Chem. Commun.* 2004, pp. 840–841.

Mehta, V., and J.S. Copper. Review and analysis of PEM fuel cell design and manufacturing. *J. Power Sources.* Vol. 114, 2003, pp. 32–53.

Mench, M.M., C.-Y. Wang, and S.T. Tynell. *An Introduction to Fuel Cells and Related Transport Phenomena.* Department of Mechanical and Nuclear Engineering, Pennsylvania State University. Draft. Available at: http://mtrl1.mne.psu .edu/Document/jtpoverview.pdf Accessed March 4, 2007.

Mench, M.M., Z.H. Wang, K. Bhatia, and C.Y. Wang. 2001. *Design of a Micro-direct Methanol Fuel Cell.* Electrochemical Engine Center, Department of Mechanical and Nuclear Engineering, Pennsylvania State University.

Mogensen, M., N.M. Sammes, and G.A. Tompsett. Physical, chemical, and electrochemical properties of pure and doped ceria. *Solid State Ionics.* Vol. 129, 2000, pp. 63–94.

Morita, H., M. Komoda, Y. Mugikura, Y. Izaki, T. Watanabe, Y. Masuda, and T. Matsuyama. Performance analysis of molten carbonate fuel cell using a Li/Na electrolyte. *J. Power Sources.* Vol. 112, 2002, pp. 509–518.

O'Hayre, R., S.-W. Cha, W. Colella, and F.B. Prinz. 2006. *Fuel Cell Fundamentals.* New York: John Wiley & Sons.

Rowe, A., and X. Li. Mathematical modeling of proton exchange membrane fuel cells. *J. Power Sources.* Vol. 102, 2001, pp. 82–96.

Silva, V.S., J. Schirmer, R. Reissner, B. Ruffmann, H. Silva, A. Mendes, L.M. Madeira, and S.P. Nunes. Proton electrolyte membrane properties and direct methanol fuel cell performance. *J. Power Sources.* Vol. 140, 2005, pp. 41–49.

Smitha, B., S. Sridhar, and A.A. Khan. Solid polymer electrolyte membranes for fuel cell applications—A review. *J. Membre. Sci.* Vol. 259, 2005, pp. 10–26.

Sousa, R., Jr., and E. Gonzalez. Mathematical modeling of polymer electrolyte fuel cells. *J. Power Sources.* Vol. 147, 2005, pp. 32–45.

Souzy, R., B. Ameduri, B. Boutevin, G. Gebel, and P. Capron. Functional fluoropolymers for fuel cell membranes. *Solid State Ionics.* Vol. 176, 2005, pp. 2839–2848.

U.S. Patent 5,211,984. Membrane Catalyst Layer for Fuel Cells. Wilson, M. The Regents of the University of California. May 18, 1993.

U.S. Patent 5,234,777. Membrane Catalyst Layer for Fuel Cells. Wilson, M. The Regents of the University of California. August 10, 1993.

U.S. Patent 6,696,382 B1. Catalyst Inks and Method of Application for Direct Methanol Fuel Cells. Zelenay, P., J. Davey, X. Ren, S. Gottesfeld, and S. Thomas. The Regents of the University of California. February 24, 2004.

Wang, C., M. Waje, X. Wang, J.M. Trang, R.C. Haddon, and Y. Yan. Proton exchange membrane fuel cells with carbon nanotube-based electrodes. *Nano Lett.* Vol. 4, No. 2, 2004, pp. 345–348.

Wong, C.W., T.S. Zhao, Q. Ye, and J.G. Liu. Experimental investigations of the anode flow field of a micro direct methanol fuel cell. *J. Power Sources.* Vol. 155, 2006, pp. 291–296.

Xin, W., X. Wang, Z. Chen, M. Waje, and Y. Yan. Carbon nanotube film by filtration as cathode catalyst support for proton exchange membrane fuel cell. *Langmuir.* Vol. 21, 2005, pp. 9386–9389.

Yamada, M., and I. Honma. Biomembranes for fuel cell electrodes employing anhydrous proton conducting uracil composites. *Biosensors Bioelectronics.* Vol. 21, 2006, pp. 2064–2069.

You, L., and H. Liu. A two-phase flow and transport model for PEM fuel cells. *J. Power Sources.* Vol. 155, 2006, pp. 219–230.

Sousa, R. Jr. and E. Gonzalez. Mathematical modeling of polymer electrolyte fuel cells. J. Power Sources Vol. 147, 2005, pp. 32-45.

Sousa, R. Jr., M. Escobar, E. Barcenas, G. Celso, and E. Capron. Proton and flow polymers for the... diesel membrane. In... water Source. Vol. 170, 2007, pp. 24-36.

U.S. Patent 4,311,864. Membrane Catalyst Layer for Fuel Cells. Wilson, M. The Regents of the University of California, May 19, 1992.

U.S. Patent 5,234,777. Membrane Catalyst Layer for Fuel Cells. Wilson, M. The Regents of the University of California, August 10, 1993.

U.S. Patent 6,716,551. Gas Water Inlet and Method of Application for PEM Membrane Fuel Cells, Zelenay P., J. Davey, S. Roh, S. Gottesfeld and S. Thomas. The Regents of the University of California, February 23, 2004.

Wang, C., M. Waje, X. Wang, J.M. Tang, R.C. Haddon and Y. Yan. Monodisperse as ultrathin fuel cells with carbon nanotube-based electrodes. Nano Lett. Vol. 4 No. 2, 2004, pp. 345-348.

Weng, G., F.S. Zhou, Q. Ye and J.T. Liu. Experimental investigations of the anode flow field in a micro direct methanol fuel cell. J. Power Sources, Vol. 135, 2006, pp. 254-264.

Xie, W. X. Wang Y., Chen M., Waje, and Y. Yan. Carbon nanotube film for filtration as cathode catalyst support for proton exchange membrane fuel cell. Langmuir Vol. 21, 2005, pp. 9386-9389.

Yamada, M. and ... Plasma Formembrane for fuel cell electrodes employing anhydrous proton conducting uniat compounds. Bioelectrochemistry, Vol. 21, 2004, pp. 2064-2066.

You, L. and H. Liu. A two-phase flow and transport model for PEM fuel cells. J. Power Sources, Vol. 155, 2006, pp. 21-34.

CHAPTER 8

Modeling the Gas Diffusion Layers

8.1 Introduction

The gas diffusion layer is between the catalyst layer and the bipolar plates. In a PEM fuel cell, the catalyst, gas diffusion, and membrane layers (the membrane electrode assembly (MEA)) are sandwiched between the flow field plates. The gas diffusion layers (GDL) are the outermost layers of the MEA. They provide electrical contact between electrodes and the bipolar plates, and distribute reactants to the catalyst layers. They also allow reaction product water to exit the electrode surface and permit the passage of water between the electrodes and the flow channels. The gas diffusion layer provides five functions for a PEM fuel cell:

- Electronic conductivity
- Mechanical support for the proton exchange membrane
- Porous media for the catalyst to adhere to
- Reactant access to the catalyst layers
- Product removal from it

Figure 8-1 illustrates the flows into and out of the GDL layer in a PEM fuel cell. Specific topics covered in this chapter include the following:

- Physical description of the gas diffusion layer
- Basics of modeling porous media
- Modes of transport in porous media
- Types of models

There have been many approaches taken in modeling the GDL layer, as shown in Table 8-1. The approach taken depends upon how the rest of the fuel cell is modeled. There are several types of flow models for porous media, such as Fick's diffusion, the Stefan-Maxwell equations, as well

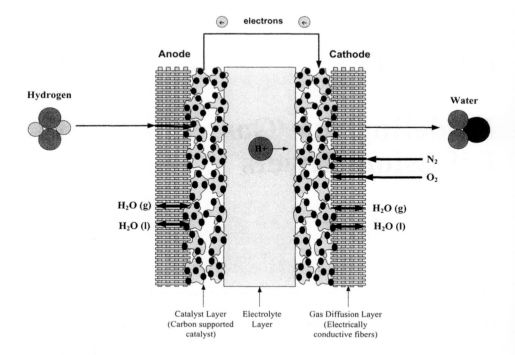

FIGURE 8-1. Flows into and out of the gas diffusions layers in the PEM fuel cell.

TABLE 8-1
Equations Used to Model the Gas Diffusion Layer

Model Characteristic	Description/Equations
No. of dimensions	1, 2, or 3
Mode of operation	Dynamic or steady-state
Phases	Gas, liquid, or a combination of gas and liquid
Mass transport	Nernst-Planck + Schlogl, Nernst-Planck + drag coefficient, or Stefan-Maxwell equation
Ion transport	Ohm's law
Energy balance	Isothermal or full energy balance

numerous other derivations based upon these equations. This chapter will describe the basic theory, and commonly used equations for modeling the fuel cell gas diffusion layer (GDL).

8.2 Physical Description of the Gas Diffusion Layer

Gas diffusion backings are made of a porous, electrically conductive material. The diffusion media are often composed of either a single gas diffusion

TABLE 8-2
Properties of Commercially Available Carbon Papers Used as Substrates for PEM
Fuel Cell Electrodes[1]

Carbon Paper	Thickness (mm)	Porosity (%)	Density (g/cm³)
Toray TGPH 090	0.30	77	0.45
Kureha E-715	0.35	60 to 80	0.35 to 0.40
Spectracarb 2050A-1041	0.25	60 to 90	0.40

layer or a composite structure of a gas diffusion layer and a microporous layer. Most models in the literature only include the gas diffusion layers. The GDL can be treated with a fluoropolymer and carbon black to improve water management and electrical properties. These material types promote effective diffusion of the reactant gases to the membrane/electrode assembly. The structure allows the gas to spread out as it diffuses to maximize the contact surface area of the catalyst layer membrane. The thicknesses of various gas diffusion materials vary between 0.0017 and 0.04 cm, the density varies between 0.21 and 0.73 g/cm², and the porosity varies between 70% and 80%. The most commonly used GDL materials are carbon cloth and carbon paper. Properties of some of the commercially available carbon papers are shown in Table 8-2.

The GDL helps to manage water in PEM fuel cells because it only allows an appropriate amount of water to contact the membrane/electrode assembly to keep the membrane humidified. In addition, it promotes the exit of liquid water from the cathode to help eliminate flooding. This layer is typically wet-proofed to ensure the pores in the carbon cloth or paper do not become clogged with water.

Many treatments exist for the gas diffusion layer. Most of these treatments are used to make the diffusion media hydrophobic to avoid flooding in the fuel cell. Either the anode or the cathode diffusion media, or both, can be PTFE treated. The diffusion material is dipped into a 5% to 30% PTFE solution, followed by drying and sintering. The interface with the catalyst layer can be fitted with a coating or microporous layer to ensure better electrical contact and efficient water transport in and out of the diffusion layer. This layer consists of carbon or graphite particles mixed with PTFE binder. The resulting pores are between 0.1 and 0.5 microns (μm) and are, therefore, much smaller than the pore size of the carbon fiber papers.

8.3 Basics of Modeling Porous Media

When considering the type of model to use for porous media, there are two main theoretical choices: (1) the motion of gas molecules through the pores of the media; or (2) the interaction of the molecules of gas and solid. If the substrate has large pores, it is intuitive to think in terms of the fraction of the media available for gas transport. Conversely, when the pore size in the substrate is very fine, and the size of the gas molecules and solid particles becomes comparable, the second option is used. These two options comprise the two main theories of modeling porous media. This section introduces the concepts such as pore structure, fluid properties, capillary and permeability needed in order to understand the basics of modeling of porous media.

8.3.1 Pore Structure

A porous medium generally consists of solid matrix and pore space. The description of the microscopic structure of pore space is difficult due to its geometrical complexity. There is typically a large network of pores communicating through relatively narrow constrictions. Since the shape of an actual pore is quite irregular, approximations of pore shape with regular geometries, such as cube, sphere, etc., are commonly made in theoretical studies to study the effect of pore structure. After the geometry of the pore structure has been specified, surface areas and volumes of the pores can be calculated.

8.3.2 Fluid Properties

The void space in the porous medium is assumed to be filled with the different phases. The volume fraction occupied by each phase is the saturation of that phase. Therefore, the total fraction of all phases is equal to one:

$$\sum_{\text{allphases}} s_i = 1 \tag{8-1}$$

The two phases that are usually considered are the liquid (l) and gaseous (g) phases. Each phase contains one or more components. The mass fraction of component i in phase k is denoted by c_{ik}. In each of the phases, the mass fractions should add up to unity, so that for N different components:

$$\sum_{i=1}^{N} c_{ig} = \sum_{i=1}^{N} c_{il} = 1 \tag{8-2}$$

If a density ρ and a viscosity μ are functions of phase pressure $p_i(i = g, l)$, and the composition of each phase, then:

$$\rho_g = \rho_g(p_g, c_g), \quad \mu_g = \mu_g(p_g, c_g) \tag{8-3}$$

The liquid density and a viscosity for the liquid plate can be written in a similar manner.

8.3.3 Capillarity

Capillary pressure is defined as the pressure difference between two immiscible fluids at equilibrium within the pore space, which can be expressed as:

$$p_c = p_L - p_G \tag{8-4}$$

where p_G is the pressure of the gas phase, and p_L is the pressure of the liquid phase. Capillary pressure involves the interfacial tension and the interfacial curvature in Laplace's equation:

$$P_c = \gamma\left(\frac{1}{r_1} + \frac{1}{r_2}\right) \tag{8-5}$$

where γ is the interfacial tension, and the curvature of the interface is characterized by two principal radii of the curved surface r_1 and r_2. The principal curvature radii at a point on the interface lie on two planes perpendicular to each other and intersect at the point.

In a cylindrical capillary tube, capillary pressure is given by

$$p_c = -\frac{2\gamma\cos\theta}{r} \tag{8-6}$$

where r is the radius of the tube and θ is the contact angle. However, the geometry of pores in porous media is more complex, and tortuous paths are often used to represent channels in porous media. An expression of capillary pressure for an interface in a tube with rectangular cross-section is:

$$p_c = \frac{\gamma}{R_t}\frac{\theta+\cos^2\theta-\dfrac{\pi}{4}-\sin\theta\cos\theta}{\cos\theta-\sqrt{\dfrac{\pi}{4}-\theta+\sin\theta\cos\theta}} \tag{8-7}$$

for i, j = g, l. Although other dependencies are reported, it is usually assumed that the capillary pressure is a function of the saturations only.

8.3.4 Permeability

The (absolute) permeability is a measure of the material's ability to transmit a single fluid at certain conditions. Since the orientation and interconnection of the pores are essential for flow, the permeability is not necessarily proportional to the porosity, but permeability (K) is normally strongly correlated to pore void fraction (φ).

In macroscale modeling, it is assumed that all phases may be present at the same location—although the phases do not actually mix. The permeability of one phase depends upon the saturation conditions and interaction with the substrate at a specific location. A property called relative permeability, denoted by kr_i, were i is g or l, describes how one phase flows in the presence of the others. The effective permeability, k, is used to define the relative permeability, k_r:

$$k = k_r k_{sat} \qquad (8\text{-}8)$$

where k_{sat} is the saturated permeability, or the permeability at complete saturation, of the medium. k_{sat} depends only on the structure of the medium.

Relative permeabilities are nonlinear functions of the saturations, therefore, the sum of the relative permeabilities at a specific location (with a specific composition) is not necessarily equal to one. Relative permeabilities can depend on the pore-size distribution, the fluid viscosity, and the interfacial forces between the fluids.

8.4 Modes of Transport in Porous Media

There are several mechanisms by which mass transport can occur in porous media. There are four main modes of transport depending upon the molecule acceleration and environment. The distinctions given here are somewhat arbitrary—but they represent an attempt to group molecules into distinct categories to help facilitate physical understanding and modeling. The four main types of transport are as follows:

- **Free Molecule or Knudsen Flow:** This occurs when the length between molecules is very small, or the species density is low. The collisions between molecules can be ignored in comparison with collisions of molecules with the walls of the porous media.
- **Viscous Flow (Bulk/Continuum Flow):** The gas acts as a continuum fluid driven by a pressure gradient, and collisions between molecules dominate over collisions between the molecules and the wall.
- **Ordinary (Continuum) Diffusion:** The different species of a mixture move relative to each other under the influence of concentration, temperature, or other external force gradients. Collisions between molecules dominate ordinary diffusion.
- **Surface Flow:** The molecules move along a solid surface in an adsorbed layer. This mechanism is assumed independent of the others. It can be integrated into a model with the three other mechanisms in order to give more accurate results.

If the mean-free-path of a molecule is less than 0.01 times the pore radius, ordinary diffusion dominates. If the mean-free-path is greater than

10 times the pore radius, Knudsen diffusion dominates. This implies that Knudsen diffusion should be considered if the pore radius is less than about 0.5 μm. The typical gas diffusion layer has pores between 0.5 and 20 μm in radius, and a microporous layer contains pores between 0.05 and 2. Therefore, depending upon the material used, Knudsen diffusion may not have to be considered in gas diffusion layers, but it should be accounted for in microporous and catalyst layers.

8.4.1 Free Molecule (Knudsen) Flow in Porous Media

Knudsen or free molecule flow is where gas molecules collide more with the walls of the container than with the other gas molecules. This occurs when the mean-free-path of the gas molecules is approximately the length scale of the container, or there are very low gas densities (which means large mean-free-paths). In free molecule flow, there is no distinction between flow and diffusion (which are continuum phenomena), and gas composition is not important since there is no interaction between gas molecules of the same or different species.

For a gas with molecular density n, the Knudsen molar flux can be expressed as:

$$J_K = -\left(\frac{2}{3}r\right)\sqrt{\frac{8RT}{\pi M}}\left(\frac{dc}{dz}\right) \tag{8-9}$$

where r is the radius, R is the ideal gas constant, T is the temperature, M is the molar mass of the gas, and dc/dz is the rate of change of gas concentration. The Knudsen diffusion coefficient D_K for flow in a cylindrical long straight pore with diffuse scattering is:

$$D_K = \left(\frac{2}{3}r\right)\sqrt{\frac{8RT}{\pi M}} \tag{8-10}$$

When geometries other than cylindrical are used, Equation 8-10 can be used with different geometrical parameters. Knudsen flow through porous media can be modeled using Equation 8-10 with the single-pore diffusion coefficient replaced with a porous medium diffusion coefficient defined as:

$$D_{K,porous\_medium} = \frac{\varepsilon}{\tau}D_{K\,single\_phase} \tag{8-11}$$

where ε is the porosity, and τ is the tortuosity, which is typically incorporated into the value of K_0. The porosity can be defined as:

$$\varepsilon = \frac{V_v}{V_v + V_s} \tag{8-12}$$

where V_v is the volume of the voids and V_s is the volume fraction of solids. If modeling the liquid water in the GDL is neglected, then ε_G is set to the value of the bulk porosity of the medium, ε_0. Tortuosity is accounted for in GDL models by using a Bruggeman expression:

$$\tau = \varepsilon^{-0.5} \qquad (8\text{-}13)$$

Although Equation 8-13 is a good estimation for tortuosity, it can sometimes underpredict the tortuosity at low porosities.

8.4.2 Viscous (Darcy) Flow in Porous Media

Viscous (Darcy) flow refers to flow in the laminar continuum regimen that is caused by a pressure gradient. The gas behavior is determined by the coefficient of viscosity, which is independent of pressure for gases. Since bulk flow does not separate the components of gas mixtures, mixtures of different gasses can be treated in the same manner as a pure gas. Under laminar flow conditions, the single-phase flow of incompressible fluids in porous media is governed by Darcy's law:

$$V = -\frac{K}{\mu}(\nabla P - \rho g) \qquad (8\text{-}14)$$

where V is the Darcy velocity, K is the absolute permeability, μ is the dynamic viscosity of the fluid, ∇P is the pressure gradient, and ρ is the fluid density.

For a porous medium with multiphase flow, the relative permeability of fluid i, K_{ri}, is defined as:

$$K_{ri} = \frac{K_i}{K} \qquad (8\text{-}15)$$

where K_i is the effective permeability of fluid i and Equation 8-14 can be written as

$$V = -\frac{K_{ri}K}{\mu}(\nabla P - \rho g) \qquad (8\text{-}16)$$

where the subscript i denotes fluid i.

A transport equation for the viscous flux of gases can be obtained by applying Newton's second law to an element of compressible fluid, as shown in Figure 8-2.

From this derivation, the total viscous flux per unit area for cylindrical passages is:

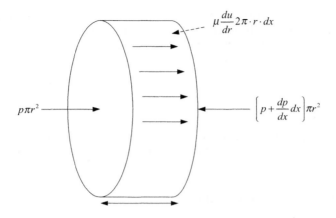

FIGURE 8-2. Forces acting on fluid element.

$$J_{visc} = -\frac{nR^2}{8\mu}\frac{dp}{dx}$$

(8-17)

where R is the radius of the cylinder, n is the compressibility of the fluid calculated using the ideal gas equation of state: $n = p/KT$, and dp/dx is the pressure change in the x direction.

When a different geometry is used, equations can be derived of the same form, but with different geometrical parameters. The general equation that can be used is often defined as:

$$J_{visc} = -\frac{nB_0}{\mu}\nabla p$$

(8-18)

where B_0 is called the "viscous flow parameter" (the Knudsen flow parameter), which must be selected to represent the geometry of a particular problem. For example, B_0 for straight circular capillaries of radius R is $B_0 = R_2/8$.

Viscous flow parameters are usually obtained from empirical data since the geometry of porous media is complex. The viscous flow through porous media can be modeled using Equation 8-18 with the single-pore viscous flow parameter replaced with a porous medium parameter defined as:

$$B_{0,pm} = \frac{\varepsilon}{\tau}B_{0,pd}$$

(8-19)

where ε is the porosity, τ is the tortuosity, and $B_{0,pd}$ is the single mean pore diameter.

In rigorous derivations, the tortuosity factor is often squared in order to correct for the path along the pore to across the medium.

Combining Equations 8-18 and 8-19 the value of B_0 for the porous medium is:

$$J_{visc} = -\frac{nB_{0,pm}}{\mu}\nabla p \tag{8-20}$$

A more rigorous method of determining the viscous flow parameter uses the concept of porosity and tortuosity, but also takes into account the variation of surface area of pores. This is critical since viscous flow is dependent on the interaction of pore walls and gas flow.

The relationship for calculating B_0 is:

$$B_{0,pm} = \frac{\varepsilon^3}{(1-\varepsilon)^2}\frac{1}{k(S/V_s)^2} \tag{8-21}$$

where S is the total surface area per unit volume, V_s is the volume fraction of solids, ε is the porosity, and $k = 2\tau$, and is called the Kozeny function (which is usually about 5). The Kozeny function allows the inclusion of the extra tortuosity factor.

8.4.3 Ordinary (Continuum) Diffusion in Porous Media

Ordinary diffusion is the most common diffusion mechanism. For binary mixtures, the species diffusive flux is directly proportional to its concentration gradient:

$$J_{1D} = -D_{12}\nabla n_1 \tag{8-22}$$

$$J_{2D} = -D_{21}\nabla n_2 \tag{8-23}$$

where $J_1D + J_2D = JD = 0$ is the zero net diffusive flux. Therefore, $D_{12} = D_{21}$. In multicomponent mixtures, the fluxes of all of the species are important to consider since they affect the diffusive transport of any one species. This is because the momentum transferred to any one species will depend on the relative motion of all other species. For multicomponent mixtures, the Stefan-Maxwell equation can be used:

$$n\nabla x_1 = \sum_{k=1}^{N}\left[\frac{(x_iJ_{jD} - x_jJ_{iD})}{D_{ij}}\right] \tag{8-24}$$

The Stefan-Maxwell equation gives the transport equation in terms of fluxes of species concentrations. However, the species fluxes in terms of concentrations are needed. Therefore, the equations must be inverted at great computational expense. Although the equation can be inverted, many

Fickian (binary diffusion) approximations have been developed for use instead of the Stefan-Maxwell (multicomponent) transport equations to avoid the computational cost.

It is possible to use Fick's binary law of diffusion to yield the flux of a single species in terms of the concentration gradients of the other species. However, the diffusion coefficients are not the same as the binary diffusion coefficients in the resulting equation:

$$J_i = -D_{i1}\nabla n_1 - D_{i2}\nabla n_2 \ldots - D_{ik}\nabla n_k \qquad (8\text{-}25)$$

This equation is actually a form of the inverse Stefan-Maxwell equation, and the Fickian multicomponent diffusion coefficients are representative of the gas mixture mole fractions and binary diffusion coefficients.

Diffusion through porous media can be modeled using Equation 8-24 with the free gas diffusion coefficient replaced with a porous diffusion coefficient defined as:

$$D_{ij,porous} = \frac{\varepsilon}{\tau} D_{ij,free} \qquad (8\text{-}26)$$

where $\frac{\varepsilon}{\tau}$ is the porosity-tortuosity factor.

8.4.4 Combining Transport Mechanisms for Binary Mixtures

In order to determine how the fluxes are related, the free molecule and continuum diffusive fluxes for one species of a binary mixture are:

$$J_{iK} = -D_{1k}\nabla n_1 \qquad (8\text{-}27)$$

$$J_{1D} = -D_{12}\nabla n_1 + x_1 J_D \qquad (8\text{-}28)$$

which can be rewritten using the ideal gas law ($p = nkT$) as:

$$J_{iK} = -\frac{D_{1k}}{kT}\nabla n_1 \qquad (8\text{-}29)$$

$$J_{1D} = -\frac{D_{12}}{kT}\nabla n_1 + x_1 J_D \qquad (8\text{-}30)$$

If the first species is not accelerating, then the average momentum transferred to it through collisions with the walls must be balanced by the force acting on molecules due to the partial pressure gradient. Since Knudsen flow describes collisions between molecules and the wall, and ordinary diffusion describes collisions between molecules, Equations 8-29 and 8-30 can be rewritten as:

$$-\nabla p_{1(wall)} = -\frac{kT}{D_{1K}} J_{1K}$$

(8-31)

$$-\nabla p_{1(molecules)} = -\frac{kT}{D_{12}} (J_{1D} - x_1 J_D)$$

(8-32)

The total partial pressure in the gas mixture is due to both Knudsen and ordinary diffusion, therefore:

$$-\nabla p_1 = -\frac{kT}{D_{1K}} J_{1K} + \frac{kT}{D_{12}} (J_{1D} - x_1 J_D)$$

(8-33)

This formula is valid for the diffusion of one component of a binary mixture, and is valid for the entire pressure range between the free molecule limit and the continuum limit.

Incorporating the viscous flow into the model is also additive because there are no viscous terms in the other diffusion equations. The independence is valid for any isotropic system, and is sometimes referred to as Curie's theorem. As first shown by Equation 8-18 the viscous flux is described by:

$$J_{visc} = -\frac{nB_0}{\mu} \nabla p$$

(8-34)

The surface flux can also be added to the total flux equation due to experimental evidence. The surface flux is described by:

$$J_{1S} = -D_{1S} \nabla n_1$$

(8-35)

Therefore, the total flux of one species is given by:

$$J_1 = J_{1D} + J_{1visc} + J_{1s} = J_{1D} + x_1 J_{1visc} + J_{1s}$$

(8-36)

and the total flux is:

$$J = J_1 + J_2 = J_D + J_{visc} + J_s$$

(8-37)

After substituting Equations 8-30, 8-33, and 8-34 into 8-37, and rearranging:

$$-\frac{1}{kT} \nabla p_1 = \left(\frac{1}{D_{1K}} + \frac{1}{D_{1K}}\right) J_1 - \frac{x_1}{D_{12}} J - \frac{x_1}{D_{1K}} J_{visc} - \left(\frac{1}{D_{1K}} + \frac{1}{D_{12}}\right) J_s - \frac{x_1}{D_{12}} J_s$$

(8-38)

$$J_1 = -D_1 \left[1 + \frac{D_{1s}}{D_1} + x_1 \frac{(D_{1s} - D_{2s})}{D_{12}} \right] \nabla n_1 + x_1 \delta_1 J - x_1 \gamma_1 \left(\frac{nB_0}{\eta} \right) \nabla p + \frac{x_1}{kT} \frac{D_{2s}}{D_{12}} \nabla p$$

$$(8\text{-}39)$$

where

$$\frac{1}{D_1} = \frac{1}{D_{1K}} + \frac{1}{D_{12}} \quad \delta_1 = \frac{D_1}{D_{12}} = \frac{D_{1k}}{D_{1k} + D_{12}}$$

$$\gamma_1 = \frac{D_1}{D_{1k}} = \frac{D_{12}}{D_{1k} + D_{12}} = 1 - \delta_1$$

$$(8\text{-}40)$$

The resulting mass transport equation contains all four mass transport mechanisms. If certain types of diffusion are dropped from this equation, one should be able to obtain many widely used mass transport formulae. If the surface diffusion is considered negligible, the equation will be simplified considerably:

$$J_{1s} = -D_1 \nabla n_1 + x_1 \delta_1 J - x_1 \gamma_1 \left(\frac{nB_0}{\mu} \right) \nabla p$$

$$(8\text{-}41)$$

If the pores of the porous media are significantly larger than the mean-free-path of the gases, the Knudsen term will be negligible, therefore:

$$J_1 = -D_{12} \nabla n_1 + x_1 J$$

$$(8\text{-}42)$$

When the Knudsen term is dropped, the second viscous flow term will also be dropped. This is because there will be no collisions with the pore wall, and no viscous shear forces transmitted through the gas from the walls.

If the pores are much smaller than the mean-free-path, continuum diffusion will be negligible. In this case:

$$J_1 = -D_{1k} \nabla n_1 - x_{11} \left(\frac{nB_0}{\eta} \right) \nabla p$$

$$(8\text{-}43)$$

In the absence of a pressure difference, the equation reduces to the definition of Knudsen diffusion. Inclusion of the pressure drop is more involved. Any pressure drop present must be due to Knudsen collisions since there is no local pressure drop due to gas–gas molecular collisions. If the pore size is intermediate, both ordinary and Knudsen diffusion will be significant. In the absence of a pressure drop, the equation can be written as:

$$J_1 = -D_1 \nabla n_1 + x_1 \delta_1 J$$

$$(8\text{-}44)$$

If there is zero net flux, then:

$$J_1 = -D_1 \nabla n_1 = -\left[\frac{1}{D_{1K}} + \frac{1}{D_{12}}\right]^{-1} \nabla n \qquad (8\text{-}45)$$

where D_1 is the multimechanism diffusion coefficient, which is valid for ordinary and Knudsen diffusion. This expression is known as the Bosanquet formula.

8.5 Types of Models

Porous media models are abundant in the literature, but there are only a few gas diffusion layer models for fuel cells that treat these layers in a rigorous manner. As mentioned in Section 8.3, GDL models focus on either the flow through the substrate (the pores), or the interaction of the solid substrate with the molecules. When the modeling focus is on the pores in the substrate, either Fick's diffusion law (for one-component diffusion through a homogeneous medium) or the Stefan-Maxwell equations (for multicomponent gas diffusion) can be used. The interaction of the gas and solid, (or more commonly known as the Dusty Gas Model) is derived by applying kinetic theory to the interaction of both gas–gas and gas–solid molecules, with the porous media treated as "dust" in the gas. The Fickian diffusion model is computationally much simpler than the Stefan-Maxwell formulation, but cannot be used for multicomponent mixtures (unless a binary mixture approximation or tertiary diffusion coefficients are used). In addition, these models either consider just the gas or liquid phase, or include both. Table 8-3 gives a brief description of the main types of GDL models in the literature.

Besides the gas and liquid transport described in the first few sections of this chapter, other important properties of the GDL layers are the electronic conduction and the evaporation/condensation of the gas/liquid in the GDL. The governing equations for the GDL are shown in Table 8-4.

TABLE 8-3
Types of Gas Diffusion Layer Models in the Literature

Type of Model	Description
Gas phase models	Gas phase models assumes that there is only the gas phase flow in the GDL.
Liquid phase models	Liquid phase models assumes that there is only the liquid phase flow in the GDL.
Two-phase flow models	Two-phase flow models describe how gas and liquid interact in a porous medium.

TABLE 8-4
Fuel Cell Gas Diffusion Layer Variables and Equations

Variable	Equation	Equation No.
Overall liquid water flux (N_L)	Mass balance	7-2, 7-10, 9-4 or Chapter 5 equations
Overall membrane water flux (N_W)	Mass balance	7-2, 7-10, 9-4 or Chapter 5 equations
Gas phase component flux ($N_{G,i}$)	Mass balance	7-2, 7-10, 9-4 or Chapter 5 equations
Gas phase component partial pressure ($p_{G,i}$)	Stefan-Maxwell	5-63
Liquid pressure (P_L)	Darcy's law	8-52
Electronic phase current density (i_1)	Ohm's law	8-46, 9-2, 9-20
Electronic phase potential (Φ_1)	Charge balance	8-46
Temperature (T)	Energy balance	Chapter 6 equations
Total gas pressure (p_G)	Darcy's law	8-50
Liquid saturation (S)	Saturation relation	8-54

8.5.1 Conductivity

Most models neglect conductivity calculations, since the GDL layer is made of carbon. However, a rigorous model should include this calculation since it can become a limiting factor due to geometry or composition. Ohm's law can be used to take this into account:

$$i_1 = -\sigma_0 \varepsilon_1^{1.5} \nabla \Phi_1 \tag{8-46}$$

where ε_1 and σ_0 are the volume fraction and electrical conductivity, respectively. The Bruggeman correction is used in Equation 8-46 to account for porosity and tortuosity. Since the GDL is often coated with Teflon to promote hydrophobicity, carbon is the conducting phase and the Teflon is insulating.

8.5.2 Evaporation/Condensation

Depending upon the local temperatures, pressures, and saturation conditions, water can evaporate or condense in the GDL layer. These reactions are often modeled by an expression that is similar to:

$$r_{evap} = k_m a_{G,L}(p_w - p_w^{vap}) \tag{8-47}$$

where r_{evap} is the molar rate of evaporation per unit volume, k_m is a mass-transfer coefficient per unit interfacial surface area, $a_{G,L}$ is the interfacial gas/liquid surface area per unit volume, p_w is the partial pressure of water in the gas phase, and p_w^{vap} is the vapor pressure of water, which can be corrected for pore effects by the Kelvin equation. There are several models in the literature that use an interfacial area that depends upon the water content of the GDL. Usually, the gas is assumed to be saturated if any liquid water exists. In a rigorous GDL model, both gas and liquid transport should be included.

8.5.3 Gas Phase Transport

Most models use the Stefan-Maxwell equations for gas phase transport in the fuel cell GDL layers. This equation is written using more common notation the Equation 8-22:

$$\nabla x_i = \sum_{j \neq i} \frac{x_i N_j - x_j N_i}{c_T D_{i,j}^{eff}} \tag{8-48}$$

where c_T is the total concentration or molar density of all of the gas species, x_i is the mole fraction of species i, and $D_{i,j}^{eff}$ is the effective binary interaction parameter between i and j, by the Onsager reciprocal relationships, $D_{i,j}^{eff} = D_{j,i}^{eff}$ for ideal gases.

Knudsen diffusion has been taken into consideration in a few models in the literature (see Section 8.4.1). Knudsen diffusion and Stefan-Maxwell diffusion can be treated as mass-transport resistances in series, and are combined to yield:

$$\nabla x_i = -\frac{N_i}{c_T D_{K_i}^{eff}} + \sum_{j \neq i} \frac{x_i N_j - x_j N_i}{c_T D_{i,j}^{eff}} \tag{8-49}$$

where the $D_{K_i}^{eff}$ is the effective Knudsen diffusion coefficient. The porous media itself constitutes another species with zero velocity with which the diffusing species interact.

While most models treat gas phase flow as purely due to diffusion, some models take into account convection in the gas phase (see Section 8.4.2). This is usually done by the addition of Darcy's law for the gas phase:

$$v_G = -\frac{k_G}{\mu_G} \nabla p_G \tag{8-50}$$

where k is the effective permeability. The above relation can be made into a flux by multiplying it by the total concentration of the gas species. One way to include the effect of gas phase pressure-driven flow is to use Equation 8-50 as a separate momentum equation. Another way to include pressure-driven flow is to incorporate the Equation 8-50 into Equation 8-49, as per the Dusty Gas model:

$$\nabla x_i = -\frac{x_i k_G}{D_{K_i}^{eff} \mu_G} \nabla p_G + \sum_{j \neq i} \frac{x_i N_j - x_j N_i}{c_T D_{i,j}^{eff}} - \frac{N_i}{c_T D_{K_i}^{eff}} \tag{8-51}$$

However, using this equation is not necessarily the best method of incorporating velocity into the model. Instead, one of the Stefan-Maxwell equations should be replaced by Equation 8-51 because it is the summation of the mass velocities of the gas species. There are many models that incorporate gas phase pressure-driven flow in the diffusion media, however, it is unknown how significant this effect is. Most modeling results show that the pressure difference through the fuel cell is minimal, and the assumption of uniform gas pressure is acceptable for most conditions. However, there have been some models that do show a small pressure difference that reduces mass transfer. In addition, the change in pressure may change more than estimated as the temperature in the fuel cell also changes.

If a model is two or three dimensional, the gas phase pressure needs to be considered. This is because the pressure difference down a gas channel is much more significant than that through the fuel cell. When an interdigitated flow field design is used (see Chapter 10), the gas-phase pressure-driven flow needs to be accounted for. In these types of fuel cells, the gas channels are not continuous through the fuel cell, and gas is forced through the channels by both convection and diffusion to reach the next gas channel.

8.5.4 Treatment of Liquid Water

Liquid water has been modeled using several methods in fuel cells. The simplest way to include liquid water in a GDL model is to treat it as a solid species that occupies a certain volume fraction. When this method is used, transport is not considered, and it just decreases the gas phase volume. This, in turn, decreases the effective diffusion coefficients of the gas species, and somewhat takes into account flooding. The models that use this approach usually use the volume fraction of water as a fitting parameter.

A more sophisticated method of including liquid water is to have a way in which the model includes the transport of water in the model. These models assume that the liquid water exists as droplets that are carried along in the gas stream. Therefore, while evaporation and condensation occur, a separate liquid phase does not have to be modeled. The liquid is assumed to be a component of the gas, which usually has little effect on the rest of the system. An advantage of this type of model over the one previously described is that it allows for the existence and location of liquid water to be noted, and to a limited extent the change in the water pressure or concentration.

In order to model liquid-water flow accurately, two-phase models are required. Liquid phase transport is similar to the gas phase pressure-driven flow described above. Since the liquid water is assumed to be pure, there is no diffusion component for the water movement. Therefore, the flux form of Darcy's law models the flow of liquid water:

$$N_{w,L} = -\frac{k}{V_w \mu} \nabla p_L \qquad (8\text{-}52)$$

where V_w is the molar volume of water and all of the properties are valid for pure water. Many models use Equation 8-52 with a liquid phase volume fraction. This is a good assumption since there is isolated gas and liquid pores in the medium. However, this can be improved by adding a type of transfer between them.

Finally, there are also models that use a phase mixture approach. The two phases are treated as a single-phase mixture, and all parameters are calculated for the mixture instead of each phase. This approach does effectively determine the mass flux, but the entire mixture moves at a certain velocity when in fact each phase may move a different velocities. Despite this, the models adequately predict water balance in the fuel cell.

8.5.5 Rigorous Two-Phase Flow Models

It is commonly known that gas and liquid interact in a porous medium. There have been many rigorous models developed in the literature for 2-phase-porous media during the last few decades. The models that have been developed for fuel cells are on the simpler end, which makes them easy to integrate into a fuel cell model, but less accurate. The interaction between liquid and gas is characterized by a capillary pressure, contact angle, surface tension, and pore radius, as first mentioned in Section 8.2:

$$p_c = p_L - p_G = -\frac{2\gamma \cos\theta}{r} \qquad (8\text{-}53)$$

where γ is the surface tension of water, r is the pore radius, and θ is the internal contact angle that a drop of water forms with a solid. Equation 8-53 relates how liquid water wets the material. For a hydrophilic pore, the contact angle is $0° < \theta < 90°$, and for a hydrophobic one, it is $90° < \theta < 180°$. An important part of the two-phase models is how the liquid saturation is predicted as a function of position. The saturation, S, is the amount of pore volume that is filled with liquid:

$$\varepsilon_G = \varepsilon_0(1 - S) \qquad (8\text{-}54)$$

From Equation 8-54 one can see that the saturation greatly affects the effective gas phase diffusion coefficients. Therefore, flooding can be characterized by saturation. In the literature, the saturation is typically calculated using an empirical relation for the capillary pressure and saturation.

In order to determine the gas and liquid pressures at each control volume in the diffusion medium, the capillary pressure must be known at every position. In typical two-phase flow models, the movement of both liquid and gas is determined by Darcy's law for each phase. Equation 8-55 relates the two pressures to each other. Many models use the capillary pressure as the driving force for the liquid-water flow:

$$N_{w,L} = -\frac{k}{V_w \mu} \nabla p_L = -\frac{k}{V_w \mu} (\nabla p_C + \nabla p_G) = -\frac{k}{V_w \mu} \nabla p_C \qquad (8\text{-}55)$$

As first mentioned in Section 8.3.4, a useful relation for calculating the effective permeability, k, is to define a relative permeability, k_r,

$$k = k_r k_{sat} \qquad (8\text{-}56)$$

where k_{sat} is the saturated permeability, or the permeability at complete saturation, of the medium. k_{sat} depends only on the structure of the medium and has been empirically determined, or estimated, using a Carman-Kozeny equation.

The addition of gas and water balances completes the set of equations. The parameters in this model are mixture parameters using capillary phenomena. Although the mixture moves at a mass-average velocity, the interfacial drag between the phases and other conditions allows each separate phase velocity to be determined. The liquid phase velocity is written as:

$$v_L = \lambda_L \frac{\rho_m}{\rho_L} v_m + \frac{k \lambda_L (1 - \lambda_L)}{\varepsilon_0 \rho_L v_m} [\nabla p_c + (\rho_L - \rho_G) g] \qquad (8\text{-}57)$$

where the subscript m stands for the mixture, ρ_k and v_k are the density and kinematic viscosity of phase k, respectively, and λ_L is the relative mobility of the liquid phase:

$$\lambda_L = \frac{k_{r,L}/v_L}{k_{r,L}/v_L + k_{r,G}/v_G} \qquad (8\text{-}58)$$

The mixture velocity is basically determined from Darcy's law using the properties of the mixture. This mixture velocity is a major improvement from the previously described approaches.

8.6 GDL Modeling Example[2]

This section presents the derivations and modeling for the cathode GDL of the fuel cell created by Beuscher et al.[3] The models in[4] are derived from multiphase flow in porous media from the hydrogeological literature. The differences between the GDL layer and the modeling of unsaturated soils are that the GDL is hydrophobic and soil is hydrophilic, the pore-size distributions are different, and the GDL is a nonhomogeneous weave of carbon fibers. Despite these differences, the hydrogeological models are quite useful for fuel cell GDL modeling. However, it sometimes may be difficult to use these models since many properties such as the temperatures, phases, pressure, and the velocity of the species in and surrounding the GDL are unknown parameters while the fuel cell is operating.

The simplified geometry is shown in Figure 8-3. The dashed lines at the top of Figure 8-3 illustrate the portion of the channel where the gas is flowing through. The bottom of the diagram is the catalyst side where heat and water are added to the system, and gas is absorbed. On the upper channel sides, gas is added, and heat and water are removed. Since half of the upper boundary is the solid cathode material, and half is open channel, the boundary conditions are mixed. The portion where there is no flux into the cathode has Neumann boundary conditions, and the portion where there is no liquid water in the channels has Dirichlet boundary conditions.

The GDL in Figure 8-3 is 4d units long, and h units high. The aspect ratio is the perturbation parameter, and can be defined as $\varepsilon = h/d \ll 1$. The lower surface abuts the cathode catalyst layer, and the upper surface is

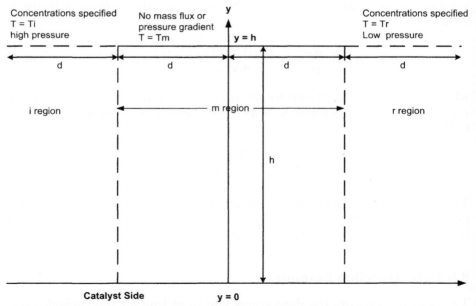

FIGURE 8-3. Division of the gas diffusion layer[5].

open to a channel on the left and right. The center region abuts a graphite cathode. The channels can be at different pressures, and all quantities are assumed to be steady-state. The pressure, P, temperature, T, oxygen concentration, u, water vapor concentration, v, and liquid-water volume fraction, θ, will be calculated. All of the variables will be functions of θ.

Since the physical process exhibited is the same as in the transport of groundwater in unsaturated porous media, the governing equation is Richard's equation, which gives the moisture velocity (V_θ) of liquid and vapor in porous media. The general form of the equation is:

$$V_\theta = -\kappa_\theta(\theta)\nabla\psi \qquad (8\text{-}59)$$

where κ_θ is the hydraulic conductivity of the GDL to the liquid water, and Ψ is the moisture potential. The total potential should also have a gravitational component, but it is disregarded from the Equation 8-59 because there is little liquid water present. The moisture potential should include all relevant properties of the GDL, such as tortuosity and wetting potential[6].

Since the nonhysterestic case is considered, θ will be a single-valued function of θ only $(\psi = \psi(\theta))$. Assuming incompressibility (the density of water is constant), the conservation equation becomes:

$$\nabla \cdot V_\theta = \Sigma \qquad (8\text{-}60)$$

where Σ is the source term introduced to incorporate condensation and evaporation:

$$\nabla \cdot (-\kappa_\theta(\theta)\nabla\psi) = \Sigma \qquad (8\text{-}61)$$

The diffusion coefficient of water can be defined by:

$$D_\theta(\theta) = \kappa_\theta(\theta)\frac{d\psi}{d\theta} \qquad (8\text{-}62)$$

The chain rule of differentiation can be used:

$$\nabla \cdot [D_\theta(\theta)\nabla\theta] + \Sigma = 0 \qquad (8\text{-}63)$$

Evaporation is a temperature-dependent process and is modeled using Arrhenius' law:

$$\text{evaporation} \propto \exp\left(-\frac{E_A}{RT}\right)\theta \qquad (8\text{-}64)$$

where E_A is the activation energy and R is the gas constant. Condensation is not a temperature-dependent process, and depends only upon the concentration of the water vapor:

$$\text{condensation} \propto \tilde{\upsilon} \tag{8-65}$$

Introducing the constant of proportionality, β_θ, for evaporation and β_v for condensation:

$$\beta_\theta \exp\left(-\frac{E_A}{RT}\right)\theta + \beta_v\tilde{\upsilon} = \Sigma \tag{8-66}$$

Therefore, this becomes:

$$\nabla \cdot [D_\theta(\theta)\nabla\theta] - \beta_\theta \exp\left(-\frac{E_A}{RT}\right)\theta + \beta_v\tilde{\upsilon} = 0 \tag{8-67}$$

For the gases (oxygen), either Fickian diffusion, or the Stefan-Maxwell equation can be used to describe the diffusion processes. Fick's equation for diffusion and transport is:

$$\nabla \cdot (D_u(\theta)\nabla\tilde{u} - \tilde{u}\tilde{V}_g) = 0 \tag{8-68}$$

where $\tilde{V}_g$ is the velocity of the gas phase, and D_u is the diffusion coefficient of oxygen.

In order to model the vapor transport, the evolution of the vapor phase of water must include convection:

$$\nabla \cdot [D_v(\theta)\nabla\tilde{\upsilon} - \tilde{\upsilon}\tilde{V}_g] + \upsilon_1\left[\beta_\theta \exp\left(-\frac{E_A}{RT}\right)\theta + \beta_v\tilde{\upsilon}\right] = 0 \tag{8-69}$$

where D_y is the diffusion coefficient of the water vapor, and v_1 is a normalization factor. The evaporation term produces vapor and the condensation term removes it.

In order to model the temperature, the following terms must be taken into account: (1) Fourier's law for heat conduction, (2) convection, (3) heat gain due to condensation, and (4) heat loss due to evaporation. Since the gas and liquid water velocities are small, it is assumed that the thermal transport from water and gas can be neglected. Therefore:

$$\nabla \cdot [\tilde{k}(\theta)\nabla\tilde{T}] + \rho_\theta L\left[\beta_\theta \exp\left(-\frac{E_A}{RT}\right)\theta + \beta_v\tilde{\upsilon}\right] = 0 \tag{8-70}$$

where $\tilde{k}$ is the thermal conductivity, ρ_θ is the density of liquid water, and L is the latent heat.

8.6.1 No Liquid Governing Equations

In the case that there is no liquid, all of the terms due to θ are neglected. Also, condensation and evaporation terms drop out of the equations. If the gas phase convects, the velocity is governed by Darcy's law:

$$\tilde{V}_g = -\frac{\kappa_g(\theta)}{\mu} \nabla \tilde{P} \tag{8-71}$$

where κ_g is the permeability of the GDL to gases and μ is the viscosity of the gas. The permeability κ_g depends upon θ because liquid water will remove the available pore space for the gas. In order to solve for pressure, the continuity equation can be used:

$$\nabla \cdot \tilde{V}_g = 0 \tag{8-72}$$

$$\nabla \cdot \tilde{V}_g = \nabla \cdot \left(-\frac{\kappa_g(\theta)}{\mu} \nabla \tilde{P} \right) = 0 \tag{8-73}$$

$$\nabla \cdot \tilde{V}_g = -\frac{1}{\mu} \nabla \cdot (\kappa_g(\theta) \nabla \tilde{P}) = 0 \tag{8-74}$$

Since θ has been dropped, $\kappa_g(\theta)$ is a constant:

$$\frac{\kappa_g(\theta)}{\mu} \nabla \cdot \nabla \tilde{P} = 0 \tag{8-75}$$

Equation 8-75 now becomes:

$$D_u \cdot \nabla \cdot (\nabla \tilde{u}) - \nabla \cdot (\tilde{u} \tilde{V}_g) = 0 \tag{8-76}$$

$$\nabla \cdot \tilde{u} + \frac{\kappa_g(\theta)}{\mu D_u} \left[\frac{\partial \tilde{u}}{\partial \tilde{x}} \cdot \frac{\partial \tilde{P}}{\partial \tilde{x}} + \frac{\partial \tilde{u}}{\partial \tilde{y}} \cdot \frac{\partial \tilde{P}}{\partial \tilde{y}} \right] = 0 \tag{8-77}$$

The condensation and evaporation term is dropped because of the no-liquid assumption.

$$\nabla \cdot \tilde{v} + \frac{\kappa_g(\theta)}{\mu D_v} \left[\frac{\partial \tilde{v}}{\partial \tilde{x}} \cdot \frac{\partial \tilde{P}}{\partial \tilde{x}} + \frac{\partial \tilde{v}}{\partial \tilde{y}} \cdot \frac{\partial \tilde{P}}{\partial \tilde{y}} \right] = 0 \tag{8-78}$$

Equation 8-78 becomes:

$$\nabla^2 \cdot \tilde{T} = 0 \tag{8-79}$$

The governing equations and boundary conditions motivate the following dimensionless parameters:

$$x = \frac{\tilde{x}}{d}, y = \frac{\tilde{y}}{d}, u = \frac{\tilde{u}}{u_1}, v = \frac{\tilde{v}}{v_1}, T(x,y) = \frac{T(\tilde{x}, \tilde{y}) - T_m}{T_1 - T_m},$$

$$P(x,y) = \frac{2\tilde{P}(\tilde{x}, \tilde{y}) - (P_1 + P_r)}{P_1 + P_r} \tag{8-80}$$

Substituting these into the previous equations:

$$\frac{P_1 - P_r}{2d^2} \frac{\partial^2 \tilde{P}}{\partial \tilde{x}^2} + \frac{P_1 - P_r}{2h^2} \frac{\partial^2 \tilde{P}}{\partial \tilde{y}^2} = 0 \quad \varepsilon^2 \frac{\partial^2 \tilde{P}}{\partial \tilde{x}^2} + \frac{\partial^2 \tilde{P}}{\partial \tilde{y}^2} = 0 \tag{8-81}$$

$$u_1 \left(\frac{1}{d^2} \frac{\partial^2 \tilde{u}}{\partial \tilde{x}^2} + \frac{1}{h^2} \frac{\partial^2 \tilde{u}}{\partial \tilde{y}^2} \right) + \frac{\kappa_g u_1 (P_1 - P_r)}{2\mu D_u} \left(\frac{1}{d^2} \frac{\partial \tilde{u}}{\partial \tilde{x}} \cdot \frac{\partial \tilde{P}}{\partial \tilde{x}} + \frac{1}{h^2} \frac{\partial \tilde{u}}{\partial \tilde{y}} \cdot \frac{\partial \tilde{P}}{\partial \tilde{y}} \right) = 0 \tag{8-82}$$

$$\varepsilon^2 \left(\frac{\partial^2 \tilde{u}}{\partial \tilde{x}^2} + \frac{\partial^2 \tilde{u}}{\partial \tilde{y}^2} \right) + Pe_u \left(\varepsilon^2 \frac{\partial \tilde{u}}{\partial \tilde{x}} \cdot \frac{\partial \tilde{P}}{\partial \tilde{x}} + \frac{\partial \tilde{u}}{\partial \tilde{y}} \frac{\partial \tilde{P}}{\partial \tilde{y}} \right) = 0 \tag{8-83}$$

$$Pe_u = \frac{\kappa_g u_1 (P_1 - P_r)}{2\mu D_u} \tag{8-84}$$

where Pe is the Peclet number for the oxygen. When v is replaced with u, then:

$$\varepsilon^2 \left(\frac{\partial^2 v}{\partial x^2} + \frac{\partial^2 v}{\partial y^2} \right) + Pe_u \left(\varepsilon^2 \frac{\partial v}{\partial x} \cdot \frac{\partial P}{\partial x} + \frac{\partial v}{\partial y} \cdot \frac{\partial P}{\partial y} \right) = 0 \quad Pe_v = \frac{\kappa_g (P_1 - P_r)}{2\mu D_v} \tag{8-85}$$

Therefore,

$$\frac{(T_1 - T_m)}{d^2} \frac{\partial^2 r}{\partial x^2} + \frac{(T_1 - T_m)}{h^2} \frac{\partial^2 r}{\partial y^2} = 0 \tag{8-86}$$

$$\left(\frac{h}{d} \right)^2 \frac{\partial^2 r}{\partial x^2} + \frac{\partial^2 T}{\partial y^2} = 0 \tag{8-87}$$

substituting into:

$$\varepsilon^2 \frac{\partial^2 T}{\partial x^2} + \frac{\partial^2 T}{\partial y^2} = 0 \tag{8-88}$$

8.6.2 No Liquid, No Convection, Constant Flux

For the case of no liquid, no convection and constant flux, the transport will now just be Fickian, and the pressure is constant. Therefore:

$$\varepsilon^2 \frac{\partial^2 u}{\partial x^2} + \frac{\partial^2 u}{\partial y^2} = 0 \tag{8-89}$$

$$\varepsilon^2 \frac{\partial^2 v}{\partial x^2} + \frac{\partial^2 v}{\partial y^2} = 0 \tag{8-90}$$

$$\varepsilon^2 \frac{\partial^2 T}{\partial x^2} + \frac{\partial^2 T}{\partial y^2} = 0 \tag{8-91}$$

When examining the boundary conditions with constant pressure, the regions of positive and negative x are symmetric about x = 0. If the region from $-2 \leq x \leq 0$ is used, the boundary conditions are as follows:

$$\frac{\partial T^m}{\partial x}(0, y) = 0 \tag{8-92}$$

$$\frac{\partial u^m}{\partial x}(0, y) = 0 \tag{8-93}$$

$$\frac{\partial v^m}{\partial x}(0, y) = 0 \tag{8-94}$$

At the cathode catalyst layer interface, constant flux is assumed. Oxygen flow out of the gas diffusion layer, therefore:

$$D_u \frac{\partial \tilde{u}}{\partial n} = -D_u \frac{\partial \tilde{u}}{\partial y}(\tilde{x}, 0) = -\tilde{q}_u \tag{8-95}$$

The water vapor and temperature fluxes are given by:

$$D_v \frac{\partial \tilde{v}}{\partial y}(\tilde{x}, 0) = -\tilde{q}_v \tag{8-96}$$

$$\tilde{k}_C \frac{\partial \tilde{T}}{\partial y}(\tilde{x}, 0) = -\tilde{q}_T \tag{8-97}$$

Substituting Equation 8-80 into Equations 8-95 through 8-97:

$$D_u \frac{u_1}{h} \frac{\partial u}{\partial y}(x, 0) = \tilde{q}_u \tag{8-98}$$

$$\frac{\partial u}{\partial y}(x, 0) = q_u \varepsilon^2, \quad q_u \varepsilon^2 = \frac{q_u h}{D_u u_1} \tag{8-99}$$

$$D_v \frac{v_1}{h} \frac{\partial v}{\partial y}(x,0) = -\tilde{q}_v \tag{8-100}$$

$$\frac{\partial v}{\partial y}(x,0) = -q_v \varepsilon^2, \quad q_v \varepsilon^2 = \frac{q_v h}{D_v v_1} \tag{8-101}$$

$$k_C \frac{T_1 - T_m}{h} \frac{\partial T}{\partial y}(x,0) = -\tilde{q}_T \tag{8-102}$$

$$\frac{\partial T}{\partial y}(x,0) = -q_T, \quad q_T = \frac{q_T h}{k_C(T_1 - T_m)} \tag{8-103}$$

Using $\varepsilon = 0.2$ as the perturbation parameter, the dependent variable can be written as:

$$T(x,y,\varepsilon) = T_0(x,y) + o(\varepsilon) \tag{8-104}$$

Now substituting Equation 8-91 into 8-104:

$$\varepsilon^2 \frac{\partial^2 T_0}{\partial x^2} + \frac{\partial^2 T_0}{\partial y^2} + o(\varepsilon) = 0 \tag{8-105}$$

Then:

$$\frac{\partial^2 T_0}{\partial y^2} = 0 \tag{8-106}$$

Using the following boundary conditions:

$$\frac{\partial T^m}{\partial x}(0,y) = 0, \quad \frac{\partial T}{\partial y}(x,0) = -q_T, \quad \frac{\partial T^1}{\partial x}(-2,y) = 0 \tag{8-107}$$

Integrating and using the boundary conditions:

$$\frac{\partial T_0}{\partial y} = -q_T, \quad T_0 = -q_T y + f(x), \quad f'(-2) = f'(0) = 0 \tag{8-108}$$

Now the problem has two different regions:

$$T_0^1(x, y) = q_T(1 - y) + 1 \quad -2 \le x \le -1 \tag{8-109}$$

$$T_0^m(x, y) = q_T(1 - y) \quad -1 \le x \le 0 \tag{8-110}$$

An interior layer variable is introduced to account for the discontinuity at $x = 0$:

$$z = \frac{x+1}{\varepsilon} \quad T(x, y) = q_T(1 - y) + T^i(z, y)$$

(8-111)

where T^i is the interior temperature.

Substituting 8-111 into 8-91, 8-103, and 8-86:

$$\frac{\partial^2 T^i}{\partial z^2} + \frac{\partial^2 T^i}{\partial y^2} = 0$$

(8-112)

$$-q_T + \frac{\partial T^i}{\partial y}(z, 0) = -q_T$$

(8-113)

$$\frac{\partial T^i}{\partial y}(z, 0) = 0$$

(8-114)

$$T^i(z, 1) = 1 \quad z < 0$$

(8-115)

$$T^i(z, 1) = 0 \quad z > 0$$

(8-116)

Now matching the outer solution:

$$q_T(1 - y) + T^i(-\infty, y) = T^l(-1^-, y) = q_T(1 - y) + 1$$

(8-117)

$$T^i(-\infty, y) = 1$$

(8-118)

$$q_T(1 - y) + T^i(-\infty, y) = T^m(-1^+, y) = q_T(1 - y)$$

(8-119)

$$T^i(\infty, y) = 0$$

(8-120)

A set of transformations are now introduced:

$$f_1 = z + iy, \quad f_2 = \exp(\pi \cdot f_1), \quad f_3 = \frac{f_2 - 1}{f_2 + 1}, \quad f_4 = \frac{1}{2} + \frac{1}{\pi} \sin^{-1}(f_3)$$

(8-121)

The solution is then:

$$T^i = \Re f_4$$

(8-122)

Example 8-1 illustrates modeling the temperature in the interior layer wing MATLAB:

EXAMPLE 8-1: Modeling the Temperature in the Interior Layer

Create a three-dimensional plot in MATLAB for the temperature of the interior layer using the equations derived in this section.

Using MATLAB to solve:

```
%%%%%%%%%%%%%%%%%%%%%%%%%%%%%%%%%%%%%%%%%%
% EXAMPLE 8-1: Modeling the Temperature in the
Interior Layer
% UnitSystem SI
%%%%%%%%%%%%%%%%%%%%%%%%%%%%%%%%%%%%%%%%%%
clear all
clc
format rat
```

% Define the parameters

```
eps = 0.2; % Perturbation Parameter
```

% Define the number of grid points in x and y direction

```
% nx = number of grid points in x direction
% ny = number of grid points in y direction
nx = 201;
ny = 201;
```

% Define the dimension of the domain

```
Lx = 1.6; % Length in x of the computation region
Ly = 1.0; % Length in y of the computation region
```

% Calculate the mesh size

```
hx = Lx/(nx-1); % Grid spacing in x
hy = Ly/(ny-1); % Grid spacing in y
```

% Generate the mesh/grid

```
x(1) = -1.8;
y(1) = 0.0;
for k = 2:nx
   x(k)=x(k-1)+hx;
end
for j = 2:ny
   y(j)=y(j-1)+hy;
end
```

$$z = \frac{x+1}{\varepsilon} \qquad T(x,y) = q_T(1-y) + T^i(z,y) \qquad \text{(8-111)}$$

where T^i is the interior temperature.

Substituting 8-111 into 8-91, 8-103, and 8-86:

$$\frac{\partial^2 T^i}{\partial z^2} + \frac{\partial^2 T^i}{\partial y^2} = 0 \qquad \text{(8-112)}$$

$$-q_T + \frac{\partial T^i}{\partial y}(z, 0) = -q_T \qquad \text{(8-113)}$$

$$\frac{\partial T^i}{\partial y}(z, 0) = 0 \qquad \text{(8-114)}$$

$$T^i(z,1) = 1 \quad z < 0 \qquad \text{(8-115)}$$

$$T^i(z,1) = 0 \quad z > 0 \qquad \text{(8-116)}$$

Now matching the outer solution:

$$q_T(1 - y) + T^i(-\infty, y) = T^l(-1^-, y) = q_T(1 - y) + 1 \qquad \text{(8-117)}$$

$$T^i(-\infty, y) = 1 \qquad \text{(8-118)}$$

$$q_T(1 - y) + T^i(-\infty, y) = T^m(-1^+, y) = q_T(1 - y) \qquad \text{(8-119)}$$

$$T^i(\infty, y) = 0 \qquad \text{(8-120)}$$

A set of transformations are now introduced:

$$f_1 = z + iy, \quad f_2 = \exp(\pi \cdot f_1), \quad f_3 = \frac{f_2 - 1}{f_2 + 1}, \quad f_4 = \frac{1}{2} + \frac{1}{\pi}\sin^{-1}(f_3) \qquad \text{(8-121)}$$

The solution is then:

$$T^i = \Re f_4 \qquad \text{(8-122)}$$

Example 8-1 illustrates modeling the temperature in the interior layer wing MATLAB:

EXAMPLE 8-1: Modeling the Temperature in the Interior Layer

Create a three-dimensional plot in MATLAB for the temperature of the interior layer using the equations derived in this section.

Using MATLAB to solve:

```
%%%%%%%%%%%%%%%%%%%%%%%%%%%%%%%%%%%%%%%%%%%%
```

% EXAMPLE 8-1: Modeling the Temperature in the Interior Layer

% UnitSystem SI

```
%%%%%%%%%%%%%%%%%%%%%%%%%%%%%%%%%%%%%%%%%%%%
```

```
clear all
clc
format rat
```

% Define the parameters

```
eps = 0.2; % Perturbation Parameter
```

% Define the number of grid points in x and y direction

```
% nx = number of grid points in x direction
% ny = number of grid points in y direction
nx = 201;
ny = 201;
```

% Define the dimension of the domain

```
Lx = 1.6; % Length in x of the computation region
Ly = 1.0; % Length in y of the computation region
```

% Calculate the mesh size

```
hx = Lx/(nx-1); % Grid spacing in x
hy = Ly/(ny-1); % Grid spacing in y
```

% Generate the mesh/grid

```
x(1) = -1.8;
y(1) = 0.0;
for k = 2:nx
   x(k)=x(k-1)+hx;
end
for j = 2:ny
   y(j)=y(j-1)+hy;
end
```

```
% Inner Layer Variable

z = (x+1.0)/eps;

[Z,Y] = meshgrid(z,y);

% Now determine the functions

f1 = complex(Z,Y);
f2 = exp((pi*f1));
f3 = (f2-1.0)./(f2+1.0);
f4 = (1.0/2.0)-(1.0/pi)*(asin(f3));

% Inner Variable

Ti = real(f4);

% Make a 3D plot of Ti

figure
surf(Z,Y,Ti,'EdgeColor','none')
set(gca,'DataAspectRatio',[1 1 1])
axis([-4.0 4.0 0 1 0 1])
view(-35,32)
xlabel(' z ')
ylabel(' y ')
zlabel(' Ti(z,y) ')
print -djpeg Tinteriorlayer
```

The plot for the temperature in the interior layer is shown in Figure 8-4.

The derivation of the oxygen concentration is analogous to the temperature derivation. Substituting Equation 8-89 into an equation similar to 8-104:

$$\frac{\partial^2 u_0}{\partial y^2} = 0 \qquad (8\text{-}123)$$

For the left region:

$$\frac{\partial u^1}{\partial y}(x, 0) = q_u \varepsilon^2 \qquad (8\text{-}124)$$

Integrating once:

$$\frac{\partial u_0}{\partial y} = f(x) \qquad (8\text{-}125)$$

Now the problem has two different regions:

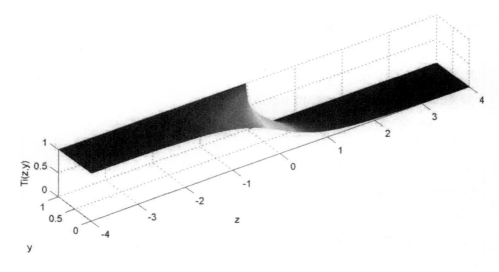

FIGURE 8-4. Three-dimensional plot of the temperature in the interior layer[7].

$$T^I(x, y) = 1 - q_u \varepsilon^2 (1 - y) \quad -2 \le x \le -1 \tag{8-126}$$

$$u^m(x, y, \varepsilon) = u_0^m(x, y) + \varepsilon^2 u_2^m(x, y) + o(1) \quad -1 \le x \le 0 \tag{8-127}$$

Substituting Equation 8-127 into 8-89 and 8-99:

$$\varepsilon^2 \frac{\partial^2 (u_0^m + u_2^m)}{\partial x^2} + \frac{\partial^2 (u_0^m + u_2^m)}{\partial y^2} = 0 \tag{8-128}$$

$$\frac{\partial^2 u_0^m}{\partial y^2} = 0 \tag{8-129}$$

$$-\frac{\partial^2 u_0^m}{\partial x^2} = \frac{\partial^2 u_2^m}{\partial y^2} \tag{8-130}$$

$$\frac{\partial u_0^m}{\partial y}(x, 0) = 0 \tag{8-131}$$

$$\frac{\partial u_2^m}{\partial y}(x, 0) = q_u \tag{8-132}$$

Now matching the outer solution:

$$u_0^m(-1, y) = 1 \tag{8-133}$$

$$\frac{\partial u_0^m}{\partial y} = 0 \tag{8-134}$$

$$u_0^m = f_0(x) \tag{8-135}$$

The boundary conditions are:

$$f_0(-1) = 1, \ f_0'(0) = 0 \tag{8-136}$$

Substituting Equation 8-135 into Equation 8-130 and 8-132:

$$\frac{\partial^2 u_2^m}{\partial y^2} = -f_0''(x) \tag{8-137}$$

$$\frac{\partial u_2^m}{\partial y} = -f_0''(x)(y-1) \tag{8-138}$$

$$\frac{\partial u_2^m}{\partial y}(x, 0) = -f_0''(x) = q_u \tag{8-139}$$

Continuing to simplify:

$$f_0' = q_u x \tag{8-140}$$

$$u_0^m(x, y) = f_0 = 1 - q_u \left(\frac{1 - x^2}{2} \right) \tag{8-141}$$

Substituting after integrating:

$$u_2^m(x, y) = q_u \left(y - \frac{y^2}{2} \right) + f_2(x) \tag{8-142}$$

where $f_2(x)$ is determined from the next order in the perturbation expansion. If $\varepsilon \to 0$ in the solutions, then:

$$u(x, y) = 1 + \varepsilon u^i(z, y) \tag{8-143}$$

Taking $\varepsilon \to 0$, and substituting 8-143 into 8-89 and 8-99:

$$\frac{\partial^2 u^i}{\partial z^2} + \frac{\partial^2 u^i}{\partial y^2} = 0 \tag{8-144}$$

$$\frac{\partial u^i}{\partial y}(z, 0) = 0 \tag{8-145}$$

$$1 + \varepsilon u^i(z, 1) = 1 \quad z < 0 \tag{8-146}$$

$$u^i(z, 1) = 0 \tag{8-147}$$

$$\frac{\partial u^i}{\partial y}(z, 1) = 0 \quad z > 0 \tag{8-148}$$

Now matching the outer solution:

$$1 + \varepsilon u^i(-\infty, y) = u^l(-1^-, y) = 1 + O(\varepsilon^2) \tag{8-149}$$

$$u^i(-\infty, y) = 0 \tag{8-150}$$

$$\frac{\partial u^i}{\partial z}(\infty, y) = \frac{\partial u^m}{\partial x}(-1^-, y) \tag{8-151}$$

$$\frac{\partial u^i}{\partial z}(\infty, y) = -q_u \tag{8-152}$$

The solutions for water vapor concentration are again analogous to those computed for T and u:

$$v^l(x, y) = 1 + q_u \varepsilon^2(1 - y) \quad -2 \leq x \leq -1 \tag{8-153}$$

$$v^m(x, y, \varepsilon) = v_0^m(x, y) + v_2^m(x, y) + o(1) \quad -1 \leq x \leq 0 \tag{8-154}$$

$$v_0^m(x, y) = 1 - q_v\left(\frac{1 - x^2}{2}\right) \tag{8-155}$$

$$v_2^m(x, y) = q_v\left(y - \frac{y^2}{2}\right) + g_2(x) \tag{8-156}$$

$$v(x, y) = 1 + \varepsilon v^i(z, y) \tag{8-157}$$

$$\frac{\partial^2 v^i}{\partial z^2} + \frac{\partial^2 v^i}{\partial y^2} = 0 \tag{8-158}$$

$$\frac{\partial v^i}{\partial y}(z, 0) = 0 \tag{8-159}$$

$$v^i(z, 1) = 1 \quad z < 0 \tag{8-160}$$

$$v^i(z, 1) = 0 \tag{8-161}$$

$$\frac{\partial v^i}{\partial y}(z, 1) = 0 \quad z > 0 \tag{8-162}$$

Now matching the outer solution:

$$v^i(-\infty, y) = 0 \qquad (8\text{-}163)$$

$$\frac{\partial v^i}{\partial z}(\infty, y) = q_v \qquad (8\text{-}164)$$

To determine the regions that are oversaturated, the following variable needs to be defined:

$$S(v, T) = \frac{v - v_{sat}(T)}{v_{sat}(T)} \qquad (8\text{-}165)$$

If $S > 0$, liquid water will be present.

Example 8-2 shows how to obtain and plot the temperature, water and oxygen concentration and saturation using the equations derived in this section.

EXAMPLE 8-2: Modeling the Gas Diffusion Layer

Create two-dimensional plots in MATLAB for the temperature, oxygen concentration, water vapor concentration, and saturation using the equations derived in this Section 8.6.2.

Using MATLAB to solve:

```
%%%%%%%%%%%%%%%%%%%%%%%%%%%%%%%%%%%%%%%%%
% EXAMPLE 8-2: Modeling the Gas Diffusion Layer
% UnitSystem SI
%%%%%%%%%%%%%%%%%%%%%%%%%%%%%%%%%%%%%%%%%
clc
clear all
format long e
% Define the parameters
eps = 0.2; % Perturbation Parameter
% Define the number of grid points in x and y direction
% nx = number of grid points in x direction
% ny = number of grid points in y direction
nx = 101;
ny = 65;
```

% SOR parameters

```
omega = 1.4; % SOR parameter
t = (2.0*cos(pi/(nx*ny)))^2;
% Calculate a optimum value of SOR parameter
omega1 = (16.0+sqrt((256.0-(64.0*t))))/(2.0*t);
omega2 = (16.0-sqrt((256.0-(64.0*t))))/(2.0*t);
oopt = min(omega1,omega2)
if ( (oopt <= 1.0) II (oopt >= 2.0 ))
   oopt = 1.0;
end
omega = oopt;
```

% Define the dimension of the domain

```
Lx = 2.0; % Length in x of the computation region
Ly = 1.0; % Length in y of the computation region
```

% Calculate the mesh size

```
hx = Lx/(nx-1); % Grid spacing in x
hy = Ly/(ny-1); % Grid spacing in y
```

% Generate the mesh/grid

```
x(1) = -1.0;
y(1) = 0.0;
for i = 2:nx
   x(i)=x(i-1)+hx;
end
for j = 2:ny
   y(j)=y(j-1)+hy;
end
```

% Solve the Temperature equation

```
%%%%%%%%%%%%%%%%%%%%%%%%%%%%%%%%%%%%%%%%%%%
```

% Initialize the temperature field as zero

```
T = zeros(nx,ny);
```

% Max-Norm on Error {L-inf Error) Initialized

```
Linf = 1.0;
iteration = 0;
while (Linf > 1.0e-5)
   iteration = iteration+1;
   % Store the old values of T in Told
   Told = T;
```

```
% Apply the boundary conditions
for i = 1:nx
    % BC for the bottom boundary
    T(i,1)=T(i,2)+hy;
    % BC for the top boundary
    if ( x(i) <= 0 )
        T(i,ny) = 1;
    else
        T(i,ny) = 0;
    end
end

for j = 1:ny
    % BC for the Left boundary
    T(1,j)=T(2,j);

    % BC for the Left boundary
    T(nx,j)=T(nx-1,j);
end

% Now compute the interior domain using 2nd order finite difference
for i = 2:nx-1
    for j = 2:ny-1
        term1 = ((eps/hx)^2)*(T(i-1,j)+T(i+1,j));
        term2 = ((1.0/hy)^2)*(T(i,j-1)+T(i,j+1));

        num = term1+term2;
        den = 2.0*(((eps/hx)^2)+((1.0/hy)^2));

        Tgs = num/den;
        T(i,j) = (omega*Tgs)+((1.0-omega)*Told(i,j));
    end
end

% Calculate the error
Terr = abs(T-Told);
Linf = norm(Terr,2);

% Print the convergence history every 100 iterations
ccheck = round(iteration/100)-(iteration/100);
if ( (iteration == 1) || ( ccheck == 0) )
    fprintf('%d \t %e \n', iteration, Linf)
end
end

% Plot the solutions
figure
[X,Y] = meshgrid(x,y);
clevel = [-1 -0.005 0 0.053 0.152 0.252 0.352 0.451 0.551 0.65 0.75 0.949 1.049
    1.148 1.248 1.348 1.447 1.547 1.646 1.746 1.848 1.945 2.0 3.0];
contourf(X',Y',T,clevel)
```

```
colorbar
axis([-1.2 1.2 -0.8 1.8])
xlabel(' x ')
ylabel(' y ')
zlabel(' T(x,y) ')
hold on
yi = [0:0.01:1];
xi = zeros(length(yi));
plot(xi,yi,'black')
print -djpeg figures\chap4\temperature4
```

% Save data to a file

```
save data\chap4\temp.dat T -ASCII -DOUBLE
```

% Solve the Oxygen concentration equation

```
%%%%%%%%%%%%%%%%%%%%%%%%%%%%%%%%%%%%%%%%%%
%omega = 1.0; % SOR parameter
```

% Initialize the oxygen concentration field as zero

```
u = zeros(nx,ny);
```

% Max-Norm on Error {L-inf Error) Initialized

```
Linf = 1.0;
iteration = 0;

while (Linf > 1.0e-5)
    iteration = iteration+1;
    % Store the old values of T in Told
    uold = u;

% Apply the boundary conditions
for i = 1:nx
    % BC for the bottom boundary
    u(i,1)=u(i,2)-(1.81*(eps^2)*hy);

    % BC for the top boundary
    if ( x(i) <= 0 )
        u(i,ny) = 1;
    else
        u(i,ny) = u(i,ny-1);
    end
end

for j = 1:ny
    % BC for the Left boundary
    u(1,j)=u(2,j);
    % BC for the Left boundary
    u(nx,j)=u(nx-1,j);
end
```

```
% Now compute the interior domain using 2nd order finite difference
for i = 2:nx-1
   for j = 2:ny-1
      term1 = ((eps/hx)^2) * (u(i-1,j)+u(i+1,j));
      term2 = ((1.0/hy)^2) * (u(i,j-1)+u(i,j+1));

      num = term1+term2;
      den = 2.0 * (((eps/hx)^2)+((1.0/hy)^2));
      ugs = num/den;
      u(i,j) = (omega * ugs)+((1.0-omega) * uold(i,j));
   end
end

% Calculate the error
uerr = abs(u-uold);
Linf = norm(uerr,2);
% Plot the convergence history every 100 iterations
ccheck = round(iteration/100)-(iteration/100);
if ( iteration == 1 )
   fighandle = figure;
   hold on
end
if ( ccheck == 0 )
   fprintf('%d \t %e \n', iteration, Linf)
   plot(iteration,Linf,'b-',iteration,Linf,'r.')
   hold on
   end
end
close(fighandle);

% Plot the solutions
figure
[X,Y] = meshgrid(x,y);
clevel = [-1 -0.5 -0.087 -0.031 0 0.025 0.08 0.136 0.192 0.247 0.303 0.359 0.415
   0.47 0.526 0.582 0.638 0.693 0.749 0.805 0.86 0.916 0.972 1 3.0];
contourf(X',Y',u,clevel)
colorbar
axis([-1.2 1.2 -0.8 1.8])
xlabel(' x ')
ylabel(' y ')
zlabel(' u(x,y) ')
hold on
yi = [0:0.01:1];
xi = zeros(length(yi));
plot(xi,yi,'black')
print -djpeg figures\chap4\oxygen-concentration4
```

% Save data to a file

```
save data\chap4\oxygen.dat u -ASCII -DOUBLE
```

% Solve the Water vapor concentration equation

% %

% Initialize the water vapor concentration field as zero

```
v = zeros(nx,ny);
```

% Max-Norm on Error {L-inf Error) Initialized

```
Linf = 1.0;
iteration = 0;

while (Linf > 1.0e-5)
   iteration = iteration+1;
   % Store the old values of T in Told
   vold = v;

% Apply the boundary conditions
for i = 1:nx
   % BC for the bottom boundary
   v(i,1)=v(i,2)+(0.755*(eps^2)*hy);

   % BC for the top boundary
   if ( x(i) <= 0 )
      v(i,ny) = 1;
   else
      v(i,ny) = v(i,ny-1);
   end
end

for j = 1:ny
   % BC for the Left boundary
   v(1,j)=v(2,j);

   % BC for the Left boundary
   v(nx,j)=v(nx-1,j);
end

% Now compute the interior domain using 2nd order finite difference
for i = 2:nx-1
   for j = 2:ny-1
      term1 = ((eps/hx)^2)*(v(i-1,j)+v(i+1,j));
      term2 = ((1.0/hy)^2)*(v(i,j-1)+v(i,j+1));

      num = term1+term2;
      den = 2.0*(((eps/hx)^2)+((1.0/hy)^2));

      vgs = num/den;
      v(i,j) = (omega*vgs)+((1.0-omega)*vold(i,j));
   end
end
```

```
% Calculate the error
verr = abs(v-vold);
Linf = norm(verr,2);
% Plot the convergence history every 100 iterations
ccheck = round(iteration/100)-(iteration/100);
if ( iteration == 1 )
   fighandle = figure;
   hold on
end
if ( ccheck == 0 )
   fprintf('%d \t %e \n', iteration, Linf)
   plot(iteration,Linf,'b-',iteration,Linf,'r.')
   hold on
   end
end
close(fighandle);

% Plot the solutions
figure
[X,Y] = meshgrid(x,y);
clevel = [0. 1. 1.012 1.035 1.058 1.082 1.105 1.128 1.152 1.175 1.198 1.222 1.245
   1.268 1.291 1.315 1.338 1.361 1.385 1.408 1.431 3.0];
contourf(X',Y',v,clevel)
colorbar
axis([-1.2 1.2 -0.8 1.8])
xlabel(' x ')
ylabel(' y ')
zlabel(' v(x,y) ')
hold on
yi = [0:0.01:1];
xi = zeros(length(yi));
plot(xi,yi,'black')
print -djpeg figures\chap4\watervapor-concentration4
```

% Save data to a file

```
save data\chap4\watervapor.dat v -ASCII -DOUBLE
```

% Calculate the water vapor saturation

```
% and the S variable from equation 4.40
for i = 1:nx
   for j = 1:ny
      term1 = 710.0/((2.0*T(i,j))+353.0);
      term2 = (7.87e-2)*T(i,j);
      term3 = (5.28e-4)*(T(i,j)^2);
      term4 = (2.65e-6)*(T(i,j)^3);
```

```
     vsat(i,j) = term1 * (exp((-0.869+term2-term3+term4)));
     S(i,j) = (v(i,j)-vsat(i,j))/vsat(i,j);
  end
end

% Plot the solutions
figure
[X,Y] = meshgrid(x,y);
clevel = [-1. 0 0.045 0.07 0.105 0.134 0.167 0.2 0.233 0.266 0.299 0.332 0.365
   0.397 0.43 0.463 0.496 0.529 0.562 0.595 0.628 0.661 0.694 0.711 1.0];
contourf(X',Y',S,clevel)
colorbar
axis([-1.2 1.2 -0.8 1.8])
xlabel(' x ')
ylabel(' y ')
zlabel(' S(v,T) ')
hold on
yi = [0:0.01:1];
xi = zeros(length(yi));
plot(xi,yi,'black')
print -djpeg figures\chap4\saturation4
```

% Save data to a file

```
save data\chap4\saturation.dat S -ASCII -DOUBLE
```

The contour plots for the temperature, oxygen concentration, water vapor concentration, and saturation are shown in Figures 8-5, 8-6, 8-7, and 8-8.

Chapter Summary

The gas diffusion layer must be a good proton conductor, chemically stable, and able to withstand the temperatures and compression forces of the fuel cell stack. There are many methods that have been used to model porous media in the literature. Some of the common methods include modeling the gas and fluid through the pores, or modeling the interaction of the gas/fluid with the solid porous media. Commonly used methods for modeling the GDL include Fick's law, Darcy's law, and the Stefan-Maxwell diffusion for the mass transport. Ohm's law is typically used for charge transport, and energy balances can be made on the system in order to obtain the most accurate flow rates, velocities, and pressure drops through the porous media layer.

Problems

- Create the MATLAB code for the case of no liquid, no convection, with a radiation condition for the problem set up in Section 8.6.

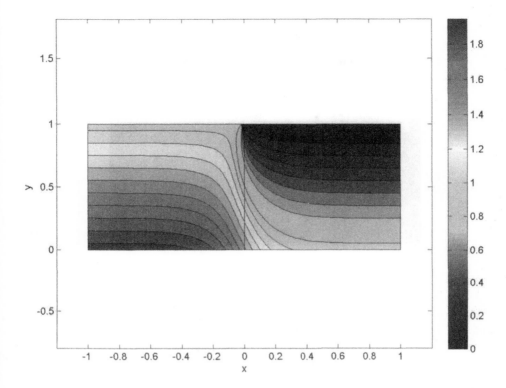

FIGURE 8-5. Contour plot of temperature.

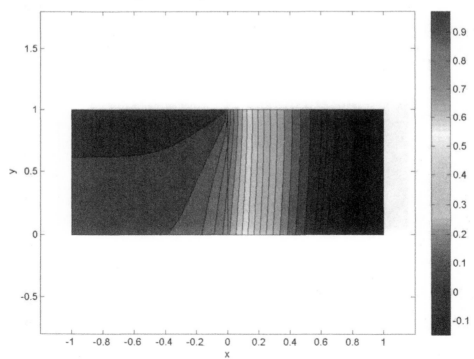

FIGURE 8-6. Contour plot of oxygen concentration.

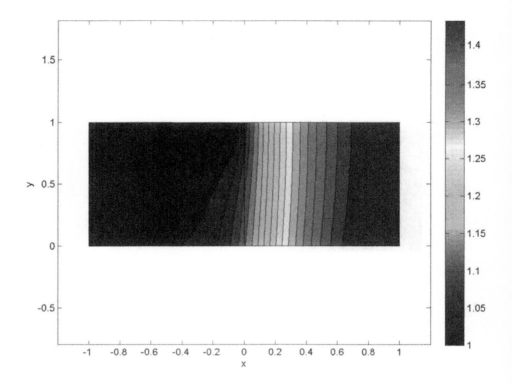

FIGURE 8-7. Contour plot of water vapor concentration.

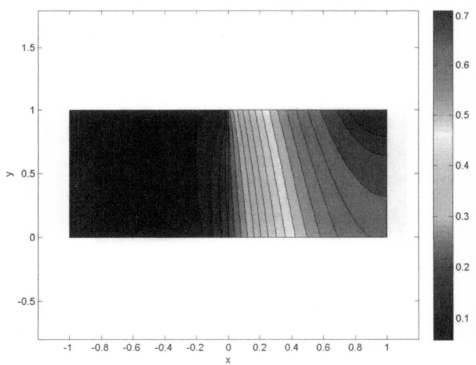

FIGURE 8-8. Contour plot of saturation.

- Create a two-dimensional contour plot in MATLAB for saturation when both liquid water and vapor are present based upon the problem presented in Section 8.6.
- Create a cathode GDL model in MATLAB using the Stefan-Maxwell equation.
- Create an anode GDL model in MATLAB using Fick's law of diffusion.

Endnotes

[1] Hinds, G. October 2005. *Preparation and Characterization of PEM Fuel Cell Electrocatalysts: A Review.* National Physical Laboratory. NPL Report DEPC MPE 019.
[2] Beuscher, U., et al. 2004. *Multiphase Flow in a Thin Porous Material.* W.L. Gore, Inc. Twentieth Annual Workshop on Mathematical Problems in Industry, June 21–25, 2004. University of Delaware. Delaware, U.S.A.
[3] Ibid.
[4] Ibid.
[5] Ibid.
[6] Ibid.
[7] Ibid.

Bibliography

Aarnes, J.E., T. Gimse, and K.-A. Lie. An introduction to the numerics of flow in porous media using MATLAB.
Antoli, E. Recent developments in polymer electrolyte fuel cell electrodes. *J. Appl. Electrochem.* Vol. 34, 2004, pp. 563–576.
Barbir, F. 2005. *PEM Fuel Cells: Theory and Practice.* Burlington, MA: Elsevier Academic Press.
Chan, K., Y.J. Kim, Y.A. Kim, T. Yanagisawa, K.C. Park, and M. Endo. High performance of cup-stacked type carbon nanotubes as a Pt-Ru catalyst support for fuel cell applications. *J. Appl. Phys.* Vol. 96, No. 10, November 2004.
Chen, R., and T.S. Zhao. Mathematical modeling of a passive feed DMFC with heat transfer effect. *J. Power Sources.* Vol. 152, 2005, pp. 122–130.
Coutanceau, C., L. Demarconnay, C. Lamy, and J.M. Leger. Development of electrocatalysts for solid alkaline fuel cell (SAFC). *J. Power Sources.* Vol. 156, 2006, pp. 14–19.
Crystal Lattice Structures Web page. Updated March 13, 2007. Center for Computational Materials Science of the United States Naval Research Laboratory. Available at: http://cst-www.nrl.navy.mil/lattice/. Accessed March 20, 2007.
Girishkumar, G., K. Vinodgopal, and P.V. Kamat. Carbon nanostructures in portable fuel cells: Single-walled carbon nanotube electrodes for methanol oxidation and oxygen reduction. *J. Phys. Chem. B.* Vol. 108, 2004, pp. 19960–19966.
Hahn, R., S. Wagner, A. Schmitz, and H. Reichl. Development of a planar micro fuel cell with thin film and micro patterning technologies. *J. Power Sources.* Vol. 131, 2004, pp. 73–78.

Haile, S.M. Fuel cell materials and components. *Acta Material.* Vol. 51, 2003, pp. 5981–6000.

Hinds, G. September 2004. *Performance and Durability of PEM Fuel Cells: A Review.* National Physical Laboratory. NPL Report DEPC MPE 002.

Kharton, V.V., F.M.B. Marques, and A. Atkinson. Transport properties of solid electrolyte ceramics: A brief review. *Solid State Ionics.* Vol. 174, 2004, pp. 135–149.

Larminie, J., and A. Dicks. 2003. *Fuel Cell Systems Explained.* 2nd ed. West Sussex, England: John Wiley & Sons.

Li, W., C. Liang, W. Zhou, J. Qui, Z. Zhou, G. Sun, and Q. Xin. Preparation and characterization of multiwalled carbon nanotube-supported platinum for cathode catalysts of direct methanol fuel cells. *J. Phys. Chem. B.* Vol. 107, 2003, pp. 6292–6299.

Li, X. 2006. *Principles of Fuel Cells.* New York: Taylor & Francis Group.

Lin, B. 1999. Conceptual design and modeling of a fuel cell scooter for urban Asia. Princeton University, master's thesis.

Lister, S., and G. McLean. PEM fuel cell electrode: A review. *J. Power Sources.* Vol. 130, 2004, pp. 61–76.

Liu, J.G., T.S. Zhao, Z.X. Liang, and R. Chen. Effect of membrane thickness on the performance and efficiency of passive direct methanol fuel cells. *J. Power Sources.* Vol. 153, 2006, pp. 61–67.

Matsumoto, T., T. Komatsu, K. Arai, T. Yamazaki, M. Kijima, H. Shimizu, Y. Takasawa, and J. Nakamura. Reduction of Pt usage in fuel cell electrocatalysts with carbon nanotube electrodes. *Chem. Commun.* 2004, pp. 840–841.

Mehta, V., and J.S. Copper. Review and analysis of PEM fuel cell design and manufacturing. *J. Power Sources.* Vol. 114, 2003, pp. 32–53.

Mench, M.M., C.-Y. Wang, and S.T. Tynell. *An Introduction to Fuel Cells and Related Transport Phenomena.* Department of Mechanical and Nuclear Engineering, Pennsylvania State University. PA, USA. Draft. Available at: http://mtrl1 .mne.psu.edu/Document/jtpoverview.pdf. Accessed March 4, 2007.

Mench, M.M., Z.H. Wang, K. Bhatia, and C.Y. Wang. 2001. *Design of a Micro-direct Methanol Fuel Cell.* Electrochemical Engine Center, Department of Mechanical and Nuclear Engineering, Pennsylvania State University. PA, USA.

Mogensen, M., N.M. Sammes, and G.A. Tompsett. Physical, chemical, and electrochemical properties of pure and doped ceria. *Solid State Ionics.* Vol. 129, 2000, pp. 63–94.

Morita, H., M. Komoda, Y. Mugikura, Y. Izaki, T. Watanabe, Y. Masuda, and T. Matsuyama. Performance analysis of molten carbonate fuel cell using a Li/Na electrolyte. *J. Power Sources.* Vol. 112, 2002, pp. 509–518.

O'Hayre, R., S.-W. Cha, W. Colella, and F.B. Prinz. 2006. *Fuel Cell Fundamentals.* New York: John Wiley & Sons.

Rowe, A., and X. Li. Mathematical modeling of proton exchange membrane fuel cells. *J. Power Sources.* Vol. 102, 2001, pp. 82–96.

Silva, V.S., J. Schirmer, R. Reissner, B. Ruffmann, H. Silva, A. Mendes, L.M. Madeira, and S.P. Nunes. Proton electrolyte membrane properties and direct methanol fuel cell performance. *J. Power Sources.* Vol. 140, 2005, pp. 41–49.

Smitha, B., S. Sridhar, and A.A. Khan. Solid polymer electrolyte membranes for fuel cell applications—A review. *J. Membr. Sci.* Vol. 259, 2005, pp. 10–26.

Sousa, R., Jr., and E. Gonzalez. Mathematical modeling of polymer electrolyte fuel cells. *J. Power Sources.* Vol. 147, 2005, pp. 32–45.

Souzy, R., B. Ameduri, B. Boutevin, G. Gebel, and P. Capron. Functional fluoropolymers for fuel cell membranes. *Solid State Ionics.* Vol. 176, 2005, pp. 2839–2848.

Springer et al. Polymer electrolyte fuel cell model. *J. Electrochem. Soc.* Vol. 138, No. 8, 1991, pp. 2334–2342.

Todd, B. 2002. *Multi-Component Gas Transport in Porous Media.* Available at: http://www.fuelcellknowledge.org/.

Wang, C., M. Waje, X. Wang, J.M. Trang, R.C. Haddon, and Y. Yan. Proton exchange membrane fuel cells with carbon nanotube-based electrodes. *Nano Lett.* Vol. 4, No. 2, 2004, pp. 345–348.

Wong, C.W., T.S. Zhao, Q. Ye, and J.G. Liu. Experimental investigations of the anode flow field of a micro direct methanol fuel cell. *J. Power Sources.* Vol. 155, 2006, pp. 291–296.

Xin, W., X. Wang, Z. Chen, M. Waje, and Y. Yan. Carbon nanotube film by filtration as cathode catalyst support for proton exchange membrane fuel cell. *Langmuir.* Vol. 21, 2005, pp. 9386–9389.

Yamada, M., and I. Honma. Biomembranes for fuel cell electrodes employing anhydrous proton conducting uracil composites. *Biosensors Bioelectronics.* Vol. 21, 2006, pp. 2064–2069.

You, L., and H. Liu. A two-phase flow and transport model for PEM fuel cells. *J. Power Sources.* Vol. 155, 2006, pp. 219–230.

Zhao, W. 2006. Mass Transfer to/from Distributed Sinks/Sources in Porous Media. University of Waterloo, PhD dissertation.

Sousa, T., Jr. and E. E. Gonzalez-Winistschi, Mathematical modeling of polymer electrolyte fuel cells, *J. Power Sources*, vol. 147, 2005, pp. 32–45.

Sousa, T., B. Franklin, B. Bouzevic, C. Cady, and R. Coppin, Functional thermofluid models for fuel cell membranes, *Fuel Cells Series*, Vol. 176, 2008, pp. 2699–2708.

Spencer, J. A., *Polymer Electrolyte Fuel Cell Model*, *J. Electrochem. Soc.*, Vol. 146, No. 8, 1991, pp. 2334–2342.

Todd, T., *Basic Matrix Computation*, *T&T Transactional Online Media*. Available at: http://www.matrixonline.org.

Wang, C., M. Wang, M. Wang, L. M. Zhou, L. P. Guttman, and Y. Yin, Proton exchange membrane fuel cells with carbon nanotube based electrodes, *Nanoletter*, Vol. 4, No. 2, 2004, pp. 345–348.

Wang, C. W., Y. F. Zhao, O. Yu, and D. O. Liu, Experimental investigation of the anode flow field in a direct methanol fuel cell, *J. Power Sources*, Vol. 158, 2006, pp. 351–358.

Xin, W., Z. Wang, Z. Chen, M. Wan, and S. Yao, Carbon nanotube film for durable cathode catalysts, *J. Power Sources*, vol. 183, catalyst distribution fuel cell, Langmuir, Vol. 27, 2009, pp. 9386–9389.

Yamada, M. and T. Honma, Bioelectric nanocomposites for fuel cell electrodes employing anhydrous proton conducting proton complexes, *Electrochim. Acta*, Electrochimica, Vol. 21, 2008, pp. 8963–8969.

You, L. X. and H. Liu, A two-phase flow and transport model for PEM fuel cells, *J. Power Sources*, Vol. 155, 2006, pp. 373–383.

Zhang, W., 2006, *Mass Transfer to/from Destabilized Slides: Studies in Porous Media*, University of Waterloo PhD dissertation.

CHAPTER 9

Modeling the Catalyst Layers

9.1 Introduction

The fuel cell electrode layers are where electrochemical reactions occur. The electrode layer is made up of the catalyst and a gas diffusion layer. When the hydrogen in the flow channels meets the electrode layer, it diffuses into the gas diffusion layer as described in Chapters 4 and 8. At the anode, the hydrogen is broken into protons and electrons. The electrons travel to the carbon cloth, flow field plate, to the contact, and then to the load. The protons travel through the polymer exchange membrane to the cathode. At the cathode catalyst layer, oxygen combines with the protons to form water. The catalyst layer must be very effective at breaking molecules into protons and electrons, have high surface area, and preferably be low cost. The catalyst layers are often the thinnest in the fuel cell (5 to 30 microns [μm]), but are often the most complex due to multiple phases, porosity, and electrochemical reactions. It is a challenge to find a low-cost catalyst that is effective at breaking the hydrogen into protons and electrons.

Figure 9-1 shows a schematic of the fuel cell catalyst layers where the chemical reactions occur at the interphase between the electrocatalyst and electrolyte. Experimental evidence supports an agglomerate-type structure where the electrocatalyst is supported on a carbon agglomerate and covered by a thin layer of membrane. Often, a layer of liquid is also assumed to be on top of the membrane layer.

There have been many approaches taken in modeling the catalyst layer, as shown in Table 9-1. The approach taken depends upon how the rest of the fuel cell is modeled. There are both microscopic and macroscopic models for the catalyst layer. The microscopic models include pore-level models and quantum models. The quantum models deal with detailed reaction mechanisms, elementary transfer reactions, and transition states.

Specific topics covered in this chapter include the following:

- Mass and species conservation
- Ion transport

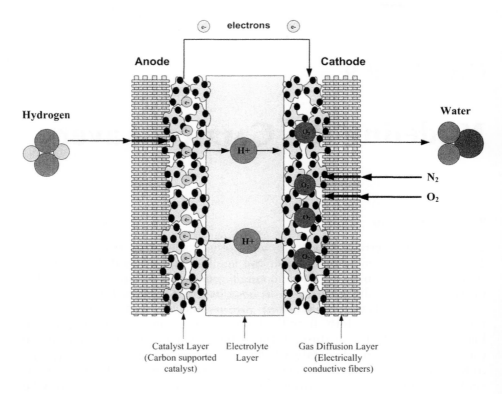

FIGURE 9-1. Catalyst transport phenomena.

TABLE 9-1
Equations Used to Model the Catalyst Layer

Model Characteristic	Description/equations
No. of dimensions	1, 2, or 3
Mode of operation	Dynamic or steady-state
Phases	Gas, liquid, or a combination of gas and liquid
Kinetics	Tafel-type expressions, Butler-Volmer equations, or complex kinetics equations
Mass transport	Nernst-Planck + Schlogl, Nernst-Planck + drag coefficient, or Stefan-Maxwell equation
Ion transport	Ohm's law
Membrane swelling	Empirical or thermodynamic models
Energy balance	Isothermal or full energy balance

- Momentum conservation
- Conservation of energy
- Other required relations

The commonly used equations for modeling the fuel cell catalyst layer will be described, along with an example for using these equations.

9.2 Physical Description of the PEM Fuel Cell Catalyst Layers

The fuel cell electrode is a thin, catalyst layer where electrochemical reactions take place. The electrodes are usually made of a porous mixture of carbon-supported platinum and isonomer. In order to catalyze reactions, catalyst particles must have contact to both protonic and electronic conductors. Furthermore, there must be passages for reactants to reach catalyst sites and for reaction products to exit. The contacting point of the reactants, catalyst, and electrolyte is conventionally referred to as the three-phase interface. In order to achieve acceptable reaction rates, the effective area of active catalyst sites must be several times higher than the geometric area of the electrode. Therefore, the electrodes are made porous to form a three-dimensional network, in which the three-phase interfaces are located. An illustration of the catalyst, electrolyte, and gas diffusion layer is shown in Figure 9-1.

The catalyst surface area is a very important characteristic of the catalyst layer; thus it is important to have small platinum particles (4 nm or smaller) with a large surface area finely dispersed on the surface of catalyst support, which is typically carbon powders with a high mesoporous area (>75 m^2/g). The typical support material is Vulcan XC72R, Black Pearls BP 2000, Ketjen Black International, or Chevron Shawinigan[1]. In order to designate the particle size distribution, the platinum particle surface area on a per-unit mass basis can be calculated by assuming all of the platinum particles are spherical:

$$A_s = \frac{\int f(D)\pi D^2 dD}{\int f(D)\rho_{Pt}\left(\frac{\pi D^3}{6}\right)dD} = \frac{6}{\rho_{Pt}D_{32}} \tag{9-1}$$

where ρ_{Pt} is the density of the platinum black and D_{32} is the volume-to-surface area mean diameter of all the particles. The active area per unit mass can be estimated from the mean D_{32}, and a typical value is 28 m^2/g Pt.

The catalyst layer is thin to help minimize cell potential losses due to the rate of proton transport and reactant gas permeation in the depth of the electrocatalyst layer. The metal active surface area should be maximized; therefore, higher Pt/C ratios should be selected (> 40% by weight).

It has been noted in the literature that the cell's performance remained unchanged as the Pt/C ratio was varied from 10% to 40% with a Pt loading of 0.4 mg/cm². When the Pt/C ratio was increased beyond 40%, the cell performance actually decreased. Fuel cell performance can be increased by better Pt utilization in the catalyst layer, instead of increasing the Pt loading.

9.3 General Equations

The catalyst layer contains many phases: liquid, gas, different solids, and the membrane. Although various models have different equations, most of these are derived from the same governing equations, regardless of the effects being modeled. The anode reaction can be described by a Butler-Volmer-type expression in most cases except for those which use a fuel other than pure hydrogen. In these cases, the platinum catalyst becomes "poisoned." The carbon monoxide adsorbs to the electrocatalytic sites and decreases the reaction rate. There are models in the literature that account for this by using a carbon monoxide site balance and examining the reaction steps involved. For the cathode, a Tafel-type expression is commonly used due to the slow kinetics of the four-electron transfer reaction.

The membrane and diffusion modeling equations apply to the same variables in the same phase in the catalyst layer. The rate of evaporation or condensation relates the water concentration in the gas and liquid phases. There are many approaches that can be used for the water content and chemical potential in the membrane. If liquid water exists, a supersaturated isotherm is used—or the liquid pressure can be assumed to be either continuous or related through a mass-transfer coefficient. In order to relate the reactant and product concentrations, potentials, and currents in the phases in the catalyst layer, kinetic expressions can be used with zero values for the total current. The kinetic expressions result in the transfer currents relate the potentials and currents in the electrode and electrolyte phases, as well as govern the production of reactant and products. To simplify the equations for an ionically and an electrically conducting phase, the following equation can be used:

$$\nabla \cdot i_2 = -\nabla \cdot i_1 = a_{1,2} i_{h,1-2} \tag{9-2}$$

where $-\nabla \cdot i_1$ represents the total anodic rate of electrochemical reactions per unit volume of electrode, $a_{1,2}$ is the interfacial area between the electrically conducting and membrane phase with no flooding, and $i_{h,1-2}$ is the transfer current for reaction h between the membrane and the electronically conducting solid. The charge balance assumes that the faradic reactions are only electrode processes. The double-layer charge is neglected under steady-state conditions. This equation is used with the conservation of mass equation to simplify it. If electroneutrality is assumed, then there is no significant charge separation compared with the volume of the domain. Since there is accumulation of charge, and electroneutrality has been assumed, the steady-state charge can be assumed to be zero:

$$\sum_k \nabla \cdot i_k = 0 \qquad (9\text{-}3)$$

A mass balance can be written for each species in each phase as first introduced in Chapter 5. The differential form for species i in phase k can be written as[2]:

$$\frac{\partial \varepsilon_k c_{i,k}}{\partial t} = -\nabla \cdot N_{i,k} - \sum a_{1,k} s_{i,k,h} \frac{i_{h,1-k}}{n_h F} + \sum_l s_{i,k,l} \sum_{p \neq k} a_{k,p} r_{1,k-p} + \sum_g s_{i,k,k} \varepsilon_k R_{g \cdot k} \qquad (9\text{-}4)$$

where the term on the left side is the total amount of species i accumulated in a certain control volume, $-\nabla \cdot N_{i,k}$ is the mass that enters or leaves the control volume by mass transport, and the last three terms account for the material that is gained or lost due to the chemical reactions. The second term on the right side accounts for electron transfer reactions that occur at the interface between phase k and the electronically conducting phase, the second summation accounts for all other interfacial reactions besides electron transfer and the final term accounts for homogeneous reactions in phase k[3].

If the reduction of oxygen is the only reaction at the cathode, the following mass balance results[4]:

$$\nabla \cdot N_{O2,g} = -\frac{1}{4F} a_{1.2} i_{0Orr} \left(\frac{p_{O2}}{p_{O2}^{ref}} \right) \exp\left(-\frac{a_c F}{RT} \eta_{ORR,1-2} \right) = -\frac{1}{4F} \nabla \cdot i_1 \qquad (9\text{-}5)$$

where i_{0Orr} is the exchange current density for the reaction, p_{O2} is the partial pressure for O_2, p_{O2}^{ref} is the reference partial pressure for O_2, a_c is the cathodic transfer coefficient, and $\eta_{ORR,1-2}$ is the cathode overpotential. Many models use catalyst loading, which is defined as the amount of catalyst in grams per geometric area. If a turnover frequency is desired, the reactive surface area of platinum can be used. This is related to the radius of the platinum particle, which assumes a roughness factor that is experimentally inferred using cyclic voltammetry measuring the hydrogen adsorption. These variables are used to calculate the specific interfacial area between the electrocatalyst and the electrolyte[5]:

$$a_{1,2} = \frac{m_{Pt} A_{Pt}}{L} \qquad (9\text{-}6)$$

where L is the thickness of the catalyst layer. Another important parameter related to the catalyst loading is the efficiency of the electrode. This tells how much of the electrode is actually being used for an electrochemical reaction. Another useful parameter is the effectiveness factor, E, which helps to examine the ohmic and mass transfer effects. This is the actual rate of reaction divided by the rate of reaction without any transport losses.

9.4 Types of Models

Catalyst layer models range from zero to three dimensions in the literature. Zero-dimensional models do not consider the actual structure of the catalyst layer. One-dimensional models account for the overall changes across the layer. There are also two- and three-dimensional models that consist of the catalyst layer and the agglomerate. Agglomerate models can be either macro- or micro-models, depending upon how they are calculated. There are many more cathode than anode models in the literature. This is due to the slower reaction rate of the cathode due to water production and mass transfer effects. The anode can almost always be modeled as a simplified cathode model—except for the case when the hydrogen is not pure, and the poisoning of the electrocatalyst is included. The common types of models for the catalyst layer are presented in Table 9-2, and include interface, microscopic, porous electrode, and agglomerate models.

TABLE 9-2
Types of Models

Type of Model	Description
Interface models	Interface models assume that the catalyst layers exist at the GDL/membrane interface. The catalyst interface layer is the location where oxygen and hydrogen are consumed and water is produced.
Microscopic and single-pore models	Microscopic and single-pore models contain cylindrical gas pores of a defined radius. The catalyst layer contains Teflon-coated pores for gas diffusion, and the rest of the electrode is flooded with liquid electrolyte.
Porous electrode models	Porous electrode models are based upon the overall reaction distribution in the catalyst layer. The agglomerates all have a uniform concentration and potential.
Agglomerate models	Agglomerate models assume a uniform reaction-rate distribution, and more accurately represent the actual structure of the catalyst layers.

9.4.1 Interface Models

There are many interface models in the literature that assume that the catalyst layer only exists at the GDL/membrane interface. This assumption means that the catalyst layers are infinitely thin, and the structure can be ignored. There are several ways to accomplish this in a model. One method is to treat the catalyst layer as a location where hydrogen and oxygen are consumed and water is produced. Models that focus exclusively on water management are set up in this manner, and use Faraday's law for the mass balance between the membrane and diffusion medium. Faraday's law is the rate at which hydrogen and oxygen are consumed and water is generated, as shown in Equations 9-7–9-9:

$$N_{H2} = \frac{I}{2F} \tag{9-7}$$

$$N_{O2} = \frac{I}{4F} \tag{9-8}$$

$$N_{H2O} = \frac{I}{2F} \tag{9-9}$$

where N is the consumption rate (mol/s), I is the current (A), and F is Faraday's constant (C/mol).

A more sophisticated method of modeling the catalyst layer is to use Equations 9-7–9-9, and then use a polarization curve equation to produce a potential for the cell at a specific current density. The general equation for the fuel cell polarization curve was first presented in Chapter 3:

$$E = E_r - \frac{RT}{\alpha F}\ln\left(\frac{i+i_{loss}}{i_0}\right) - \frac{RT}{nF}\ln\left(\frac{i_L}{i_L - i}\right) - iR_i \tag{9-10}$$

The Nernst equation from Chapter 2 is used to determine the theoretical electrical potential of the reaction. Equation 9-11 shows the potential for hydrogen electrochemically reacting with oxygen.

$$E_r = \frac{\Delta G}{2F} + RT\ln\frac{P_{H2} \cdot P_{O2}^{1/2}}{P_{H2O}} \tag{9-11}$$

The Nernst equation is used to find the potential at the active locations, and the local potential using the half reactions. To obtain a good approximation of the actual fuel cell potential, the voltage losses described in Chapters 2 through 5 can be utilized.

As mentioned in Chapters 2–5, the activation over-potential and the rate of species consumption and generation are determined by the electrochemical kinetics and the current density, i. Activation losses and the current density are solved using the appropriate boundary conditions. The reaction rate depends upon the current density, and the mass flow rates are related to the electric current through Faraday's law:

$$m_{H2} = \frac{I}{2F} M_{H2} \tag{9-12}$$

$$m_{O2} = \frac{I}{4F} M_{O2} \tag{9-13}$$

$$m_{H2O} = \frac{I}{2F} M_{H2O} \tag{9-14}$$

where M_{H2}, M_{O2}, and M_{H2O} are the molecular weights of the hydrogen, oxygen, and water, respectively.

Another commonly used approach for modeling the catalyst layers is to use the Butler-Volmer equation from Chapter 3. The relationship between the current density and the activation losses for the anode is:

$$i_a = i_{0,a} \exp\left(-\alpha_a \frac{F v_{act,a}}{RT}\right) - \exp\left((1-\alpha_a)\frac{F v_{act,a}}{RT}\right) \tag{9-15}$$

where i_a is the transfer current density (A/m^3), v_{act} is the activation electrode losses, i_o is the exchange current density, and α_a is the anodic charge transfer coefficient. For the cathode:

$$i_c = i_{0,c} \exp\left(-\alpha_c \frac{F v_{act,c}}{RT}\right) - \exp\left((1-\alpha_c)\frac{F v_{act,c}}{RT}\right) \tag{9-16}$$

From Chapter 3, the exchange current density depends on the local partial pressure of reactants and the local temperature. As the partial pressure of the reactants decreases, the exchange current density will also decrease—which decreases performance. This illustrates how activation and diffusion limitations affect each other, and why the mass flux must be solved precisely. The exchange current density for the anode and cathode is:

$$i_{0,c} = i_{0,c}^0 \left(\frac{P_{O2}}{P_{O2}^0}\right)^{\gamma} \exp\left[\frac{-E_{A,c}}{R}\left(\frac{1}{T} - \frac{1}{T^0}\right)\right] \tag{9-17}$$

$$i_{0,a} = i_{0,a}^0 \left(\frac{P_{H2}}{P_{H2}^0}\right)^{\gamma1} \left(\frac{P_{H2O}}{P_{H2O}^0}\right)^{\gamma2} \exp\left[\frac{-E_{A,a}}{R}\left(\frac{1}{T} - \frac{1}{T^0}\right)\right] \tag{9-18}$$

where $i_{0,c}^0$ and $i_{0,a}^0$ are the reference exchange current density, γ_1 and γ_2 is the reaction order, T^0 is the reference temperature (303 K) and E_A is the activation energy.

One assumption that is sometimes problematic when modeling the exchange current density is that it is expressed in terms of geometric area. The actual reaction occurs at the active sites, which are a strong function of the shapes and volume of the actual particles. These microstructural parameters are difficult to control and can vary drastically depending on the processing techniques and conditions. Therefore, as one would expect, two electrodes may have vastly different numbers of active sites in the same geometric area. The average current density is the total current generated in a fuel cell divided by the geometric area:

$$i_{avg} = \frac{1}{A} \int_{V_a} j_a dV = \frac{1}{A} \int_{V_c} j_c dV \qquad (9\text{-}19)$$

Another method for modeling the catalyst layer is to incorporate kinetics equations at the interfaces. This enables the models to account for multidimensional effects, where the current density or potential changes. This allows for nonuniform current density distributions, since the potential is constant in the cell. Although interface models adequately predict catalyst layer performance, modeling the catalyst layers in more detail allows relevant interactions to be accounted for—which creates more accurate results.

9.4.2 Microscopic and Single-Pore Models

There are many early models of fuel cell catalyst layers that are microscopic, single-pore models. The catalyst layer typically contains Teflon-coated pores for gas diffusion, with the rest of the electrode being flooded with electrolyte. These models provide a little more detail about the microstructure of the catalyst layers than the interface models. There are two main types of single-pore models: gas pore and flooded agglomerate models. In the gas pore model, the pores are assumed to be straight, cylindrical gas pores of a certain radius. They extend the length of the catalyst layer, and reactions occur at the surface. The second type of model also uses gas pores, but the pores are filled with electrolyte and catalyst. In these pores, reaction, diffusion and migration occur. The equations that were previously introduced (Equations 9-2–9-14) are primarily used in these models.

In the flooded agglomerate models, diffusion along with the use of equilibrium for the dissolved gas concentration in the electrolyte is used. The flooded agglomerate model shows better agreement with experimental data than the single-pore model, which is expected because it models the actual microstructure better. A disadvantage of the single-pore models is that they do not take into account the actual structure of the catalyst layer—which has multiple pores that are tortuous. However, the single-pore models have helped to form some of the later, more complicated models that provide more realistic simulation results. It is currently unknown how accurately these equations model PEM fuel cells, since these models were originally used for phosphoric acid fuel cells, and the polymer electrolyte membrane does not necessarily penetrate the pores.

The other types of microscopic models in the literature are spherical agglomerate models, which were introduced by Antoine et al.[6]. The spherical agglomerates in these models are assumed to exist in three-dimensional hexagonal arrays. Between the agglomerates, there are either gas pores or the region is flooded with electrolytes. These models examine the interactions between agglomerate placement. The equations that are solved are Ohm's law and Fick's law with kinetic expressions. The results of these models show the concentration around an electrocatalyst particle, and the placement of these particles helps to enhance or reduce the efficiency of the catalyst layer.

9.4.3 Porous Electrode Models

The porous electrode models calculate the overall reaction distribution in the catalyst layer without including the exact geometry details. The porous electrode models consider the agglomerate structure, but the layer has a uniform concentration and potential. This theory is concerned with the overall reaction distribution in the catalyst layer. Therefore, the main effects do not occur in the agglomerates, and the agglomerates have a uniform concentration and potential. The effect of concentration is accounted for in the calculation of the charge transfer resistance, which is from the kinetic expressions, and likely to be nonlinear. The charge transfer resistances should be in parallel with a capacitor, which represents double-layer charging. This can be neglected for the steady-state operation of the fuel cells, and introduced if transients or impedance is studied. The governing equations for porous electrodes are shown in Table 9-3.

The next level of models treat the catalyst layers using a complete simple porous electrode modeling approach. Therefore, the catalyst layers have a finite thickness, and all variables are determined as in Table 9-3 Some of these models assume that the gas phase reactant concentration is uniform in the catalyst layers; most allow the diffusion to occur in the gas phase.

The final porous electrode models are similar to thin film models. Instead of gas diffusion in the catalyst layer, the reactant gas dissolves in the electrolyte and moves by diffusion and reaction[7]. The governing equations are the same, but a concentration instead of a partial pressure appears in the kinetic expressions, and the governing equations for mass transport become one of diffusion in the membrane or water. These models are simple because they only consider the length scale of the catalyst layer,

TABLE 9-3
Fuel Cell Catalyst Layer Variables and Equations

Variable	Equation	Equation No.
Overall liquid water flux (N_L)	Mass balance	7-2, 7-10, 9-4 or Chapter 5 equations
Overall membrane water flux (N_W)	Mass balance	7-2, 7-10, 9-4 or Chapter 5 equations
Gas phase component flux ($N_{G,i}$)	Mass balance	7-2, 7-10, 9-4 or Chapter 5 equations
Gas phase component partial pressure ($p_{G,i}$)	Stefan-Maxwell	5-63
Liquid pressure (P_L)	Darcy's law	8-52
Membrane water chemical potential (μ_w)	Schlogl's equation	7-6 or 7-7
Electronic phase current density (i_1)	Ohm's law	8-46
Membrane current density (i_2)	Ohm's law	7-5, 8-46, 9-2 or 9-20
Electronic phase potential (Φ_1)	Charge balance	8-46
Temperature (T)	Energy balance	Chapter 6 equations
Total gas pressure (p_G)	Darcy's law	8-50
Liquid saturation (S)	Saturation relation	8-54

and the concentrations of the other species are assumed to be in equilibrium with their respective gas phase partial pressures[8].

9.4.4 Agglomerate Models

Agglomerate models only consider effects that occur on the agglomerate scale. They assume a uniform reaction rate distribution. These models more accurately represent the structure of the catalyst layers than the simple porous electrode models. These are similar to the microscopic models, except the geometric arrangement is averaged and each phase exists in each control volume. The characteristic length scale of the agglomerate is assumed to be the same size and shape. In the model, the reactant or product diffuses through the electrolyte film surrounding the particle and agglomerate where it diffuses and reacts. The equations again are similar to those listed in Table 9-3, except that either spherical or cylindrical coordinates are used for the gradients.

The equation for the porous catalysts has been used in the literature, and is known to match experimental results. The porous catalyst equations can be used for both the anode and cathode, but this section reviews the equations for the cathode reaction. Equation 9-2 has been modified by the addition of an effectiveness factor, which allows for the actual rate of reaction to be written as:

$$\nabla \cdot i_2 = a_{1,2} i_{h,1-2} E \qquad (9\text{-}20)$$

Since the cathode reaction is a first-order reaction following Tafel kinetics, the solution of the mass conservation equation in spherical agglomerate yields an analytical expression for the effectiveness factor:

$$E = \frac{1}{3\phi^2}(3\phi\coth(3\phi) - 1) \tag{9-21}$$

where ϕ is the Thiele modulus for the system, and can be expressed as:

$$\phi = \zeta\sqrt{\frac{k'}{D_{O2,agg}^{eff}}} \tag{9-22}$$

where ζ is the characteristic length of the agglomerate (volume per surface area), $R_{agg}/3$ for spheres, $R_{agg}/2$ for cylinders, δ_{agg} for slabs, and k' is a rate constant given by:

$$k' = \frac{a_{1,2}i_{0,Orr}}{4Fc_{O2}^{ref}}\exp\left(-\frac{\alpha_c F}{RT}(\eta_{ORR,1-2})\right) \tag{9-23}$$

where the reference concentration is that concentration in the agglomerate that is in equilibrium with the reference pressure:

$$c_{O2}^{ref} = p_{O2}^{ref}H_{O2,agg} \tag{9-24}$$

where $H_{O2,agg}$ is Henry's constant for oxygen in the agglomerate. If external mass transfer limitations can be neglected, then the surface concentration can be set equal to the bulk concentration, which is assumed uniform throughout the catalyst layer in simple agglomerate models. Otherwise, the surface concentration is unknown and must be calculated. An expression for the diffusion of oxygen to the surface of the agglomerate can be written:

$$W_{O2}^{diff} = A_{agg}D_{O2,film}\frac{c_{O2}^{bulk} - c_{O2}^{surf}}{\delta_{film}} \tag{9-25}$$

where W_{O2}^{diff} is the molar flow rate of oxygen to the agglomerate, A_{agg} is the specific external surface area of the agglomerate, and the film can be either membrane or water. This expression uses Fick's law and a linear gradient, which should be valid due to the low solubility of oxygen, steady-state conditions, and thinness of the film. At steady-state, the above flux is equal to the flux due to reaction and diffusion in the agglomerate; therefore, the unknown concentrations can be replaced. Using the resultant expression in the conservation equation yields:

$$\nabla \cdot i_1 = 4Fc_{O2}^{bulk} \left(\cfrac{1}{\cfrac{\delta_{film}}{A_{agg}D_{O2,film}} + \cfrac{1}{k'E}} \right) \tag{9-26}$$

Equation 9-26 is the governing equation for the agglomerate models for the cathode catalyst layer, and without external mass transfer limitations.

9.5 Heat Transport in the Catalyst Layers

In order to accurately predict rates of reaction and species transport, the temperature and heat distribution need to be determined accurately. This was first introduced in Chapter 6, Section 6.3. Solving for heat transfer in the electrodes is a challenge because convective, radial, and conductive heat transfers all exist. There are three main differences with heat transfer equations in the catalyst layer in comparison to the other fuel cell layers:

- Conduction plays a dominant role in the solid part of the catalyst layers, whereas convection dominates the heat transfer in the species transport.
- The porous nature of the layer complicates the heat transfer model. In addition, the heat transfer from the gas phase to the solid phase may be difficult to model.
- The heat source in the catalyst is difficult to model in comparison with the other fuel cell layers (where the heat source is large and a known parameter).

Heat is generated in the electrodes through several different methods:

- Ohmic heat is generated due to the irreversible resistance to current flow.
- Heat is generated due to the potential loss of activation and transport losses. The energy not transformed into current ends up as heat, and this heat is released in the electrodes.
- The change in entropy due to the electrochemical reaction generates heat. Entropic heat effects can be endothermic or exothermic and are generated at the two electrodes in unequal amounts.

A critical parameter in modeling the heat transfer in fuel cells is determining where the heat is released in the electrode. Most researchers ignore radiative transfer, but it is known that the electrodes absorb, emit, and transmit radiation. Modeling this radiative transfer will aid in developing better fuel cell catalyst layer models.

TABLE 9-4
Parameters for Example 9-1

Parameter	Value
Temperature	348.15 K
O_2 permeation in agglomerate	1.5e-11
H_2 permeation in agglomerate	2e-11
Agglomerate radius in anode and cathode	110e-5
Total gas pressure	1 atm
Hydrogen pressure	1 atm
Air pressure	1 atm
Saturation	0.6e-12
Anode transfer coefficient	1
Cathode transfer coefficient	0.9
Constant ohmic resistance	0.02 ohm-cm^2
Limiting current density	1.4 A/cm2
Mass transport constant	1.1
Amplification constant	0.085
Gibbs function in liquid form	−228,170 J/mol
Electrode-specific interfacial area	10,000
Current density	1 to 1.2 A/cm^2

EXAMPLE 9-1: Modeling the Catalyst Layer

Model the PEM fuel cell anode and cathode catalyst layer using the equations for porous catalysts introduced in this chapter. Create expressions for the anode and cathode activation losses, liquid water rate of reaction, and hydrogen rate of reaction. Create the following plots: (1) Current density versus the effectiveness factor, (2) current density versus activation losses, (3) current density versus voltage (polarization curve), and (4) current density versus the hydrogen flux density. All of the parameters required for this example are listed in Table 9-4.

The first step is to calculate the Nernst voltage and voltage losses. To calculate the Nernst voltage for this example, the partial pressures of water, hydrogen, and oxygen will be used. First calculate the saturation pressure of water:

$$\log P_{H_2O} = -2.1794 + 0.02953 * T_c - 9.1837 \times 10^{-5} * T_c^2 + 1.4454 \times 10^{-7} * T_c^3$$

$$\log P_{H_2O} = -2.1794 + 0.02953 * 60 - 9.1837 \times 10^{-5} * 60^2 + 1.4454 \times 10^{-7} * 60^3 = 0.467$$

Calculate the partial pressure of hydrogen:

$$p_{H_2} = 0.5 * (P_{H_2}/\exp(1.653 * i/(T_K^{1.334}))) - P_{H_2O} = 1.265$$

Calculate the partial pressure of oxygen:

$$p_{O_2} = (P_{air}/\exp(4.192 * i/(T_K^{1.334}))) - P_{H_2O} = 2.527$$

The voltage losses will now be calculated. The activation losses are estimated using the Butler-Volmer equation. For the anode:

$$\nabla \cdot i_2 = a_{1,2}(1 - S)i_{anode}$$

$$i_{anode} = \left[\frac{p_{H2}}{p_{H2}^{ref}} \exp\left(\frac{\alpha_a F}{RT}(\Phi_1 - \Phi_2) \right) - \exp\left(\frac{-\alpha_c F}{RT}(\Phi_1 - \Phi_2) \right) \right]$$

For the cathode:

$$\nabla \cdot i_2 = a_{1,2}(1 - S)i_{cathode}$$

$$i_{cathode} = \left[\frac{p_{O2}}{p_{O2}^{ref}} \exp\left(\frac{-\alpha_c F}{RT}(\Phi_1 - \Phi_2 - E_r) \right) \right]$$

The ohmic losses (see Chapter 4) are estimated using Ohm's law:

$$V_{ohmic} = -(i * r)$$

The mass transport (or concentration losses—see Chapter 5) can be calculated using the following equation:

$$V_{conc} = alpha1 * i^k * \ln\left(1 - \frac{i}{i_L} \right)$$

To insure that there are no negative values calculated for V_{conc} for the MATLAB program, the mass transport losses will only be calculated if $1 - \left(\dfrac{i}{i_L} \right) > 0$, else $V_{conc} = 0$.

The Nernst voltage can be calculated using the following equation:

$$E_{Nernst} = -\frac{G_{f,liq}}{2*F} - \frac{R*T_k}{2*F} * \ln\left(\frac{P_{H2O}}{P_{H2} * p_{O2}^{1/2}} \right)$$

Since all of the voltage losses had a (−) in front of each equation, the actual voltage is the addition of the Nernst voltage plus the voltage losses:

$$V = E_{Nernst} + V_{act} + V_{ohmic} + V_{conc}$$

The hydrogen oxidation reaction rate at the anode can be written as:

$$\nabla \cdot i_2 = a_{1,2}i_{h,1-2}E$$

$$\nabla \cdot N_{H_2,G} = -\frac{1}{2F}a_{1,2}(1-S)i_{anode}E$$

$$i_{anode} = \left[\frac{p_{H2}}{p_{H2}^{ref}}\exp\left(\frac{\alpha_a F}{RT}(\Phi_1 - \Phi_2)\right) - \exp\left(\frac{-\alpha_c F}{RT}(\Phi_1 - \Phi_2)\right)\right]$$

The liquid water cathode catalyst reaction can be written as:

$$\nabla \cdot N_{H2O,L} = -\frac{1}{4F}a_{1,2}(1-S)i_{cathode}E$$

$$i_{cathode} = \left[\frac{p_{O2}}{p_{O2}^{ref}}\exp\left(\frac{-\alpha_c F}{RT}(\Phi_1 - \Phi_2 - E_r)\right)\right]$$

where the effectiveness factor is:

$$E = \frac{1}{3\phi^2}(3\phi\coth(3\phi) - 1)$$

The Thiele modulus is expressed by:

$$\phi = \zeta\frac{k'}{D_{O2,agg}^{eff}}$$

The kinetic portion of the Thiele modulus is:

$$k' = \frac{a_{1,2}i_{0,Orr}}{4Fc_{O2}^{ref}}\exp\left(-\frac{\alpha_c F}{RT}(\eta_{cathode})\right)$$

Using MATLAB to solve:

```
%%%%%%%%%%%%%%%%%%%%%%%%%%%%%%%%%%%%%%%%%
% EXAMPLE 9-1: Modeling the Catalyst Layer
% UnitSystem SI
%%%%%%%%%%%%%%%%%%%%%%%%%%%%%%%%%%%%%%%%%
% Handling homogeneous reactions
% Parameters
F = 96 487;            % Faraday's constant
R = 8.314 34;          % Ideal gas constant
```

```
R2 = 83.1434;              % Ideal gas constant
T = 348.15;                % Temperature (K)
Tc = T - 273.15;           % Temperature (degrees C)
Psi_O2_agg = 1.5e-11;      % O2 permeation in agglomerate
Psi_H2_agg = 2e-11;        % H2 permeation in agglomerate
R_agg_an = 110e-5;         % Agglomerate radius in anode
R_agg_cat = 110e-5;        % Agglomerate radius in cathode
P_gas = 1;                 % Total gas pressure
P_H2 = 1;                  % Hydrogen pressure in atm
P_air = 1;                 % Air pressure in atm
S = 0.6e-12;               % Saturation
x_O2_g = 0.21;             % Mole fraction of O2 in the gas phase
x_H2_g = 1;                % Mole fraction of H2 in the gas phase
alpha_a = 1;               % Anode transfer coefficient
alpha_c = 0.9;             % Cathode transfer coefficient
R_ohm = 0.02;              % Constant ohmic resistance (ohm-cm^2)
il = 1.4;                  % Limiting current density (A/cm2)
k = 1.1;                   % Constant k used in mass transport
Alpha1 = 0.085;            % Amplification constant
Gf_liq = -228170;          % Gibbs function in liquid form (J/mol)
a120 = 10000;              % Electrode specific interfacial area (1/cm)
volt = 0:0.01:1.2;         % Voltage
i = 0:0.01:1.2;            % Current Density (A/cm^2)
```

%%

% Calculation of Partial Pressures

% Calculations of saturation pressure of water

```
x = -2.1794 + 0.02953.*Tc-9.1837.*(10.^-5).*(Tc.^2) + 1.4454.*(10.^-7).*(Tc.^3);
P_H2O = (10.^x);
```

% Calculation of partial pressure of hydrogen

```
pp_H2 = 0.5.*((P_H2)./(exp(1.653.*i./(T.^1.334)))-P_H2O);
```

% Calculation of partial pressure of oxygen

```
pp_O2 = (P_air./exp(4.192.*i/(T.^1.334)))-P_H2O;
```

%%

% Reaction 1: H2O generation as liquid

% Exchange current density (A/cm^2)

```
i_orr = 1.0e-7.*exp((73269./R).*((1./303)-(1./T)));
```

% Kinetic portion of the Thiele modulus

```
k_O2 = a120.*i_orr./(4*F).*exp(((-alpha_c.*F)./(R.*T)).*(-volt));
```

% Thiele modulus

```
phi_O2 = R_agg_cat.*sqrt(k_O2 ./ Psi_O2_agg);
```

% Effectiveness factor due to mass transfer & reaction

```
E_O2 = 3 ./ phi_O2.^2.*(phi_O2 ./ tanh(phi_O2) - 1)
```

% Reaction rate of liquid water at cathode catalyst layer

```
rate_rx_H2Ol = k_O2.*x_O2_g.*P_gas.*(1 - S).*E_O2;
```

```
%%%%%%%%%%%%%%%%%%%%%%%%%%%%%%%%%%%%%%%%
```

% Reaction 2: Hydrogen oxidation

% Exchange current density (A/cm^2)

```
i_hor = 1e-3.*exp((9500./R).*((1./303)-(1./T)));
```

% Kinetic portion of the Thiele modulus

```
k_h = a120.*i_hor ./(2.*F).*exp((alpha_a*F)./(R.*T).*volt);
```

% Thiele modulus

```
phi_H2 = R_agg_an.*sqrt(k_h./Psi_H2_agg);
```

% Effectiveness factor due to mass transfer & reaction

```
E_H2 = 3 ./ phi_H2.^2.*(phi_H2 ./ tanh(phi_H2) - 1);
i_h = exp(-(alpha_c.*F)./(R.*T).*volt)./ exp((alpha_a.*F)./(R.*T).*volt);
```

% Reaction rate of hydrogen at anode catalyst layer

```
rate_rx_H2 = k_h.*(x_H2_g.*P_gas - i_h).*(1 - S).*E_H2;
```

```
%%%%%%%%%%%%%%%%%%%%%%%%%%%%%%%%%%%%%%%%
```

% Calculate activation losses from Butler-Volmer equation

% Activation loss at the anode

```
V_act_anode = ((R.*T)./((alpha_a + alpha_c).*F)).*log(i./(i_hor.*a120.*(1 -
    S).*(x_H2_g.*P_gas)));
```

% Activation loss at the cathode

```
V_act_cathode = log(i./(-a120.*(1 - S).*i_orr.*(x_O2_g.*P_gas)))*((R.*T)./
    (-alpha_c.*F));
```

% Total activation loss

```
V_act = V_act_anode + V_act_cathode;
```

% Ohmic Losses

```
V_ohmic = -(i.*R_ohm);
```

% Mass Transport Losses

```
term = (1-(i./il));
if term > 0
V_conc = Alpha1.*(i.^k).*log(1-(i./il));
else
V_conc = 0;
end
```

% Calculation of Nernst voltage

```
E_nernst = -Gf_liq./(2.*F)-((R.*T).*log(P_H2O./(pp_H2.*(pp_O2.^0.5))))./(2.*F)
```

% Calculation of output voltage

```
V_out = E_nernst + V_ohmic + V_act + V_conc;
if term < 0
V_conc = 0;
end
if V_out < 0
V_out = 0;
end
```

% Plot the cell current versus the effectiveness factor

```
figure1 = figure('Color',[1 1 1]);
hdlp=plot(i,E_O2,i,E_H2);
title('Cell Current vs. Effectiveness Factor','FontSize',14,'FontWeight','Bold')
xlabel('Cell Current (A/cm^2)','FontSize',12,'FontWeight','Bold');
ylabel('Effectiveness Factor','FontSize',12,'FontWeight','Bold');
set(hdlp,'LineWidth',1.5);
grid on;
```

% Plot the cell current versus voltage

```
figure2 = figure('Color',[1 1 1]);
hdlp=plot(i,V_act);
title('Cell Current vs. Voltage','FontSize',14,'FontWeight','Bold')
xlabel('Cell Current (A/cm^2)','FontSize',12,'FontWeight','Bold');
ylabel('Voltage (Volts)','FontSize',12,'FontWeight','Bold');
set(hdlp,'LineWidth',1.5);
grid on;
```

```
% Plot the polarization curve

figure3 = figure('Color',[1 1 1]);
hdlp=plot(i,V_out);
title('Cell Current vs. Voltage','FontSize',14,'FontWeight','Bold')
xlabel('Cell Current (A/cm^2)','FontSize',12,'FontWeight','Bold');
ylabel('Voltage (Volts)','FontSize',12,'FontWeight','Bold');
set(hdlp,'LineWidth',1.5);
grid on;

% Plot the flux density of hydrogen

figure4 = figure('Color',[1 1 1]);
hdlp=plot(i,rate_rx_H2);
title('Superficial flux density of hydrogen','FontSize',14,'FontWeight','Bold')
xlabel('Cell Current (A/cm^2)','FontSize',12,'FontWeight','Bold');
ylabel('Flux density of H2 (mol/cm^2-s)','FontSize',12,'FontWeight','Bold');
set(hdlp,'LineWidth',1.5);
grid on;
```

Figures 9-2 through 9-5 show the figures plotted from Example 9-1.

One of the main challenges in modeling the catalyst layer is finding reliable parameters. The reference exchange current density, the transfer coefficients, and the reaction order are all dependent on the rate determining step(s) of the complex electrochemical reaction, as well as the electrode microstructure. The model becomes more difficult when one has to consider the oxidation of various gases (such as CH_4, CO, and H_2 simultaneously). The reaction order has not been extensively studied, and the experimental kinetic data are still scarce. Thus, there is a need to establish exactly how different fuels are simultaneously oxidized.

Chapter Summary

Modeling the catalyst layer is very complex because it has properties of all of the other fuel cell layers combined. Some of the important phenomena that need to be included in a rigorous catalyst layer model include mass, energy, and charge balances along with a relation that accounts for the contact between the porous GDL and polymer membrane layer. In addition, knowing how the catalyst agglomerates are distributed along the GDL is a challenge. The kinetics equations are the most important when modeling the catalyst layer. Commonly used equations are the Tafel and the Butler-Volmer equations. There are many choices for how the catalyst layer is modeled, and the complexity of the model needs to be determined by the level of accuracy required and the resources available to help create the model.

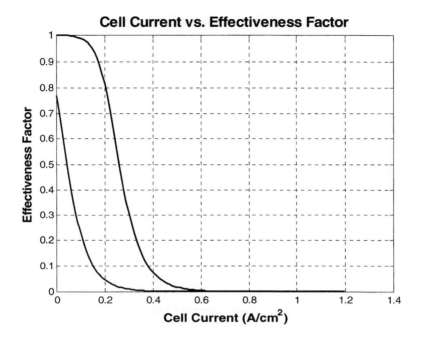

FIGURE 9-2. Cell current as a function of effectiveness factor.

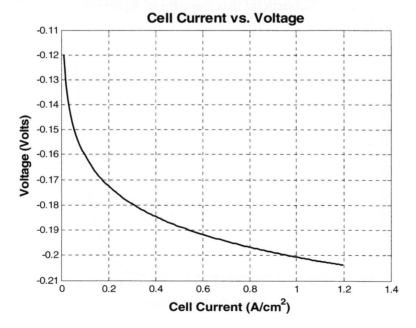

FIGURE 9-3. Butler-Volmer activation losses.

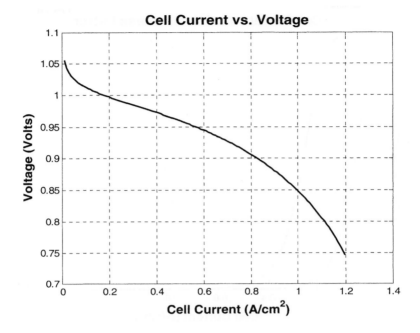

FIGURE 9-4. Polarization curve for Example 9-1.

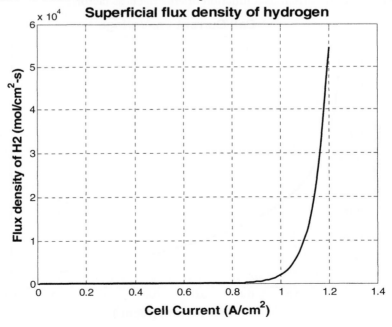

FIGURE 9-5. Superficial flux density of hydrogen.

Problems

- Explain the difference between interface, single-pore, porous electrode, and agglomerate models.
- Calculate the flux density of oxygen for Example 9-1.
- What would make the polarization curve for Example 9-1 more representative of an actual polarization curve?
- Complete the model started in Example 9-1 with the rest of the equations listed in Table 9-3.
- Two 100-cm² fuel cells are operating at 75°C and 2 atm with 100% humidity. They are both generating the same current, and the only difference between the two cells is the catalyst loading, which is 0.35 mg/cm² Pt in the first cell. What is the platinum loading of the second cell?

Endnotes

[1] Barbir, F. *PEM Fuel Cells: Theory and Practice.* 2005. Burlington, MA: Elsevier Academic Press.
[2] Weber, A.Z., and J. Newman. Modeling transport in polymer electrolyte fuel cells. *Chem. Rev.* 104, 2004, pp. 4679–4726.
[3] Ibid.
[4] Ibid.
[5] Ibid.
[6] Antoine, O., Y. Bultel, R. Durand, P. Ozil. Electrocatalysis, diffusion and ohmic drop in PEMFC: Particle size and spatial discrete distribution effects *Electrochim. Acta.* Vol. 43, 1998, pp. 3681–3691.
[7] Weber, Modeling Transport in Polymer Electrolyte Fuel Cells.
[8] Ibid.

Bibliography

Antoli, E. Recent developments in polymer electrolyte fuel cell electrodes. *J. Appl. Electrochem.* Vol. 34, 2004, pp. 563–576.
K. Chan, Y.J. Kim, Y.A. Kim, T. Yanagisawa, K.C. Park, and M. Endo. High performance of cup-stacked type carbon nanotubes as a Pt-Ru catalyst support for fuel cell applications. *J. Appl. Phys.* Vol. 96, No. 10, November 2004.
Chen, R., and T.S. Zhao. Mathematical modeling of a passive feed DMFC with heat transfer effect. *J. Power Sources.* Vol. 152, 2005, pp. 122–130.
Coutanceau, C., L. Demarconnay, C. Lamy, and J.M. Leger. Development of electrocatalysts for solid alkaline fuel cell (SAFC). *J. Power Sources.* Vol. 156, 2006, pp. 14–19.
Crystal Lattice Structures Web page. Updated March 13, 2007. Center for Computational Materials Science of the United States Naval Research Laboratory. Available at: http://cst-www.nrl.navy.mil/lattice/. Accessed March 20, 2007.
EG&G Technical Services. November 2004. *The Fuel Cell Handbook.* 7th ed. Washington, DC: U.S. Department of Energy

Fuel Cell Scientific. Available at: http://www.fuelcellscientific.com/. Provider of fuel cell materials and components.

Girishkumar, G., K. Vinodgopal, and Prashant V. Kamat. Carbon nanostructures in portable fuel cells: Single-walled carbon nanotube electrodes for methanol oxidation and oxygen reduction. *J. Phys. Chem. B*. Vol. 108, 2004, pp. 19960–19966.

Hinds, G. Preparation and characterization of PEM fuel cell electrocatalysts: A review. National Physical Laboratory. NPL Report DEPC MPE 019, October 2005.

Kharton, V.V., F.M.B. Marques, and A. Atkinson. Transport properties of solid electrolyte ceramics: A brief review. *Solid State Ionics*. Vol. 174, 2004, pp. 135–149.

Larminie, J., and Andrew D. 2003. *Fuel Cell Systems Explained*. 2nd ed. West Sussex, England: John Wiley & Sons.

Li, W., C. Liang, W. Zhou, J. Qui, Z. Zhou, G. Sun, and Q. Xin. Preparation and characterization of multiwalled carbon nanotube-supported platinum for cathode catalysts of direct methanol fuel cells. *J. Phys. Chem. B*. Vol. 107, 2003, pp. 6292–6299.

Li, X. *Principles of Fuel Cells*. 2006. New York: Taylor & Francis Group.

Lin, B. Conceptual design and modeling of a fuel cell scooter for urban Asia. 1999. Princeton University, masters thesis.

Liu, J.G., T.S. Zhao, Z.X. Liang, and R. Chen. Effect of membrane thickness on the performance and efficiency of passive direct methanol fuel cells. *J. Power Sources*. Vol. 153, 2006, pp. 61–67.

Hahn, R., S. Wagner, A. Schmitz, and H. Reichl. Development of a planar micro fuel cell with thin film and micro patterning technologies. *J. Power Sources*. Vol. 131, 2004, pp. 73–78.

Haile, S.M. Fuel cell materials and components. *Acta Material*. Vol. 51, 2003, pp. 5981–6000.

Hinds, G. Performance and durability of PEM fuel cells: A review. National Physical Laboratory. NPL Report DEPC MPE 002, September 2004.

Lister, S., and G. McLean. PEM fuel cell electrode: A review. *J. Power Sources*. Vol. 130, 2004, pp. 61–76.

Matsumoto, T., T. Komatsu, K. Arai, T. Yamazaki, M. Kijima, H. Shimizu, Y. Takasawa, and J. Nakamura. Reduction of Pt usage in fuel cell electrocatalysts with carbon nanotube electrodes. *Chem. Commun.* 2004, pp. 840–841.

Mehta, V., and J.S. Copper. Review and analysis of PEM fuel cell design and manufacturing. *J. Power Sources*. Vol. 114, 2003, pp. 32–53.

Mench, M.M., C.-Y. Wang, and S.T. Tynell. *An Introduction to Fuel Cells and Related Transport Phenomena*. Department of Mechanical and Nuclear Engineering, Pennsylvania State University. PA, USA. Draft. Available at: http://mtrl1.mne.psu.edu/Document/jtpoverview.pdf. Accessed March 4, 2007.

Mench, M.M., Z.H. Wang, K. Bhatia, and C.Y. Wang. 2001. *Design of a Micro-direct Methanol Fuel Cell*. Electrochemical Engine Center, Department of Mechanical and Nuclear Engineering, Pennsylvania State University. PA, USA.

Mogensen, M., N.M. Sammes, and G.A. Tompsett. Physical, chemical, and electrochemical properties of pure and doped ceria. *Solid State Ionics*. Vol. 129, 2000, pp. 63–94.

Morita, H., M. Komoda, Y. Mugikura, Y. Izaki, T. Watanabe, Y. Masuda, and T. Matsuyama. Performance analysis of molten carbonate fuel cell using a Li/Na electrolyte. *J. Power Sources*. Vol. 112, 2002, pp. 509–518.

Rowe, A., and X. Li. Mathematical modeling of proton exchange membrane fuel cells. *J. Power Sources.* Vol. 102, 2001, pp. 82–96.

O'Hayre, R., S.-W. Cha, W. Colella, and F.B. Prinz. 2006. *Fuel Cell Fundamentals.* New York: John Wiley & Sons.

Silva, V.S., J. Schirmer, R. Reissner, B. Ruffmann, H. Silva, A. Mendes, L.M. Madeira, and S.P. Nunes. Proton electrolyte membrane properties and direct methanol fuel cell performance. *J. Power Sources.* Vol. 140, 2005, pp. 41–49.

Smitha, B., S. Sridhar, and A.A. Khan. Solid polymer electrolyte membranes for fuel cell applications—A review. *J. Membr. Sci.* Vol. 259, 2005, pp. 10–26.

Sousa, R., Jr., and E. Gonzalez. Mathematical modeling of polymer electrolyte fuel cells. *J. Power Sources.* Vol. 147, 2005, pp. 32–45.

Souzy, R., B. Ameduri, B. Boutevin, G. Gebel, and P. Capron. Functional fluoropolymers for fuel cell membranes. *Solid State Ionics.* Vol. 176, 2005, pp. 2839–2848.

U.S. Patent 5,211,984. Membrane Catalyst Layer for Fuel Cells. Wilson, M. The Regents of the University of California. May 18, 1993.

U.S. Patent 5,234,777. Membrane Catalyst Layer for Fuel Cells. Wilson, M. The Regents of the University of California. August 10, 1993.

U.S. Patent 6,696,382 B1. Catalyst Inks and Method of Application for Direct Methanol Fuel Cells. Zelenay, P., J. Davey, X. Ren, S. Gottesfeld, and S. Thomas. The Regents of the University of California. February 24, 2004.

Wang, C., M. Waje, X. Wang, J.M. Trang, R.C. Haddon, and Y. Yan. Proton exchange membrane fuel cells with carbon nanotube-based electrodes. *Nano Lett.* Vol. 4, No. 2, 2004, pp. 345–348.

Wong, C.W., T.S. Zhao, Q. Ye, and J.G. Liu. Experimental investigations of the anode flow field of a micro direct methanol fuel cell. *J. Power Sources.* Vol. 155, 2006, pp. 291–296.

Xin, W., X. Wang, Z. Chen, M. Waje, and Y. Yan. Carbon nanotube film by filtration as cathode catalyst support for proton exchange membrane fuel cell. *Langmuir.* Vol. 21, 2005, pp. 9386–9389.

Yamada, M., and I. Honma. Biomembranes for fuel cell electrodes employing anhydrous proton conducting uracil composites. *Biosensors Bioelectronics.* Vol. 21, 2006, pp. 2064–2069.

You, L., and H. Liu. A two-phase flow and transport model for PEM fuel cells. *J. Power Sources.* Vol. 155, 2006, pp. 219–230.

CHAPTER 10

Modeling the Flow Field Plates

10.1 Introduction

After the actual fuel cell layers (membrane electrode assemble (MEA)) have been assembled, the cell(s) must be placed in a fuel cell stack to evenly distribute fuel and oxidant and collect the current to power the desired devices. In a fuel cell with a single cell, there are no bipolar plates (only single-sided flow field plates), but in fuel cells with more than one cell, there is usually at least one bipolar plate (flow fields on both sides of the plate). Bipolar plates perform many roles in fuel cells. They distribute fuel and oxidant within the cell, separate the individual cells in the stack, collect the current, carry water away from each cell, humidify gases, and keep the cells cool. In order to simultaneously perform these functions, specific plate materials and designs are used. Commonly used designs can include straight, serpentine, parallel, interdigitated, or pin-type flow fields. Materials are chosen based upon chemical compatibility, resistance to corrosion, cost, density, electronic conductivity, gas diffusivity/impermeability, manufacturability, stack volume/kW, material strength, and thermal conductivity. The materials most often used are stainless steel, titanium, nonporous graphite, and doped polymers. Several composite materials have been researched and are beginning to be mass produced.

Most PEM fuel cell bipolar plates are made of resin-impregnated graphite. Solid graphite is highly conductive, chemically inert, and resistant to corrosion but expensive and costly to manufacture. Flow channels are typically machined or electrochemically etched into the graphite or stainless steel bipolar plate surfaces. However, these methods are not suitable for mass production, which is why new bipolar materials and manufacturing processes are currently being researched.

Figure 10-1 shows an exploded view of a fuel cell stack. The stack is made of repeating cells of MEAs and bipolar plates. Increasing the number

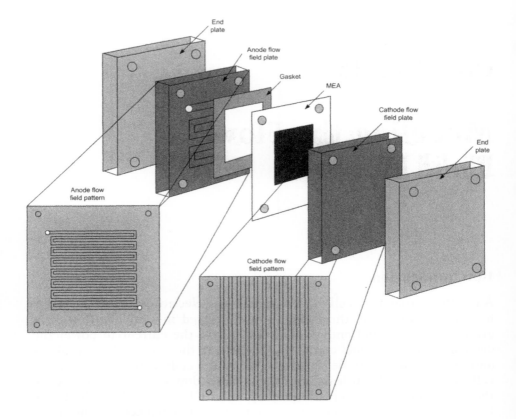

FIGURE 10-1. An exploded view of a fuel cell stack.

of cells in the stack increases the voltage, while increasing the surface area increases the current.

Fuel cell bipolar plates account for most of the stack weight and volume; therefore, it is desirable to produce plates with the smallest dimensions possible (< 3-mm width) for portable and automotive fuel cells[1]. Flow channel geometry has an effect on reactant flow velocities and mass transfer, and therefore, on fuel cell performance. Therefore, modeling the flow field channels is helpful when deciding on optimal mass transfer, pressure drop, and fuel cell water management.

Specific topics covered in this chapter include the following:

- Flow field plate materials
- Flow field design
- Channel shape, dimensions, and spacing

- Pressure drop in flow channels
- Heat transfer from the plate channels to the gas

This chapter covers the modeling required for bipolar plate modeling and optimization. An efficient design for the bipolar plates or cell interconnects is necessary for creating the most efficient fuel cell stack possible for the desired application.

10.2 Flow Field Plate Materials

There are many types of materials that have been used for flow field plates. As mentioned previously, graphite and stainless steel are the most common, but other materials such as aluminum, steel, titanium, nickel, and polymer composites are also used. Metallic plates are suitable for mass production and also can be made into very thin layers, which results in lightweight and portable stacks. The bipolar plates are exposed to a corrosive environment, and dissolved metal ions can diffuse into the membrane, which lowers ionic conductivity and reduces fuel cell life. A coating or coatings are needed to prevent corrosion while promoting conductivity. Some commonly used coatings are graphite, gold, silver, palladium, platinum, carbon, conductive polymer, and other types. Some of the issues with protective coatings include (1) the corrosion resistance of the coating, (2) micropores and microcracks in the coating, and (3) the difference between the coefficient of thermal expansion and the coating.

Graphite–carbon composite plates have been made using thermoplastics or thermosets with conductive fillers. These materials are usually chemically stable in fuel cells, and are suitable for mass production techniques, such as compression molding, transfer molding, or injection molding. Often, the construction and design of these plates are a trade-off between manufacturability and functional properties. Important properties that need to be considered when designing these plates are tolerances, warping, and the skinning effect. Some issues associated with these plates are that they are slightly brittle and bulky. Although the electrical conductivity is several orders of magnitude lower than the conductivity of the metallic plates, the bulk resistivity losses are only on the order of a magnitude of several millivolts.

One of the most important properties of the fuel cell stack is the electrical conductivity. The contact resistance from interfacial contacts between the bipolar plate and the gas diffusion layer is a very important consideration. The interfacial contact resistivity losses can be determined by putting a bipolar plate between two gas diffusion layers, and then passing an electrical current through the sandwich and measuring voltage drop. The total voltage drop is a strong function of clamping pressure. Bulk resistance of the bipolar plate and the gas diffusion media is a strong function of the clamping force.

Interfacial contact resistance depends not only upon the clamping pressure, but the surface characteristics of the bipolar plate and Gas Diffusion Layer (GDL) in contact. The relationship between the contact resistance and the clamping pressure between the GDL and a bipolar plate is as follows:

$$R = \frac{A_a K G^{D-1}}{\kappa L^D} \left[\frac{D}{(2-D)p^*} \right]^{D/2}$$ (10-1)

where R is the contact resistance, Ωm^2, A_a is the apparent contact area at the interface, m^2, K is the geometric constant, G is the topothesy of a surface profile, m, D is the fractal dimension of a surface profile, and κ is the effective electrical conductivity of two surfaces, S/m, described by:

$$\frac{1}{\kappa} = \frac{1}{2} \left(\frac{1}{\kappa_1} + \frac{1}{\kappa_2} \right)$$ (10-2)

L is the scan length, m, and p^* is the dimensionless clamping pressure (ratio of actual clamping pressure and comprehensive modulus of gas diffusion layer). Chapter 12 discusses fuel cell stack design and clamping pressure in more detail.

10.3 Flow Field Design

In fuel cells, the flow field should be designed to minimize pressure drop (reducing parasitic pump requirements), while providing adequate and evenly distributed mass transfer through the carbon diffusion layer to the catalyst surface for reaction. The three most popular channel configurations for PEM fuel cells are serpentine, parallel, and interdigitated flow. Serpentine and parallel flow channels are shown in Figures 10-2 through 10-4. Some small-scale fuel cells do not use a flow field to distribute the hydrogen and/or air, but rely on diffusion processes from the environment. Because the hydrogen reaction is not rate limiting, and water blockage in the humidified anode can occur, a serpentine arrangement is typically used for the anode in smaller PEM fuel cells.

The serpentine flow path is continuous from start to finish. An advantage of the serpentine flow path is that it reaches the entire active area of the electrode by eliminating areas of stagnant flow. A disadvantage of serpentine flow is the fact that the reactant is depleted through the length of the channel, so that an adequate amount of the gas must be provided to avoid excessive polarization losses. The pressure drop is high in serpentine channels because flow velocity scales with the square of the feature size, and the channel length is inversely proportional to the feature size. For

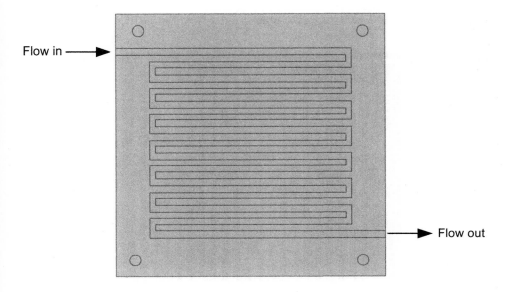

FIGURE 10-2. A serpentine flow field design.

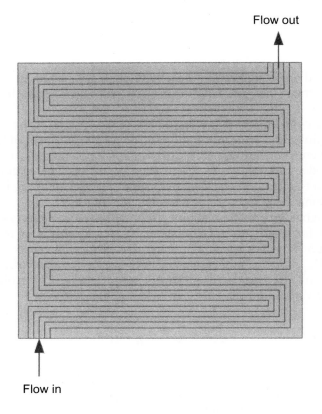

FIGURE 10-3. Multiple serpentine flow channel design.

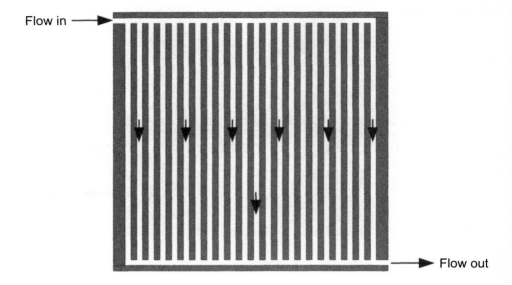

FIGURE 10-4. A parallel flow field design.

high current density operation, very large plates, or when air is used as an oxidant, alternate designs have been proposed based upon the serpentine design[2].

Several continuous flow channels can be used to limit the pressure drop, and reduce the amount of power used for pressurizing the air through a single serpentine channel. This design allows no stagnant area formation at the cathode surface due to water accumulation. The reactant pressure drop through the channels is less than the serpentine channel, but still high due to the long flow path of each serpentine channel[3].

Although some of the reactant pressure losses can increase the degree of difficulty for hydrogen recirculation, they are helpful in removing the product water in vapor form. The total reactant gas pressure is $P_T = P_{vap} + P_{gas}$, where P_{vap} and P_{gas} are the partial pressures of the partial pressure and reactant gas in the reactant gas stream, respectively. The molar flow rate of the water vapor and reactant can be related as follows:

$$\frac{N_{vap}}{N_{gas}} = \frac{P_{vap}}{P_{gas}} = \frac{P_{vap}}{P_T - P_{vap}} \tag{10-3}$$

The total pressure loss along a flow channel will increase the amount of water vapor that can be carried and taken away by the gas flow if the relative humidity is maintained[4]. This can help remove water in both the anode and cathode flow streams.

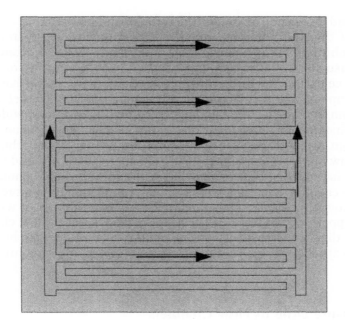

FIGURE 10-5. An interdigitated flow field design.

The reactant flow for the interdigitated flow field design is parallel to the electrode surface. Often, the flow channels are not continuous from the plate inlet to the plate outlet. The flow channels are dead-ended, which forces the reactant flow, under pressure, to go through the porous reactant layer to reach the flow channels connected to the stack manifold. This design can remove water effectively from the electrode structure, which prevents flooding and enhances performance. The interdigitated flow field pushes gas into the active layer of the electrodes where forced convection avoids flooding and gas diffusion limitations. This design is sometimes noted in the literature as outperforming conventional flow field design, especially on the cathode side of the fuel cell. The interdigitated design is shown in Figure 10-5.

10.4 Channel Shape, Dimensions, and Spacing

Flow channels are typically rectangular in shape, but other shapes such as trapezoidal, triangular, and circular have been demonstrated. The change in channel shape can have an affect upon the water accumulation in the cell, and, therefore, the fuel and oxidant flow rates. For instance, in rounded flow channels, the condensed water forms a film at the bottom of the channel, and in tapered channels, the water forms small droplets. The shape and size of the water droplets are also determined by the hydrophobicity and hydrophilicity of the porous media and channel walls. Channel dimensions are usually around 1 mm, but a large range exist for micro- to large-scale fuel cells (0.1 mm to 3 mm). Simulations have found that optimal channel dimensions for macrofuel cell stacks (not MEMS fuel cells) are 1.5, 1.5, and 0.5 mm for the channel depth, width, and land width (space between channels), respectively. These dimensions depend upon the total stack design and stack size. The channels' dimensions affect the fuel and oxidant flow rates, pressure drop, heat and water generation, and the power generated in the fuel cell. Wider channels allow greater contact of the fuel to the catalyst layer, have less pressure drop, and allow more efficient water removal. However, if the channels are too wide, there will not be enough support for the MEA layer. If the spacing between flow channels is also wide, this benefits the electrical conductivity of the plate but reduces the area exposed to the reactants, and promotes the accumulation of water[5].

10.5 Pressure Drop in Flow Channels

In many fuel cell types, the flow fields are usually arranged as a number of parallel flow channels; therefore, the pressure drop along a channel is also the pressure drop in the entire flow field. In a typical flow channel, the gas moves from one end to the other at a certain mean velocity. The pressure difference between the inlet and outlet drives the fluid flow. By increasing the pressure drop between the outlet and inlet, the velocity is increased. The flow through bipolar plate channels is typically laminar, and proportional to the flow rate. The pressure drop can be approximated using the equations for incompressible flow in pipes[6].

$$\Delta P = f \frac{L_{chan}}{D_H} \rho \frac{\bar{v}^2}{2} + \Sigma K_L \rho \frac{\bar{v}^2}{2} \qquad (10\text{-}4)$$

where f is the friction factor, L_{chan} is the channel length, m, D_H is the hydraulic diameter, m, ρ is the fluid density, kg/m^3, $\bar{v}$ is the average velocity, m/s, and K_L is the local resistance.

The hydraulic diameter for a circular flow field can be defined by:

$$D_H = \frac{4 \times A_c}{P_{cs}} \qquad (10\text{-}5)$$

where A_c is the cross-sectional area, and P_{cs} is the perimeter. For the typical rectangular flow field, the hydraulic diameter can be defined as:

$$D_H = \frac{2w_c d_c}{w_c + d_c} \qquad (10\text{-}6)$$

where w_c is the channel width, and d_c is the depth.

The channel length can be defined as:

$$L_{chan} = \frac{A_{cell}}{N_{ch}(w_c + w_L)} \qquad (10\text{-}7)$$

where A_{cell} is the cell active area, N_{ch} is the number of parallel channels, w_c is the channel width, m, and w_L is the space between channels, m.

The friction factor can be defined by:

$$f = \frac{56}{Re} \qquad (10\text{-}8)$$

The velocity at the fuel cell entrance is:

$$v = \frac{Q_{stack}}{N_{cell} N_{ch} A_{ch}} \qquad (10\text{-}9)$$

where v is the velocity in the channel (m/s), Q_{stack} is the air flow rate at the stack entrance, m^3/s, N_{cell} is the number of cells in the stack, N_{ch} is the number of parallel channels in each cell, and A_{ch} is the cross-sectional area of the channel.

The total flow rate at the stack entrance is:

$$Q_{stack} = \frac{I}{4F} \frac{S_{O2}}{r_{O2}} \frac{RT_{in}}{P_{in} - \varphi P_{sat(T_{in})}} N_{cell} \qquad (10\text{-}10)$$

where Q is the volumetric flow rate (m^3/s), I is the stack current, F is the Faraday's constant, S_{O2} is the oxygen stoichiometric ratio, r_{O2} is the oxygen content in the air, R is the universal gas constant, T_{in} is the stack inlet temperature, P_{in} is the pressure at the stack inlet, Φ is the relative humidity, P_{sat} is the saturation pressure at the given inlet temperature, and N_{cell} is the number of cells in the stack[7].

By combining the previous equations, the velocity at the stack inlet is:

$$v = \frac{i}{4F} \frac{S_{O2}}{r_{O2}} \frac{(w_c + w_L)L_{chan}}{w_c d_c} \frac{RT}{P - \phi P_{sat}} \qquad (10\text{-}11)$$

Liquid or gas flow confined in channels can be laminar, turbulent, or transitional and is characterized by an important dimensionless number

known as the Reynold's number (Re). This number is the ratio of the iner-
tial forces to viscous forces and is given by:

$$Re = \frac{\rho v_m D_{ch}}{\mu} = \frac{v_m D_{ch}}{v} \qquad (10\text{-}12)$$

where v_m is the characteristic velocity of the flow (m/s), D_{ch} is the flow
channel diameter or characteristic length (m), ρ is the fluid density (kg/m³),
μ is the fluid viscosity [kg/(m∗s or N∗s/m²], and v is the kinematic viscosity (m²/s). When Re is small (<2300), the flow is laminar. When Re is greater
than 4000, the flow is turbulent, which means that it has random fluctua-
tions. When Re is between 2300 and 4000, it is know to be in the "transitional" range, where the flow is mostly laminar, with occasional bursts of
irregular behavior. It is found that regardless of channel size or flow velocity, $f*Re = 16$ for circular channels[8].

The effective Reynold's number for rectangular channels is:

$$Re_h = \frac{\rho v_m D_h}{\mu} \quad \text{where} \quad D_h = \frac{4A}{P} \qquad (10\text{-}13)$$

where D_h is equal to $4*$(cross-sectional area)/perimeter.

A relationship in the literature for rectangular channels can be approximated by:

$$fRe = 24(1 - 1.3553 \times \alpha^* + 1.9467 \times \alpha^{*2} - 1.7012 \times \alpha^{*3} +$$
$$0.9564 \times \alpha^{*4} - 0.2537 \times \alpha^{*5}) \qquad (10\text{-}14)$$

where α^* is the aspect ratio of the cross-section, and $\alpha^* = {}^b/_a$ where 2a and
2b are the lengths of the channels' sides.

$$Re = \frac{\rho v D_H}{\mu} = \frac{1}{\mu} \frac{i}{2F} \frac{S_{O2}}{r_{O2}} \frac{(w_c + w_L)L_{chan}}{w_c + d_c} M_{air} + M_{H2O} \frac{\phi P_{sat(Tin)}}{P_{in} - \phi P_{sat(Tin)}} \qquad (10\text{-}15)$$

The velocity profile remains the parabolic shape, and the pressure
gradient is constant throughout the region once the fluid enters the fully
developed region. The flow rate for laminar flow in a circular pipe is given
by the Hagen-Poiseuille equation:

$$Q = \frac{\pi r^4}{8\mu \ell} \Delta p \qquad (10\text{-}16)$$

where r is the radius of the pipe, l is the length, Δp is the applied pressure
difference, and μ is the viscosity of the fluid.

The flow rate at the stack outlet is usually different than the inlet. If it is assumed that the outlet flow is saturated with water vapor, the flow rate is:

$$Q_{stack} = \frac{I}{4F}\left(\frac{S_{O2}}{r_{O2}} - 1\right)\frac{RT_{out}}{P_{in} - \Delta P - \varphi P_{sat(Tout)}}N_{cell} \qquad (10\text{-}17)$$

where ΔP is the pressure drop in the stack.

The variation in viscosity varies with temperature. For dilute gases, the temperature dependence of viscosity can be estimated using a simple power law:

$$\frac{\mu}{\mu_0} \approx \left(\frac{T}{T_0}\right)^n \qquad (10\text{-}18)$$

where μ_0 is the viscosity at temperature T_0. In these equations, n, μ_0, and T_0 can be obtained from experiments or calculated through kinetic theory.

Fuel cell gas streams are rarely composed of a single species. Usually, they are gas mixtures, such as oxygen and nitrogen from the air. The following expression provides a good estimate for the viscosity of a gas mixture:

$$\mu_{mix} = \sum_{i=1}^{N}\frac{x_i\mu_i}{\displaystyle\sum_{j=1}^{N}x_j\Phi_{ij}} \qquad (10\text{-}19)$$

where Φ_{ij} is a dimensionless number obtained from:

$$\Phi_{ij} = \frac{1}{\sqrt{8}}\left(1 + \frac{M_i}{M_j}\right)^{-1/2}\left[1 + \left(\frac{\mu_i}{\mu_j}\right)^{1/2}\left(\frac{M_i}{M_j}\right)^{1/4}\right]^2 \qquad (10\text{-}20)$$

where N is the total number of species in the mixture, x_i and x_j are the mole fractions of species i and j, and M_i and M_j are the molecular weight (kg/mol) of species i and j.

For porous flow fields, the pressure drop is determined by Darcy's law:

$$\Delta P = \mu\frac{Q_{cell}}{kA_c}L_{chan} \qquad (10\text{-}21)$$

where μ is the viscosity of the fluid, Q_{cell} is the geometric flow rate through the cell, m³/s, K is the permeability, m², A_c is the cross-sectional area of the flow field, m², and L_{chan} is the length of the flow field.

When using this set of equations, there are a few assumptions that are made that will cause a slight deviation from the actual values[9,10].

- The channels are typically smooth on one side of the "pipe," but the GDL side has a rough surface.
- The gas is not simply flowing through the channels. It is also reacting with the catalyst.
- The temperature may not be uniform through the channels.
- There are a number of bends or turns that should be accounted for in the channels.

EXAMPLE 10-1: Calculating the Pressure Drop

Calculate the pressure drop through a PEM fuel cell cathode flow field of a single graphite plate with 100-cm² cell area. The stack operates at 3 atm at 60°C with 100% saturated air. The flow field consists of 24 parallel serpentine channels 1 mm wide, 1 mm deep, and 1 mm apart. The cell operates at 0.7 A/cm² at 0.65 V.

The pressure drop is:

$$\Delta P = f \frac{L}{D_H} \rho \frac{\bar{v}^2}{2} + \Sigma K_L \rho \frac{\bar{v}^2}{2}$$

The hydraulic diameter is:

$$D_H = \frac{2 w_c d_c}{w_c + d_c} = \frac{2*0.1*0.1}{0.1+0.1} = 0.1\,cm$$

The channel length is:

$$L = \frac{A_{cell}}{N_{ch}(w_c + w_L)} = \frac{100}{24(0.1+0.1)} = 20.83\,cm$$

The flow rate at the stack entrance is:

$$Q_{stack} = \frac{I}{4F}\left(\frac{S}{r_{O2}}\right)\frac{RT_{in}}{P_{in} - \varphi P_{sat(Tin)}} N_{cell} = \frac{0.7*100}{4*96,485} * \frac{1}{0.21} *$$

$$\frac{8.314*333.15}{303,975.03 - 19,944} *1 = 8.423*10^{-7}\frac{m^3}{s} = 8.423\frac{cm^3}{s}$$

The velocity in a fuel cell channel near the entrance of the cell is:

$$v = \frac{Q_{stack}}{N_{cell}N_{ch}A_{ch}} = \frac{8.423}{1*24*0.1*0.1} = 35.096\frac{cm}{s}$$

The Reynold's number at the channel entrance is:

$$Re = \frac{\rho \bar{v} D_H}{\mu}$$

$$\rho = \frac{(P - P_{sat})M_{air} + P_{sat}M_{H2O}}{RT} = \frac{(303,975.03 - 19,944)*29 + 19,944*18}{8314*333.15}$$

$$= 3.10 \, \text{kgm}^3 = 0.0031 \, \text{gcm}^3$$

$$\mu = 2*10^{-5} \, \text{kg/ms} = 0.0002 \frac{g}{cms}$$

$$Re = \frac{\rho \bar{v} D_H}{\mu} = \frac{0.00123*35.096*0.1}{0.0002} = 21.584$$

For rectangular channels:

$$Ref \approx 55 + 41.5 \exp\left(\frac{-3.4}{w_c/d_c}\right) = 56$$

$$f = \frac{56}{Re} = \frac{56}{21.584} = 2.594$$

The pressure drop is:

$$\Delta P = f \frac{L}{D_H} \rho \frac{\bar{v}^2}{2} + \Sigma K_L \rho \frac{\bar{v}^2}{2} = 2.594 * \frac{0.2083}{0.001} 1.23 * \frac{0.351^2}{2} +$$

$$1.23 * \frac{0.351^2}{2} = 33.36 \, \text{Pa}$$

Using MATLAB to solve:

%%

% EXAMPLE 10-1: Calculating the Pressure Drop

% UnitSystem SI

%%

% Inputs

```
F = 96485;          % Faraday's constant
R = 8.314;          % Universal gas constant
T_in = 333.15;      % Inlet temperature (K)
P_in = 101325;      % Inlet pressure (Pa)
psi = 1;            % Relative humidity
```

```
P_sat = 19944;      % Saturation pressure (Pa)
N_cell = 1;         % Number of cells in the stack
wc = 0.1;           % Channel width (cm);
dc = 0.1;           % Channel depth (cm);
A_cell = 100;       % Active cell area (cm^2)
N_ch = 1:50;        % Number of parallel channels
wl = 0.1;           % Space between channels
be = 0;             % # of bends
n = 4;              % Cathode
i = 0.7;            % Cell current (A/cm^2)
I = i*A_cell;       % Stack current (A/cm^2)
S_O2 = 1;           % Stoichiometric ration for O2
x_O2 = 0.21;        % O2 content in the air
A_ch = wc*dc;       % Channel area
M_air = 29;         % Molecular weight of air
M_H2O = 18;         % Molecular weight of water
u = 0.0002;         % Viscosity (g/cms)
```

%%%

% Calculate Pressure drop in flow channels

% Hydraulic diameter

```
%Dh=(4*Ac)./P_cs;           % circular flowfield
Dh =(2*wc.*dc) ./ (wc + dc);  % rectangular flowfield
```

% Channel length

```
Lc = A_cell ./(N_ch*(wc + wl));
```

% Flow rate at the fuel cell stack entrance (m^3/s)(anode)

```
Q_stack =(I ./(n*F)).*(S_O2 ./ x_O2).*((R.*T_in) ./(P_in – (psi.*P_sat))).*N_
   cell; % m^3/s
Q_stack1 = Q_stack.*1000000;  % convert to cm^3/s
```

% Velocity in a fuel cell channel at the channel entrance

```
v = Q_stack1 ./ ((N_cell).*N_ch.*A_ch);% cm/s
```

% Reynold's number at the channel entrance

```
R1 = 8314;
den = (((P_in – P_sat)*M_air)+(P_sat*M_H2O)) ./(R1*T_in);% kgm^3
den1 = den ./1000;    % convert to gcm^3
Re = (den1.*v.*Dh) ./u;
```

% Friction Factor

```
f = 56 ./ Re; % for rectanglar fields
% Pressure Drop
```

```
Lc1 = Lc ./ 100;
Dh1 = Dh ./ 100;
v1 = v ./ 100;
Kl = be.*30.*f;
P = (f.*(Lc1 ./Dh1).*den.*((v1.^2) ./ 2))+(Kl.*den.*((v1.^2) ./2))
P_atm = P.*9.86923e-1 % Convert from Pa to atm
```

10.6 Heat Transfer from the Plate Channels to the Gas

Another important consideration when modeling the flow channels is the gas temperature and the associated heat transfer from the plate to the gas. The temperature of the gas affects the phase change in the channels and the GDL layer, which ultimately affects the reaction rate.

There are N nodes distributed uniformly in the y-direction across the channel, as shown in Figure 10-6. The control volumes are set up in this manner because the velocity at the wall is zero according to the no slip condition. This will be modeled in a similar manner as the heat transfer through the fuel cell layer in Chapter 6.

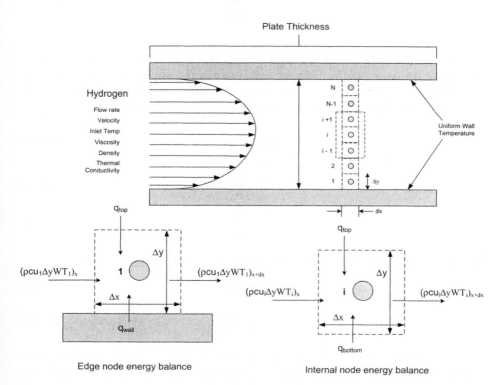

FIGURE 10-6. Heat transfer from plate channels to gas.

The distance between adjacent nodes is:

$$\Delta y = \frac{H}{N} \tag{10-22}$$

and the location of each of the nodes is given by:

$$y_i = \Delta y \left(i - \frac{1}{2} \right) \quad \text{for } i = 1 \ldots N \tag{10-23}$$

The velocity distribution in the duct is parabolic, therefore, the velocity at each nodal location is:

$$u_i = 6u_m \left[\frac{y_i}{H} - \left(\frac{y_i}{H} \right)^2 \right] \quad \text{for } i = 1 \ldots N \tag{10-24}$$

The hydraulic diameter associated with the channel is:

$$D_h = 2H \tag{10-25}$$

The Reynold's number that characterizes the flow is:

$$Re_{D_h} = \frac{\rho D_h u_m}{\mu} \tag{10-26}$$

If the Reynold's number implies that the flow is laminar, the conductive heat transfer can be approximated using the molecular conductivity, k, rather than a turbulent conductivity. The thermal diffusivity, kinematic viscosity, and Prandtl number associated with the fluid are:

$$\alpha = \frac{k}{\rho c} \tag{10-27}$$

$$v = \frac{\mu}{\rho} \tag{10-28}$$

$$Pr = \frac{v}{\alpha} \tag{10-29}$$

The Peclet and Brinkman numbers that characterize the flow are:

$$Pe = Pr \, Re_{D_H} \tag{10-30}$$

$$Br = \frac{\mu u_m^2}{k(T_s - T_{in})} \tag{10-31}$$

A control volume can be defined around each of the nodes. This definition is consistent with the approach that was used in Chapter 6 to derive the state equations for the time rate of change of temperature. As illustrated in Figure 10-6, the internal nodes are treated separately from the boundary nodes. The node has conduction in the y-direction from the adjacent nodes, and is also influenced by the heat transferred by the fluid, as well as energy carried by fluid entering the control volume at x and leaving at x + dx. This creates the following energy balance:

$$(\rho c \mu_i \Delta y W T_i)_x + \dot{q}_{top} + \dot{q}_{bottom} = (\rho c \mu_i \Delta y W T_i)_{x+dx} \quad (10\text{-}32)$$

where W is the depth of the channel. The conduction heat transfer rates are approximated with:

$$\dot{q}_{top} = \frac{k dx W}{\Delta y}(T_{i+1} - T_i) \quad (10\text{-}33)$$

$$\dot{q}_{bottom} = \frac{k dx W}{\Delta y}(T_{i-1} - T_i) \quad (10\text{-}34)$$

Equations 10-33 and 10-34 are substituted into Equation 10-32, and the x + dx term is expanded:

$$(\rho c u_i \Delta y W T_i)_x + \frac{k dx W}{\Delta y}(T_{i+1} - T_i) + \frac{k dx W}{\Delta y}(T_{i-1} - T_i) =$$

$$(\rho c u_i \Delta y W T_i)_x + \frac{d}{dx}(\rho c u_i \Delta y W T_i) dx \quad \text{for } i = 2 \ldots (N-1) \quad (10\text{-}35)$$

Note that the only term in the derivative that changes with x is the temperature, therefore Equation 10-35 can be rewritten as:

$$\frac{k dx W}{\Delta y}(T_{i+1} - T_i) + \frac{k dx W}{\Delta y}(T_{i-1} - T_i) = \rho c u_i \Delta y W \frac{dT_i}{dx} dx \quad \text{for } i = 2 \ldots (N-1)$$

$$(10\text{-}36)$$

Solving for the rate of change of T_i with respect to x:

$$\frac{dT_i}{dx} = \frac{k}{\rho c \Delta y^2 u_i}(T_{i+1} + T_{i-1} - 2T_i) \quad \text{for } i = 2 \ldots (N-1) \quad (10\text{-}37)$$

An energy balance for the control volume around node 1 leads to:

$$\dot{q}_{top} + \dot{q}_{wall} = \rho c u_1 \Delta y W \frac{dT_1}{dx} dx \quad (10\text{-}38)$$

The conductive heat transfer from node 2 is approximated with:

$$\dot{q}_{top} = \frac{k dx W}{\Delta y}(T_2 - T_1) \quad (10\text{-}39)$$

and the conductive heat transfer from the wall is:

$$\dot{q}_{wall} = \frac{2kdxW}{\Delta y}(T_s - T_1)$$ (10-40)

Substituting Equations 10-39 and 10-40 into Equation 10-38 leads to:

$$\frac{kdxW}{\Delta y}(T_2 - T_1) + \frac{2kdxW}{\Delta y}(T_s - T_1) = \rho c u_1 \Delta y W \frac{dT_1}{dx} dx$$ (10-41)

Solving for the rate of change of the temperature of node 1 is:

$$\frac{dT_1}{dx} = \frac{k}{\rho c \Delta y^2 u_1}(T_2 + 2T_s - 3T_1)$$ (10-42)

A similar process applied to node N leads to:

$$\frac{dT_N}{dx} = \frac{k}{\rho c \Delta y^2 u_N}(T_{N-1} + 2T_s - 3T_N)$$ (10-43)

EXAMPLE 10-2: Heat Transfer from Plate to Gas in the Channel

Plot the two dimensional temperature distribution of the gas in a channel as a function of axial position. The channel radius is 0.001 m, channel height is 0.002 m, and length of the channel is 0.00635. The plate temperature is 352 K, and has the following properties: density: 0.08988 kg/m³, viscosity: 8.6e-6 Pa-s, conductivity: 0.1805 J/kg-K, and specific heat: 14,304 J/kg-K. Set up a grid with 6 nodes (slices) in the y-direction. Using MATLAB to solve:

```
%%%%%%%%%%%%%%%%%%%%%%%%%%%%%%%%%%%%%%%%%%%%%

% EXAMPLE 10-2: Heat Transfer From Plate to Gas in Channel

%%%%%%%%%%%%%%%%%%%%%%%%%%%%%%%%%%%%%%%%%%%%%

% Inputs

v_H2_in = 0.000000017;      % Volumetric flow rate (m3/s)
r = 0.001;                  % Channel radius (m)
A = 0.5*pi*r^2;             % Area of flow channel
```

```
u_m = v_H2_in/A;          % Mean velocity (m/s)
H = 0.002;                % Channel height (m)
T_in = 295;               % Inlet gas temperature (K)
rho = 0.08988;            % Density (kg/m^3)
mu = 8.6e-6;              % Viscosity (Pa-s)
k = 0.1805;               % Conductivity (W/mK)
c = 14304;                % Specific heat capacity (J/kg-K)
T_s = 352;                % Plate temperature (K)
L = 0.00635;              % Length of channel (m)
W = 0.00635;              % Unit length (depth)(m)
```

%%%%%%%%%%%%%%%%%%%%%%%%%%%%%%%%%%%%%%%

% Setup y grid

```
N = 6;             % Number of nodes in y direction (-)
Dy = H/N;          % Distance between nodes (m)

for i=1:N
   y(i)=Dy*(i-1/2);     % Position of each node (m)
end
```

% Velocity

```
for i=1:N
   u(i)=6*u_m*(y(i)/H-(y(i)/H)^2);     % Velocity at each node (m/s)
end

OPTIONS=odeset('RelTol',1e-6);
[x,T]=ode45(@(x,T) dTdx_functionv(x,T,Dy,k,rho,c,u,T_s),[0,L],T_in*ones(N,1),
   OPTIONS);

[M,g]=size(T);                 % Determine number of length steps used

for j=1:M
   T_mean(j)=sum(T(j,:).*u)*    % The mean temperature is the velocity weighted
      Dy/(H*u_m);               average temperature at each axial position
   qf(j)=k*(T_s-T(j,1))/(Dy/2); % The heat flux is obtained from the thermal
                                resistance of the node at the wall
   htc(j)=qf(j)/(T_s-T_mean(j)); % Heat transfer coefficent
   Nusselt(j)=htc(j)*2*H/k;      % Nusselt number
end
```

% Plot heat distribution in the channel

```
plot(y,T);
xlabel('Axial Position (m)');
ylabel('Temperature (K)');
```

%%%%%%%%%%%%%%%%%%%%%%%%%%%%%%%%%%%%%%%

```
function[dTdx]=dTdx_functionv(x,T,Dy,k,rho,c,u,T_s)
```

```
[N,g]=size(T);        % Determine number of nodes
dTdx=zeros(N,1);    % Initialize dTdx
dTdx(1)=k*(T(2)+2*T_s-3*T(1))/(rho*c*Dy^2*u(1));
for i=2:(N-1)
dTdx(i)=k*(T(i+1)+T(i-1)-2*T(i))/(rho*c*Dy^2*u(i));
end
dTdx(N)=k*(T(N-1)+2*T_s-3*T(N))/(rho*c*Dy^2*u(N));
end
```

Figure 10-7 shows the graph of the temperature as a function of axial position at various y-locations for Example 10-2.

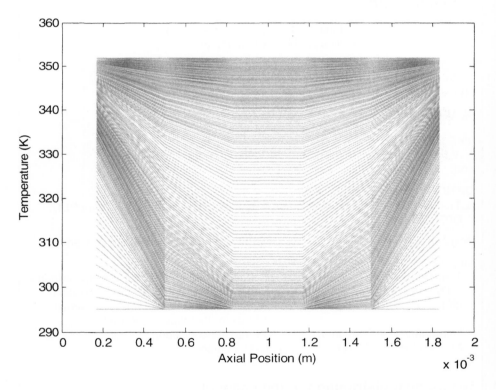

FIGURE 10-7. Temperature as a function of axial position at various y-locations.

Example 10-3 uses the concepts from Chapter 5 for determing the mass flow rates into and out of the layers in the fuel cell stack (such as the flow field plates) that have convective mass transport. The basic concepts introduced can be expanded to model the flow through the fuel cell stack.

EXAMPLE 10-3: Mass Flow Rates into Fuel Cell Layers

Create a transient MATLAB program that will calculate the mass flow rates and mole fractions of liquid water, water vapor, and hydrogen going into and out of six fuel cell layers with corrective mass transport. Assume that the hydrogen coming into the stack is fully saturated, with a volumetric flow rate of 1.7e-8 m^3/s.

The code created in this example can act as a start for a program that calculates the flow rates into and out of the bipolar plates and other layers in a fuel cell stack. Plot the flow rates after a 20 and 120 second simulation time.

As shown in Chapter 5, the mass balances into and out of each fuel cell layer needs to be calculated.

First, the volumetric flow rate needs to be converted to a molar flow rate using the ideal gas law:

$$n_{H2\_in} = \frac{PV}{RT}$$

Since the model is transient, the total molar accumulation can be written as:

$$\frac{dn_{tot}}{dt} = n_{tot\_in} - n_{tot\_out}$$

The rate of H_2 accumulation is:

$$\frac{d}{dt}(x_{H2}n_{tot}) = x_{H2\_in}n_{tot\_in} - x_{H2\_out}n_{tot\_out}$$

The rate of H_2O accumulation is:

$$\frac{d}{dt}(x_{H2O}n_{tot}) = x_{H2O\_in}n_{tot\_in} - x_{H2O\_out}n_{tot\_out}$$

The inlet molar flow rates can be calculated using the following equations:

The vapor pressure of the inlet water vapor is:

$$P_{H2Ov\_in} = \theta_{in}P_{sat}(T_{H2O\_in})$$

The mole fraction of the water vapor is:

$$X_{H2Ov\_in} = \frac{P_{H2Ov\_in}}{P_{tot}} \quad \text{where} \quad P_{tot} = 1$$

The mole fraction of the liquid water is:

$$X_{H2Ol\_in} = \frac{X_{H2Ov\_in} * P_{sat}(T_{H2O\_in})}{P_{tot}}$$

The total mole fraction of water is:

$$X_{H2O\_in} = X_{H2Ov\_in} + X_{H2Ol\_in}$$

The mole fraction of hydrogen is:

$$X_{H2\_in} = 1 - X_{H2O\_in}$$

The inlet molar flow rate of hydrogen is:

$$n_{H2\_in} = x_{H2\_in} n_{tot\_in}$$

The total inlet molar flow rate of water is:

$$n_{H2O\_in} = x_{H2O\_in} n_{tot\_in}$$

The inlet molar flow rate of water vapor is:

$$n_{H2Ov\_in} = x_{H2Ov\_in} n_{H2O\_in}$$

The inlet molar flow rate of liquid water is:

$$n_{H2Ol\_in} = x_{H2Ol\_in} n_{H2O\_in}$$

The outlet mole fractions and molar flow rates can be calculated using the same equations.

Using MATLAB to solve:

%%

% EXAMPLE 10-3: Mass flow rates into fuel cell layers

%%

% Inputs

```
const.N = 6;              % Number of layers
const.current = 0.6;      % current (amp)
const.tfinal = 20;        % Simulation time (s)
const.F = 96485.3383;     % Faraday's Constant (coulomb/mole)
```

```
const.R = 8.314472;        % Ideal gas constant (J/K-mol)
const.P_tot = 1;           % Total pressure (bar)
const.A_active = 0.03;     % Active area (cm^2)—only used for current (Amps)
const.phi = 1;                 calculations, and for energy calculations of the
const.mw_H2O = 18;             channels
const.v_H2_in = 1.7e-8;    % Molecular weight of water
const.v_air_in = 1.e-8;    % Volumetric flow rate of wet hydrogen (m^3/s)
const.T_in = 293.2;        % Volumetric flow rate of air (m^3/s)
const.Tf_in = 353.2;       % Initial temperature (K)
const.Tf_air = 273.5;      % Initial fluid temperature (K)
const.phi_air = 1;         % Initial air temperature (K)
const.airO2 = 0.21;        % Inlet humidity of air
const.airN2 = 0.79;        % Fraction of O2 in air
const.damp = 0.6;          % Fraction of N2 in air
                           % ODE solver damping factor (to avoid ringing of the
                             solution)
```

% Convert volumetric flow rate to molar flow rate using ideal gas law

```
const.n_air_in = const.v_air_in * (const.P_tot./const.Tf_air) * (1/0.0831) * 1000; %
    mol/s
```

% Convert volumetric flow rate to molar flow rate using ideal gas law

```
const.n_H2_in = const.v_H2_in * (const.P_tot./const.Tf_in) * (1/0.0831) * 1000;
    % mol/s
```

% Layers

```
% 1 – Left end plate
% 2 – Gasket
% 3 – Contact (Copper)
% 4 – Contact
% 5 – Gasket
% 6 – End plate
```

% Parameters defined at the layer boundaries

```
% 1 2 3 4 5 6 7
```

% Gas temperature

```
param.T_f = [353.2, 353.2 353.2 353.2, 353.2, 353.2, 353.2];
```

% Humidity of gas

```
param.phi = [1, 1, 1, 1, 1, 1, 1];
```

% Parameters defined at the layer centers

```
% 1 2 3 4 5 6
% Area (m^2)
param.A = [0.0367, 0.0367, 0.0367, 0.0367, 0.0367, 0.0367];

% Thickness (m)
param.thick = [0.025, 0.025, 0.002, 0.002, 0.025, 0.025];
```

% Number of slices within layer

```
param.M = [1, 1, 1, 1, 1, 1];
```

% Channel radius

```
param.r = [0.000625, 0.000625, 0.000625, 0.000625, 0.000625, 0.000625];
```

% Channel width (rectangular channels)

```
param.wc = [0.000625, 0.000625, 0.000625, 0.000625, 0.000625, 0.000625];
```

% Channel depth (rectangular channels)

```
param.dc = [0.000625, 0.000625, 0.000625, 0.000625, 0.000625, 0.000625];
```

% Channel length

```
param.L = [0.000625, 0.000625, 0.000625, 0.000625, 0.000625, 0.000625];
%%%%%%%%%%%%%%%%%%%%%%%%%%%%%%%%%%%%%%%%%%%
```

% Grid definition

```
% x – interslice coordinates
% n – molar flow rates at slice boundary
%%%%%%%%%%%%%%%%%%%%%%%%%%%%%%%%%%%%%%%%%%%
```

% Grid up mass flows. Assume each layer abuts the

```
% next one. The mass flow rate is at the boundary of each slice. x is at the
% edge of each slice (like a stair plot).
x = 0;
layer = [];
for i=1:const.N,
x = [x, x(end) + (1:param.M(i))*param.thick(i)/param.M(i)]; %Boundary points
   layer = [layer, i*ones(1,param.M(i))];
end
```

% Slice thicknesses

```
dx = diff(x); %gives approximate derivatives between x's
```

% Initial mass flows (defined at layer boundaries)

```
n_tot = zeros(size(x)); % Total molar flow
```

% Convert volumetric flow rate to molar flow rate using ideal gas law

```
n_tot(1) = const.v_H2_in * (const.P_tot./const.Tf_in) * (1/0.0831) * 1000; % mol/s
n_tot(end) = const.v_air_in * (const.P_tot./const.Tf_air) * (1/0.0831) * 1000; % mol/s

f = @ (t,n_tot) mass(t,n_tot,x,layer,param,const,dx');

options = odeset('OutputFcn', @ (t,n_tot,opt) massplot(t,n_tot,opt,x));
[t,n_tot] = ode45(f, [linspace(0,const.tfinal,100)], n_tot, options);
end % of function
%
function dndt = mass(t,n_tot,x,layer,param,const,dx)
```

% Mass flow rates for fuel cell

% Make a convenient place to set a breakpoint

```
if (t > 30)
    s = 1;
end
```

% Preallocate output

```
dndt = zeros(size(n_tot));

% Fluid flows from left to right for layers 1–3 and
% right to left for layers 4–6
inlet = [find(layer<4) find(layer>3)+1];
outlet = [find(layer<4)+1 find(layer>3)];

% Treat n_tot as a row vector
n_tot = n_tot(:)';
```

% Inlet Mole Fractions

```
P_H2Ov_inlet = param.phi(inlet) .* psat          % Calculate the vapor pressure
    (param.T_f(inlet));                               of water vapor
x_H2Ov_inlet = P_H2Ov_inlet ./ const.P_tot;      % mole fraction of water vapor
x_H2Ol_inlet = (x_H2Ov_inlet .* psat             % mole fraction of water
    (param.T_f(inlet))) ./ const.P_tot;
x_H2O_inlet = x_H2Ov_inlet + x_H2Ol_inlet;       % mole fraction of H2O
x_H2_inlet = 1 – x_H2O_inlet;                     % mole fraction of Hydrogen
```

% Inlet Molar flows

```
n_H2_inlet = x_H2_inlet .* n_tot(inlet);          % Total molar flow of H2 coming in
n_H2O_inlet = x_H2O_inlet .* n_tot(inlet);        % Total molar flow of H2O coming in
```

```
n_H2Ov_inlet = x_H2Ov_inlet.*          % Calculate the H2O vapor molar
   n_H2O_inlet;                           flow rate going in
n_H2Ol_inlet = x_H2Ol_inlet.*n_H2O_inlet;  % Calculate the H2O liquid molar
                                          flow rate going in
```

% Replace hydrogen parts

%Outlet Mole Fractions

```
P_H2Ov_outlet = param.phi(outlet).*psat    % Calculate the vapor pressure
   (param.T_f(outlet));                       of water vapor
x_H2Ov_outlet = P_H2Ov_outlet ./ const.P_tot;  % mole fraction of water vapor
x_H2Ol_outlet = (x_H2Ov_outlet.*psat       % mole fraction of water
   (param.T_f(outlet))) ./ const.P_tot;
x_H2O_outlet = x_H2Ov_outlet +             % mole fraction of H2O
   x_H2Ol_outlet;
x_H2_outlet = 1 – x_H2O_outlet;            % mole fraction of Hydrogen
```

% Outlet Molar flows

```
n_H2_outlet = x_H2_outlet.*n_tot(inlet);      % Total molar flow of H2
                                                coming out
n_H2O_outlet = x_H2O_outlet.*n_tot(inlet);    % Total molar flow of H2O
                                                coming out
n_H2Ov_outlet = x_H2Ov_outlet.*n_H2O_outlet;  % Calculate the H2O vapor
                                                molar flow rate going out
n_H2Ol_outlet = x_H2Ol_outlet.*n_H2O_outlet;  % Calculate the H2O liquid
                                                molar flow rate going out
n_outlet = n_H2_outlet + n_H2O_outlet;

%%%%%%%%%%%%%%%%%%%%%%%%%%%%%%%%%%%%%%%%%%%%%
%
% Combine into rate of change of n
%
% Do this in a loop since more than one layer outlet may contribute to
% a given element of dndt.
for i=1:length(outlet)
   dndt(outlet(i)) = dndt(outlet(i)) + n_outlet(i) – n_tot(outlet(i));
end
```

% Use damping factor to help numerical convergence

```
dndt = const.damp*dndt;

end % of function
%
function status = massplot(t,n_tot,opt,x)
if isempty(opt)
   plot(t,n_tot), title(['t = ',num2str(t)]), hold on
   status = 0;
```

```
    drawnow
  end
  end % of function

%%%%%%%%%%%%%%%%%%%%%%%%%%%%%%%%%%%%%%%%%

  function Psat = psat(T)
  % PSAT Saturation pressure
  % PSAT(T) returns the saturation pressure in bars. T is in degrees K.

  Tc=T-273;                 % Conversion to Celcius for use in Psat
  Psat_Pa=-2846.4+411.24.*Tc - 10.554.*Tc.^2 + 0.16636.*Tc.^3; % calculation
      of saturation pressure
  Psat=Psat_Pa./100000;   % Convert to bar

  end % of function
```

Figures 10-8 and 10-9 both show the transient flow rates of hydrogen and water into and out of six fuel cell layers at 20 and 120 seconds of simulation time.

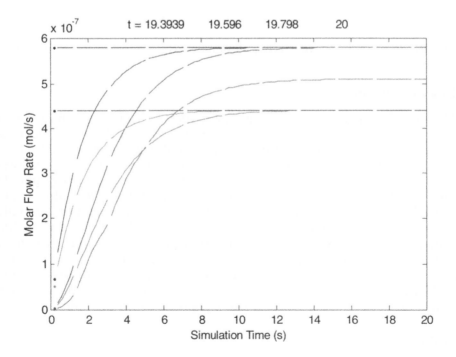

FIGURE 10-8. Hydrogen and water flow rates after 20 seconds of simulation time.

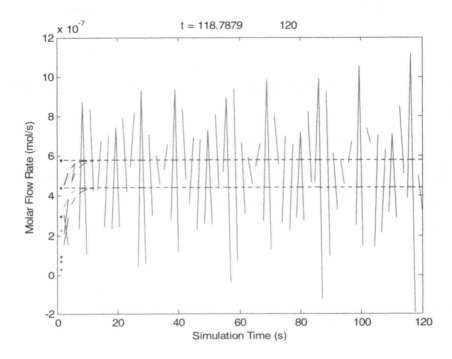

FIGURE 10-9. Hydrogen and water flow rates after 120 seconds of simulation time.

Chapter Summary

The flow field plates have multiple jobs, such as evenly distributing fuel and oxidant to the cells, collecting the current to power the desired devices, and evenly distributing or discarding heat and water products. The flow field design is critical for optimal fuel cell performance because it ensures even distribution of the reactants and products through the cell. Commonly used materials for flow field plates are graphite, stainless steel, aluminum, and polymer composites. The flow field designs that have been traditionally used are the serpentine, parallel, and interdigitated designs. The width, depth, and length of the channels in the flow field plate should be carefully considered to ensure proper flow rates, mass transfer, and pressure drop. Another consideration when designing flow field plates is the temperature of the gases in the channels. All of these factors contribute to the mass and heat transfer in the fuel cell, and can be optimized through modeling.

Problems

- A fuel cell has a 50-cm² active area and a current density of 1 A/cm² with nine parallel channels on the cathode. Each channel is 1 mm wide and 1 mm deep with 1 mm of spacing between channels. Air at the inlet is 100% humidified at 60°C. The pressure is 3 atm, and there is a 0.3 atm pressure drop through the flow field. The oxygen stoichiometric ratio is 1.5. Calculate the velocity and Reynold's number at the air inlet and outlet.

- Calculate the pressure drop through a PEM fuel cell cathode flow field of a single graphite plate with a 100-cm² cell area. The stack operates at 1 atm at 60°C with 100% saturated air. The flow field consists of 18 parallel serpentine channels 0.8 mm wide, 1 mm deep, and 1 mm apart.
- A fuel cell has a 100-cm² active area and a current density of 0.8 A/cm² with 20 parallel channels on the cathode. Each channel is 1.5 mm wide and 1.5 mm deep with 1 mm spacing between channels. Air at the inlet is 100% humidified at 70°C. The pressure is 3 atm, and there is a 0.3 atm pressure drop through the flow field. The oxygen stoichiometric ratio is 3. Calculate the velocity, Reynold's number, and pressure drop at the air inlet and outlet.

Endnotes

[1] Li, X., and I. Sabir. Review of bipolar plates in PEM fuel cells: Flow-field designs. *Int. J. Hydrogen Energy.* Vol. 30, 2005, pp. 359–371.
[2] Ibid.
[3] Ibid.
[4] Ibid.
[5] Spiegel, C.S. Designing and Building Fuel cells. 2007. New York: McGraw-Hill.
[6] Ibid.
[7] Ibid.
[8] Ibid.
[9] Barbir, F. *PEM Fuel Cells: Theory and Practice.* 2005. Burlington, MA: Elsevier Academic Press.
[10] Barbir, *PEM Fuel Cells: Theory and Practice.*

Bibliography

Cha, S.W., R. O'Hayre, Y. Saito, and F.B. Prinz. The scaling behavior of flow patterns: A model investigation. *J. Power Sources.* Vol. 134, 2004, pp. 57–71.
Chen, X., N.J. Wu, L. Smith, and A. Ignatiev. Thin film heterostructure solid oxide fuel cells. *Appl. Phys. Lett.* Vol. 84, No. 14, April 2004.
EG&G Technical Services. November 2004. *The Fuel Cell Handbook.* 7th ed. Washington, DC: U.S. Department of Energy.
Feindel, K., W. Logan, P.A. LaRocque, D. Starke, S.H. Bergens, and R.E. Wasylishen. *J. Am. Chem. Soc.* Vol. 126, 2004, pp. 11436–11437.
Gulzow, E., M. Schulze, and U. Gerke. Bipolar concept for alkaline fuel cells. *J. Power Sources.* Vol. 156, 2006, pp. 1–7.

He, S., M.M. Mench, and S. Tadigadapa. Thin film temperature sensor for real-time measurement of electrolyte temperature in a polymer electrolyte fuel cell. *Sensors Actuators A*. Vol. 12, 2006, pp. 170–177.

Hermann, A., T. Chaudhuri, and P. Spagnol. Bipolar plates for PEM fuel cells: A review. *Int. J. Hydrogen Energy*. Vol. 30, 2005, pp. 1297–1302.

Hsieh, S.S., C.-F. Huang, J.-K. Kuo, H.-H. Tsai, and S.-H. Yang. SU-8 flow field plates for a micro PEMFC. *J. Solid State Electrochem*. Vol. 9, 2005, pp. 121–131.

Hsieh, S.-S., S.-H. Yang, J.-K. Kuo, C.-F. Huang, and H.-H. Tsai. Study of operational parameters on the performance of micro PEMFCs with different flow fields. *Energy Conversion Manage*. Vol. 47, 2006, pp. 1868–1878.

Lee, S.-J., Y.-P. Chen, and C.-H. Huang. Electroforming of metallic bipolar plates with micro-featured flow field. *J. Power Sources*. Vol. 145, 2005, pp. 369–375.

Mehta, V., and J.S. Copper. Review and analysis of PEM fuel cell design and manufacturing. *J. Power Sources*. Vol. 114, 2003, pp. 32–53.

Motokawa, S., M. Mohamedi, T. Momma, S. Shoji, and T. Osaka. MEMS-based design and fabrication of a new concept micro direct methanol fuel cell. *Electrochem. Comm*. Vol. 6, 2004, pp. 562–565.

Muller, M.A., C. Muller, R. Forster, and W. Menz. Carbon paper flow fields made by WEDM for small fuel cells. *Microsystem Technol*. Vol. 11, 2005, pp. 280–281.

Muller, M.C., F. Gromball, M. Wolfle, and W. Menz. Micro-structured flow fields for small fuel cells. *Microsystem Technol*. Vol. 9, 2003, pp. 159–162.

Nguyen, N.-T., and S.H. Chan. Micromachined polymer electrolyte membrane and direct methanol fuel cells—A review. *J. Micromech. Microeng*. Vol. 16, 2006, pp. R1–R12.

O'Hayre, R., S.-W. Cha, W. Colella, and F.B. Prinz. 2006. *Fuel Cell Fundamentals*. New York: John Wiley & Sons.

U.S. Patent 6,551,736 B1. Fuel Cell Collector Plates with Improved Mass Transfer Channels. Gurau, V., F. Barbir, and J.K. Neutzler. Teledyne Energy Systems, Inc., Hunt Valley, MD. April 22, 2003.

Wang, C.Y., M.M. Mench, S. Thynell, Z.H. Wang, and S. Boslet. Computational and experimental study of direct methanol fuel cells. *Int. J. Transport Phenomena*. Vol. 3, August 2001.

CHAPTER 11

Modeling Micro Fuel Cells

11.1 Introduction

This chapter is concerned with the change in magnitude of proportion when fuel cells go from large to small. This is called "scaling," and the exact manner with which the particular quantity changes with respect to another quantity is called a "scaling law." Scaling can be examined in terms of systems of forces rather than a single force. When a device also has fixed proportions, the surface area-to-volume ratio always increases as the length scale decreases. Therefore, as objects become smaller, surface effects become relatively more important. While the point at which surface effects matter more than volume effects depends upon the system under consideration; a good rule of thumb is that millimeter-scale devices are small enough for surface effects to be important, but these effects will be dominant in the micron regimen. Some of the differences between macroscopic and microscopic systems include the following:

- Surface effects matter more than bulk effects
- Very small dead volumes
- Issues with bubbles
- No unwanted turbulent flow

The classification of microchannels varies in the literature, but a good guideline can be found in Table 11-1[1]. Table 11-2 shows the different flow regimes for various channel dimensions for air and hydrogen[2].

There are certain parameters that can be ignored when modeling macro-scale fuel cells, that need to be included when modeling micro fuel cells. Some of the performance considerations with microdevices are minimal dead volume, low leakage, good flow control, and rapid diffusion. The most commonly used stack configuration for macro and micro fuel cells is the bipolar configuration, which has been described in previous chapters and is shown in Figure 11-1. There are many alternative stack

TABLE 11-1
Classification of Microchannels

Classification	Hydraulic Diameter Range
Convectional	Dh > 3 mm
Minichannel	3 mm > Dh > 200 μm
Microchannel	200 μm > Dh > 10 μm

TABLE 11-2
Channel Dimensions (microns [μm])

	Continuum Flow	Slip Flow
Air	>67	0.67 to 67
H$_2$	>123	1.23 to 123

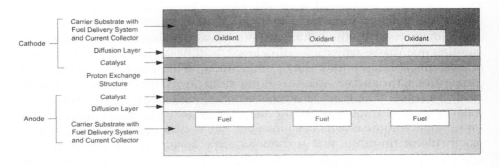

FIGURE 11-1. Basic micro fuel cell based upon traditional fuel cell design.

configurations for micro fuel cells, and the design and modeling of these are in their infancy. Important modeling parameters to include are:

- Size, weight, and volume at the desired power
- Temperature
- Humidification and water management
- Fuel and oxidant pressures

As shown in Figure 11-1, the membrane electrode assembly (MEA) is separated by a plate with flow fields to distribute the fuel and oxidant. The majority of fuel cell stacks, regardless of size and fuels used, is of this configuration.

The specific topics that will be covered in this chapter include:

- Micro fuel cells in the literature
- Microfluidics

- Flow rates and pressures
- Bubbles and particles
- Capillary effects
- Single- and two-phase pressure drop

This chapter explains the differences and potential issues between micro and macro fuel cell stacks and introduces microfluidics for modeling the micro-electro mechanical systems (MEMS) fuel cell bipolar plates.

11.2 Micro PEM Fuel Cells in the Literature

Micro fuel cells have been documented in the literature for several years. There are many companies currently working on this technology for cellular phones and other small, portable devices. However, many of the advantages of MEMS technology have not been applied yet to fuel cells, therefore, the optimization of MEMS fuel cells is in its infancy. The next few sections present an overview and comparison of the MEMS fuel cell technology recently documented in the literature.

11.2.1 The Electrodes

The thickness of the electrodes in traditional fuel cells is typically 250 to 2000 angstroms (Å) with a catalyst loading of at least 0.5 mg/cm^2. For micro fuel cells, the typical platinum loading is from 5 to 60 nm in thickness, with a platinum–ruthenium loading for the anode between 2.0 and 6.0 mg/cm^2, and a platinum loading for the cathode between 1.3 and 2.0 mg/cm^2.[3-6] An adhesion layer is deposited before the catalyst layer, and it is typically 25 to 300 Å in thickness. As mentioned previously, the catalyst loading is a cost-prohibitive factor for the PEM fuel cell. The cost of the PEM fuel cell stack would be lowered if the amount of platinum is reduced, another (cheaper) element is combined with it, the platinum is replaced with another element, or the fuel cell stack is miniaturized to the point where the required catalyst loading is not as cost prohibitive. Chapter 8 covers the details of modeling the catalyst layer. When modeling the micro fuel cell system, it is important to use a catalyst model that takes into account microscopic effects, such as an agglomerate model. Due to the small areas of the micro fuel cell, homogeneous catalyst distribution and placement are very important. Taking this into consideration is also very important for obtaining an accurate electrode model for a MEMS fuel cell.

Diffusion Layer
The diffusion layer is made of electrically conductive porous materials such as carbon or Toray paper. The thickness of the diffusion layer is

usually 0.25 to 0.40 mm. The conductivity of the paper can be improved by filling it with electrically conductive powder such as carbon black. To help remove water from the pores of the carbon paper, the diffusion layer can be treated with PTFE. Some micro fuel cell developers forgo the diffusion layer altogether, and platinum is sputtered directly on the proton exchange structure. There are several new studies that are helping to improve fuel cell performance by creating highly aligned diffusion layers from carbon nanotubes. Several studies have shown an increase in fuel cell current density from fuel cells made with carbon nanotube diffusion layers[7-12]. Depending upon the micro dimensions of the fuel cell, the GDL layer may not be as advantageous as in larger fuel cells. This layer is extremely helpful in creating an even flow rate to the catalyst layer. In micro fuel cells, the MEMS bipolar plates can be altered appropriately to provide even flow without using the GDL. This also depends highly on the fuel cell design. The GDL for micro fuel cells can be modeled using the same methods presented in Chapter 9. However, depending upon the GDL design, it may improve the model accuracy to rigorously include the geometric details for a micro fuel cell.

11.2.2 Bipolar Plates

Most traditional bipolar plates (in large fuel cells) are made from stainless steel or graphite. Stainless steel plates are heavy components for a portable or micropower system. Solid graphite plates are highly conductive, chemically inert, and resistant to corrosion, but are expensive, brittle, and costly to manufacture. Flow channels are traditionally machined or electrochemically etched to the graphite or stainless steel bipolar plate surfaces. These materials are not suitable for mass production, and would not work for MEMS-based fuel cell system. Typical materials that have been used in MEMS fuel cells are silicon wafers, carbon paper, PDMS, SU-8, and copper and stainless steel metal foils. Traditional photolithography and microfabrication techniques have begun to be used with MEMS fuel cells during the last few years.

Flow Channels

In PEM fuel cells, the flow field should be designed to minimize pressure drop while providing adequate and evenly distributed mass transfer through the carbon diffusion layer to the catalyst surface for reaction, as discussed in Chapter 10. As discussed previously, the three most popular channel configurations for traditional fuel cells are (1) serpentine, (2) parallel, and (3) interdigitated flow. Most MEMS fuel cell studies in the literature also use the same flow field patterns. Some small-scale fuel cells do not use a flow field to distribute the hydrogen and/or air but rely on diffusion processes from the environment. Since the hydrogen reaction is not rate limiting, and water blockage in the humidified anode can occur, a serpentine arrangement is typically used for the anode in smaller PEM fuel cells. Figure 11-2 illustrates interdigitated, serpentine, and spiral interdigitated flow patterns for MEMS fuel cells.

There are some MEMS fuel cells that have been fabricated with spiral interdigitated channels. Combining the advantages of the serpentine and the interdigitated flow patterns yields the spiral-interdigitated channel. Figure 11-2c shows an example of this flow field type. The peak power density of the spiral interdigitated cell decreases as the feature size decreases from 1000 to 5 μm[13]. The scaling behavior is slightly similar to interdigitated channels; however, the flow path short circuits are highly prominent in the smaller channels.

Fuel cell performance improves as the channel gas flow velocity increases because the increased flow velocity enhances mass transport. When investigating the effect of fuel cell geometry, the following geometric parameters need to be considered: the flow channel pattern, the channel and rib shape, and the diffusion layer thickness, as well as many other factors. The velocity in the flow channel will increase as the feature size decreases. However, one drawback of the smaller feature size is the increased pressure drop in the flow channels. The feature sizes for flow channels in the literature range from $100 \times 200 \times 20$ μm to 500×500 μm to 750×750 $\times 12.75$ mm, with many length, widths, and depths in between with various rib widths[14-22]. A recent study that was conducted verified that the fuel cell performance improves with the decrease in flow channel dimension, as shown in Figure 11-3.

11.2.3 Stack Design and Configuration

In the traditional fuel cell stack, the cathode of one cell is connected to the anode of the next cell. The main components of the fuel cell stack are the membrane electrode assemblies (MEAs), gaskets, bipolar plates with electrical connections, and end plates. The stack is connected together by bolts,

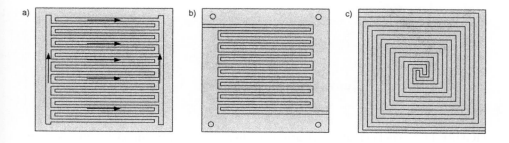

FIGURE 11-2. Interdigitated, serpentine, and spiral-interdigitated flow field designs.

a)

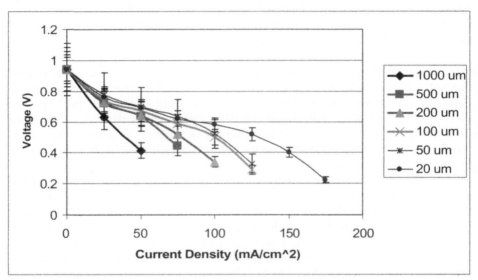

b)

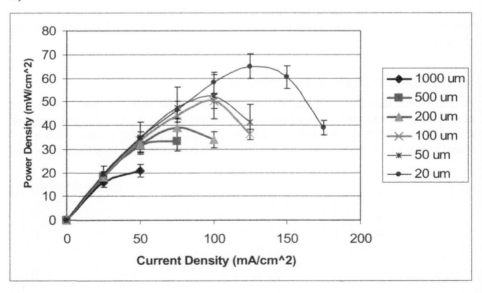

FIGURE 11-3. Fuel cell a) polarization and b) power density curves for 20–1000 μm channel widths and depths.

rods, or other method to clamp together the cells. The key aspects of fuel cell design are:

- Uniform distribution of reactants to the cell
- Uniform distribution of reactants inside the cell
- Maintenance of required temperature inside each cell
- Minimum resistive losses
- No gas leakage
- Mechanical sturdiness

Most MEMS systems use silicon as the preferred material because of the availability, low cost, and various processing technologies available. Some of the processes normally used to create micro fuel cells are anisotropic etching, deep reactive ion etching (DRIE), and CVD and PVD for depositing various materials. Polymers are being used, but silicon/glass systems are mechanically more stable, resist high temperatures/pressures, and are basically chemically inert. However, silicon is brittle, and polymers allow configurations and alternative processing techniques. Some of the polymers that are being researched include PMMA and PDMS using ion etching, polymeric surface micromachining, hot embossing, soft lithography, and laser machining. Stainless steel foil and copper films are also being researched as materials for fuel delivery/current collector plates[23,24].

There is much more variability in fuel cell design and configuration with MEMS fuel cells (1 cm² or less in area) than with the larger fuel cell stacks. An interesting design is shown in Figure 11-4, which was first pro-

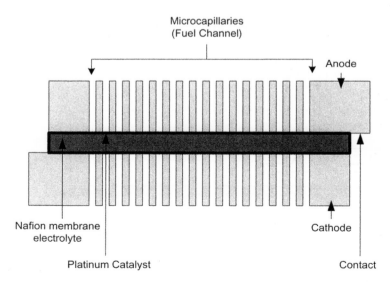

FIGURE 11-4. Cross-sectional view of the porous silicon-based stack[25,26].

posed by Aravamudhan et al.[27]. The flow fields are made of silicon, and the proton exchange membrane is wedged between the two sets of flow fields. Platinum is deposited on both microcolumns to act as an electrocatalyst and current collector. The flow field pore diameter was carefully controlled to use capillary pressure in order to distribute the fuel correctly and minimize methanol crossover. The area of each electrode is 1 cm[2].[28]

Figure 11-5 shows the planar design, which is the most common stack design used in micro fuel cells besides the traditional design shown in Figure 11-1. The planar design is two-dimensional and requires a large surface area to deliver similar performance to the bipolar configuration. The fuel and oxidant are delivered through a single side of the fuel cell[29].

Another very interesting micro fuel cell design is shown in Figure 11-6. This fuel cell structure is usually made of silicon and the channels are fabricated at small enough dimensions to allow the fuel and oxidant to flow in the laminar flow regimen without mixing[30–32]. The protons travel from one stream to the next without the aid of a proton exchange

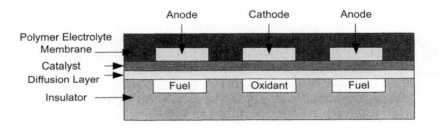

FIGURE 11-5. Cross-sectional view of planar micro fuel cell stack[33,34].

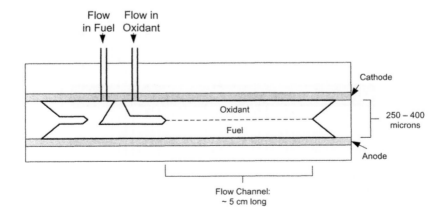

FIGURE 11-6. Cross-sectional view of a membraneless laminar flow micro fuel cell stack[35,36].

membrane. There are separate entrances for the fuel and oxidant. The electrode materials are deposited on the silicon structure by sputtering or evaporation.

Laminar flow is a new concept in developing micro fuel cells. The most widely known design is a Y-shaped microchannel system where two fuels flow side-by-side with the help of a large control and monitoring system outside the fuel cell. When considering how this system will be able to be actualized, one important concept to keep in mind is the size of the interface between the two fuels, which is defined by the depth and the length of the channel. The width is not considered as important because the interface remains the same regardless of the width of the channels.

11.3 Microfluidics

As in continuum mechanics, microfluidics uses the Navier-Stokes equations for liquids and gases. The equations are valid for liquids and gases, with the exception that gases are compressible while liquids are not. As the dimensions become smaller, the differences between gases and liquids become more apparent. The first difference is that liquids have interfaces, and there are definite boundary liquids flowing in a channel. On the other hand, gases readily mix together. The second difference becomes apparent when Navier-Stokes equations are analyzed for MEMS systems. A Knudsen number that is less than 0.01 indicates that the equations of the continuum theory should provide a good approximation, while a Knudsen number approaching unity means that the gas must be treated as a collection of particles rather than continuum. The useful forms of the Navier-Stokes equations for MEMS systems are introduced in this section.

11.3.1 Navier-Stokes Equation

The Navier-Stokes equation describes the behavior of a fluid in terms of stress and strain. In fluids, in addition to conservation of momentum, there is also an equation derived from the principal of conservation of mass:

$$\frac{\partial \rho}{\partial t} + u_j \frac{\partial \rho}{\partial x_j} + \rho \frac{\partial u_i}{\partial x_i} = 0 \qquad (11\text{-}1)$$

where ρ denotes the density of fluid and u_i is a vector of fluid velocities whose ith component is fluid velocity in direction i. The strain rate tensor:

$$\frac{\partial \varepsilon_{ij}}{\partial t} = \frac{1}{2}\left(\frac{\partial u_i}{\partial x_j} + \frac{\partial u_j}{\partial x_i} \right) \qquad (11\text{-}2)$$

The stress tensor can be thought of as a 3×3 matrix; therefore, it is written as:

$$\begin{pmatrix} \sigma_{11} & \sigma_{12} & \sigma_{13} \\ \sigma_{21} & \sigma_{22} & \sigma_{23} \\ \sigma_{31} & \sigma_{32} & \sigma_{33} \end{pmatrix} \qquad (11\text{-}3)$$

The ijth element of this matrix is the force per unit area in the direction i exerted on a surface element with normal in the j direction. The stress tensor is related to the strain tensor through:

$$\sigma_{ij} = -p\delta_{ij} + 2\mu\dot{\varepsilon}_{ij} + \lambda\dot{\varepsilon}_{kk}\delta_{ij} \qquad (11\text{-}4)$$

In the equation 11-4, the dot implies differentiation with respect to time, p is the pressure in the fluid, μ is the dynamic viscosity, and λ is a second viscosity coefficient. The equation of conservation can now be written as:

$$\rho\frac{\partial u_i}{\partial t} + \rho u_j \frac{\partial u_i}{\partial x_j} = \rho F_i + \frac{\partial \sigma_{ij}}{\partial x_j} \qquad (11\text{-}5)$$

where F_i represents body forces, while the stress tensor captures the internal stresses. Equation 11-5 is a statement of Newton's second law, $F = ma$. Using Equation 11-4 in Equation 11-5:

$$\rho\frac{\partial u_i}{\partial t} + \rho u_j \frac{\partial u_i}{\partial x_j} = \rho F_i + \frac{\partial p}{\partial x_i} + \frac{\partial}{\partial x_j}(2\mu\varepsilon_{ij} + \lambda\varepsilon_{kk}\delta_{ij}) \qquad (11\text{-}6)$$

Equation 11-6 is usually called the Navier-Stokes equation of motion and Equations 11-1 and 11-6 are called Navier-Stokes equations. These can be rewritten in vector form:

$$\frac{\partial \rho}{\partial t} + \nabla\cdot(\rho u) = 0 \qquad (11\text{-}7)$$

$$\rho\frac{\partial u_i}{\partial t} + \rho(u\cdot\nabla)u + \nabla p - \mu\nabla^2 u - (\lambda + \mu)\nabla(\nabla\cdot u) = f \qquad (11\text{-}8)$$

The Navier-Stokes equations for a viscous, compressible fluid are a system of four nonlinear partial differential equations. However, the system contains five unknown functions: pressure, density, and the three components of the velocity vector. In order to solve for the unknowns, the conservation of energy equation is usually added to the Navier-Stokes equations. This introduces one more equation and one more unknown variable, the temperature, T. A final equation relating the ρ, p, and T needs to be introduced to solve for the six unknowns.

11.3.2 Incompressible Flow

If the fluid is assumed to be incompressible, the Navier-Stokes equations may be simplified. The assumption of incompressibility implies that density is a constant. Therefore, Equations 11-7 and 11-8 reduce to:

$$\nabla \cdot u = 0 \tag{11-9}$$

$$\frac{\partial u}{\partial t} + (u \cdot \nabla)u = -\frac{1}{\rho}\nabla p + v\nabla^2 u \tag{11-10}$$

where $v = \mu/p$ is called the kinematic viscosity. The assumption of constant density reduces the number of equations and unknowns to four.

11.3.3 The Euler Equations

In addition to the assumption of constant density, if it is assumed that the fluid is inviscid as well as incompressible, the Navier-Stokes equations may be further simplified to obtain the Euler equations. This assumption means that $v = 0$, and therefore Equations 11-9 and 11-10 reduce to:

$$\nabla \cdot u = 0 \tag{11-11}$$

$$\frac{\partial u}{\partial t} + (u \cdot \nabla)u = -\frac{1}{\rho}\nabla p \tag{11-12}$$

Equations 11-11 and 11-12 are called the incompressible inviscid Navier-Stokes equations.

11.3.4 The Stokes Equations

If the incompressible Navier-Stokes equations for a flow assume a characteristic velocity, U, in a spatial region with characteristics length, l, the following equations can be introduced:

$$u = \frac{u}{U} \quad x = \frac{x}{1} \quad y = \frac{y}{1} \quad z = \frac{z}{1} \quad t = \frac{Ut}{1} \quad p = \frac{p - p_\infty}{pU^2} \tag{11-13}$$

where p_∞ is a reference pressure in the system to be thought of as the pressure in the fluids at infinity. The dimensionless system is obtained:

$$\nabla \cdot u = 0 \tag{11-14}$$

$$\frac{\partial u}{\partial t} + (u \cdot \nabla)u = -\nabla p + \frac{1}{Re}\nabla^2 u \tag{11-15}$$

As introduced previously, the dimensionless parameter Re is the Reynold's number for the flow and is given by:

$$Re = \frac{Ul}{v} \tag{11-16}$$

when the Reynold's number is low, the incompressible Navier-Stokes equations are replaced with the Stokes equations (Equations 11-17 and 11-18). The pressure is rescaled with Re where $p = p/Re$, and the limit as Re $\rightarrow$ 0 is used to obtain:

$$\nabla \cdot u = 0 \tag{11-17}$$

$$\nabla p = \nabla^2 u \tag{11-18}$$

11.3.5 Boundary and Initial Conditions

In order to formulate the boundary and initial conditions, two types of interfaces are examined: a solid–fluid interface and a fluid–fluid interface. For the boundary between a fluid and solid, the no-penetration and no-slip boundary conditions are generally used. If the fluid–solid interface is used as shown in Figure 11-7a the boundary conditions would be formulated in terms of normal, n, and the velocity, u, as:

$$u \cdot n = 0 \tag{11-19}$$

The no-slip boundary condition comes from experimental evidence. It has been observed that the fluid is moving tangentially to the solid surface at the interface between a fluid and solid. The no-slip boundary condition is stated as:

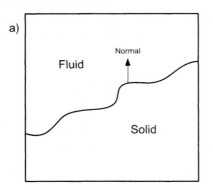
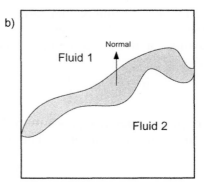

FIGURE 11-7. A a) fluid–solid interface, and b) a fluid–fluid interface.

$$u \times n = 0 \qquad (11\text{-}20)$$

When the two conditions are combined, the no-slip and no-penetration at the boundary between a solid and a fluid, they can be stated as:

$$u = 0 \qquad (11\text{-}21)$$

When dealing with inviscid fluid flow, it is only necessary to specify no-penetration into a solid, and not the full no-slip condition.

At a fluid–fluid interface, the no-slip boundary condition is used. The interface is described by the equation f(x, y, z, t) = 0, as shown in Figure 11-7b. The location of the interface between the two fluids is not usually known, determining the function f is part of the problem. The change in momentum across the interface is balanced by the tensile force of the interface, as shown in Figure 11-7:

$$(\sigma^1 - \sigma^2) \cdot n = -\gamma \left(\frac{1}{R_1} + \frac{1}{R_2} \right) n \qquad (11\text{-}22)$$

where γ is the surface tension at the interface, R_i is the radii of curvature of the interface, and the σ^i are the stress tensors in each fluid. If the fluid is not moving, the stress in the fluid becomes hydrostatic, and the equation reduces to the familiar Laplace-Young law:

$$(p_1 - p_2) = \gamma \left(\frac{1}{R_1} + \frac{1}{R_2} \right) \qquad (11\text{-}23)$$

where the p_i are the pressures of each fluid.

Since the location of the interface is unknown, an additional condition is needed to determine its position. This is called a kinematic condition, and it says the fluid that starts on the boundary remains on the boundary. In terms of the interface f as shown, this condition may be stated as:

$$\frac{\partial f}{\partial t} + u^i \cdot \nabla f = 0 \qquad (11\text{-}24)$$

Here u^i is the velocity vector of the ith fluid.

The study of MEMS sometimes requires the researcher to confront unfamiliar parameter regimes, and there are cases where the fluids do sometimes slip along a solid surface. The Knudsen number provides a measure of how close a particular system is to the slip regime:

$$Kn = \frac{\lambda}{1} \qquad (11\text{-}25)$$

where λ is the molecular mean-free-path, and l is the characteristic length of the system under consideration. As mentioned previously, the mean-free-path is the average distance traveled by a molecule between collisions. When the mean path becomes large and the system becomes comparable in size, the Knudsen number approaches one. In microfluidics, the Knudsen number becomes large, not because the mean-free-path is large, but because the system size becomes small. If Kn is less that 10^{-4}, the no-slip boundary condition can be applied. If Kn becomes larger than 10^{-4}, fluid will slip along an interface. A modified slip boundary condition is often used. For example, the wall coinciding with the x-axis and moving with velocity V_w in the direction of the x-axis can be expressed as:

$$u_1 - V_w = \left(\frac{2-\sigma}{\sigma}\right)\left(\frac{Kn}{1-bKn}\right)\frac{\partial u_1}{\partial y} \quad (11\text{-}26)$$

where σ is called the accommodation coefficient, and b is called the slip coefficient. These coefficients are typically determined experimentally.

11.3.6 Poiseuille Flow

A simple solution of the incompressible Navier-Stokes equations can be solved for flow in a pipe. If a cylinder has a radius R, and a constant pressure gradient in the z direction, then:

$$p_z = -A \quad (11\text{-}27)$$

where A is a constant. If the velocity vector has flow only in the z direction, then the incompressible Navier-Stokes equations are:

$$-\frac{1}{\rho} = p_z = \frac{v}{r}\frac{\partial}{\partial r}\left(r\frac{\partial u_3}{\partial r}\right) \quad (11\text{-}28)$$

If Equation 11-27 is substituted in Equation 11-28, and the resulting equation is integrated twice with respect to *r*:

$$u_3(r) = -\frac{Ar^2}{4\mu} + c_0 \log(r) + c_1 \quad (11\text{-}29)$$

where c_0 and c_1 are integration constants. If c_0 is zero, and c_1 is found by applying the no-slip boundary condition at r = R, then:

$$c_1 = \frac{AR^2}{4\mu} \quad (11\text{-}30)$$

$$u_3(r) = \frac{A}{4\mu}(R^2 - r^2) \tag{11-31}$$

11.3.7 Poiseuille Flow with Slip

If the slip coefficient, b, from Equation 11-26 is zero, then the boundary condition for Equation 11-28 is:

$$u_3(R) = Kn\left(\frac{2-\sigma}{\sigma}\right)\frac{\partial u_3}{\partial r}\bigg|_{r=R} \tag{11-32}$$

It is required that $c_0 = 0$ in order to keep the velocity bounded at the origin. However, the slip boundary condition implies that:

$$c_1 = \frac{AR^2}{4\mu} - \frac{AR^2}{2\mu}Kn\left(\frac{2-\sigma}{\sigma}\right) \tag{11-33}$$

Therefore, the solution for velocity is:

$$u_3(r) = \frac{A}{4\mu}(R^2 - r^2) - \frac{AR^2}{2\mu}Kn\left(\frac{2-\sigma}{\sigma}\right) \tag{11-34}$$

Equation 11-34 reduces to the no-slip solution when Kn goes to zero.

11.4 Flow Rates and Pressures

For a pipe of length, l, and circular cross-section with radius, r, the Navier-Stokes equations can be solved for a steady-state, incompressible fluid. The volumetric flow rate is given by the Hagen-Poiseuille law:

$$Q = \frac{\pi r^4 \Delta P}{8vl} \tag{11-35}$$

where ΔP is the pressure drop over the length of the pipe and v is the dynamic viscosity of the liquid. Using the no-slip boundary condition, the average fluid velocity is:

$$v = \frac{Q}{\pi r^2} = \frac{r^2 \Delta P}{8vl} \tag{11-36}$$

Therefore, the pressure drop over the length l is:

$$\Delta P = \frac{8v\upsilon l}{r^2} \qquad (11\text{-}37)$$

Now let $r = _{el}$, and suppose the velocity, v, is scaled so that:

$$\upsilon = \alpha r^n \qquad (11\text{-}38)$$

where n = 0 for the case of constant velocity, and n = 1 for the case of constant time of a particle of the fluid to transverse the pipe. Then:

$$\Delta P = \frac{8v\alpha r^{n-1}}{\varepsilon} \qquad Re = \frac{\rho\alpha r^{n+1}}{v}, \qquad Q = \pi\alpha r^{n+2} \qquad (11\text{-}39)$$

These equations illustrate that maintaining a constant velocity of the fluid requires very large pressures. Constant time results in a constant pressure drop across the pipe, however, this requires extremely high pressure gradients since:

$$\Delta P \approx 8v\alpha r^{n-2} \qquad (11\text{-}40)$$

Note that in both cases the Reynold's number decreases with decreasing r. Large pressures and large pressure gradients are unavoidable in microsystems.

The fluidic resistance is the pressure drop over the flow rate, and is independent of the average velocity of the fluid:

$$R = \frac{\Delta P}{Q} = \frac{8vl}{\pi r^4} = \frac{8v}{\pi \varepsilon r^3} \qquad (11\text{-}41)$$

which shows that small pipes have very high resistance.

11.5 Bubbles and Particles

Bubbles are more important in MEMS systems than in macroscopic systems. Since the channel size is very small, the bubbles can sometimes block entire channels, inhibit flow, create large void fractions, or introduce many other issues in microsystems. Small volumes of one fluid in another fluid have spherical shapes due to surface tension. If the fluids have different densities, the droplets will move upward or downward due to the buoyant forces acting upon them.

The buoyant force, F_B, on a spherical air bubble of radius r in a liquid density is given by:

$$F_B = \rho g \frac{4}{3} \pi r^3 \qquad (11\text{-}42)$$

The force acting to hold a bubble in place on a surface is the interfacial force, F is:

$$F_I = \pi d \gamma \qquad (11\text{-}43)$$

where γ is the interfacial tension, and d is the diameter of the contact area of the bubble.

For bubbles in flow channels, the pressure drop across a liquid–gas interface and the pressure difference needed to move the bubbles are given by:

$$\Delta P = \frac{2 \gamma \cos \theta}{r} \qquad (11\text{-}44)$$

$$P_d = \frac{2 \gamma_f}{r} \qquad (11\text{-}45)$$

where r is the channel radius and γ is a frictional surface tension parameter.

Depending upon how a bubble is positioned in a microchannel, the pressure drop and pressure required to move the bubble can vary. If the bubble impedes the flow in a capillary, the pressure may be low. If the bubble ends up in a region with different curvatures, the pressure drop may be significant, and large pressures may be required to remove the bubble.

When considering the movement of a particle in a fluid, the friction coefficient is given by Stokes' law:

$$f = 6 \pi r \mu \qquad (11\text{-}46)$$

Like bubbles, particles in microfluidic systems are important because they are of comparable size to flow channels. In order to prevent issues with particles in MEMS systems, careful filtration of fluids and gases is required.

11.6 Capillary Effects

As first discussed in Chapter 8, the surface tension force that draws liquid into a small flow channel or capillary is:

$$F_l = 2\pi r \gamma \cos(\theta) \tag{11-47}$$

where θ is the contact angle between the liquid and surface. For a vertical capillary, the gravitational force on the rising column of liquid is given by:

$$F_g = \rho g \pi r^2 h \tag{11-48}$$

When these forces are made equal, the maximum rise in height of a fluid in a capillary against gravity is:

$$h = \frac{2\gamma \cos(\theta)}{\rho g r} \tag{11-49}$$

Therefore, the height of the fluid column will greatly increase as the size of the channel is decreased. Capillary forces are very useful in microfluidics because very long lengths of channel can be filled with fluid using this force alone—as long as the capillary force is not opposed by gravity.

11.7 Single- and Two-Phase Pressure Drop

Single- and two-phase pressure drop calculations for microchannels are slightly different than the calculations for conventional channel sizes. There are several models in the literature for pressure drop in microchannels for different sizes of microchannels for gas and liquid phases. A pressure drop model proposed by Garimella et al.[37] is used in this section. The first type of pressure drop is contraction pressure drop is due to reduction in the flow area. The homogeneous flow model is used to calculate the contraction pressure drop:

$$\Delta P_{con} = \frac{G_{chan}^2}{2\rho_l} \left[\left(\frac{1}{C_c} - 1 \right)^2 + 1 - \frac{1}{\gamma_{con}^2} \right] \Psi_H \tag{11-50}$$

$$C_c = \frac{1}{0.639 * \left(1 - \dfrac{1}{\gamma_{con}} \right)^{0.5} + 1} \tag{11-51}$$

$$\Psi_{H,x\_out} = 1 + x_{out} \left(\frac{\rho_l}{\rho_v} - 1 \right) \tag{11-52}$$

$$\gamma_{con} = \frac{A_{inlet}}{A_{bipolar}} \tag{11-53}$$

The multiple flow regimen pressure drop model of Garimella et al.[38] for condensing flows of refrigerant R134a in tubes with $0.5 < D < 4.9$ mm can also be used for other types of fluids. Although this model consists of separate submodels for the intermittent flow regimen and the annular/mist/disperse flow regimens, in the current study, the annular flow portion is used for all data for ease of implementation.

In the annular flow model, the interfacial friction factor is computed from the corresponding liquid phase Re and friction factor, the Martinelli parameter, and a surface tension–related parameter:

$$\frac{f_i}{f_l} = A \cdot X^a Re_l^b \psi^c \tag{11-54}$$

where the laminar region is <2300 and A = 1.308×10^{-3}, a = 0.4273, b = 0.9295, and c = −0.1211

The Martinelli parameter X is given by:

$$X = \left(\frac{\left(\frac{dP}{dz} \right)_l}{\left(\frac{dP}{dz} \right)_v} \right)^{0.5} \tag{11-55}$$

The liquid phase Re is defined in terms of the annular flow area occupied by the liquid phase:

$$Re_l = \frac{GD(1-x)}{(1+\sqrt{a})\mu_l} = \frac{GD(1-x)}{(1+\sqrt{a})\mu_l} \tag{11-56}$$

and the gas phase Re is:

$$Re_v = \frac{GDx}{\mu_v \sqrt{a}} = \frac{GDx}{\mu_v \sqrt{a}} \tag{11-57}$$

The surface tension parameter ψ is given by:

$$\psi = \frac{j_l \mu_l}{\sigma} \tag{11-58}$$

where $j_l = \frac{G(1-x)}{\rho_l(1-\alpha)}$ is the liquid superficial velocity. The interfacial friction factor is related to the pressure drop through the void fraction model using the following equation:

$$\frac{\Delta P}{L} = \frac{1}{2} f_i \rho_v V_v^2 \cdot \frac{1}{D_i} = \frac{1}{2} \cdot f_i \frac{G^2 x^2}{\rho_v \alpha^{2.5}} \frac{1}{D} \tag{11-59}$$

Two-phase pressure drops in bends are calculated using the homogeneous flow model:

$$\Delta P_{bend,in} = k_B \frac{G^2}{2\rho_l} \Psi_{H,x\_out} \tag{11-60}$$

The mass flux required for calculating the pressure drop at an abrupt turn in the channel is determined based on the minimum flow area in the channel when the gas goes into the channel

$$G_{chan} = \frac{\dot{m}}{\pi \cdot D_{TS,tube\_ID} \cdot d_{TS}} \tag{11-61}$$

where d_{TS} is the depth of the channels. The deceleration pressure gain is calculated using:

$$\Delta P_{deceleration} = \left[\frac{G^2 x^2}{\rho_v \alpha} + \frac{G^2(1-x^2)}{\rho_l(1-\alpha)} \right]_{x=x_{in}} - \left[\frac{G^2 x^2}{\rho_v \alpha} + \frac{G^2(1-x^2)}{\rho_l(1-\alpha)} \right]_{x=x_{out}} \tag{11-62}$$

where

$$\alpha|_x = \left[1 + \left(\frac{1-x}{x} \right)^{0.74} \left(\frac{\rho_v}{\rho_l} \right)^{0.65} \left(\frac{\mu_l}{\mu_v} \right)^{0.13} \right]^{-1} \tag{11-63}$$

The deceleration pressure gain and contraction and expansion losses are proportional to the square of the mass flux, and therefore, the increase or decrease at the same rate with a change in mass flux. The contraction and expansion pressure drop decreases with an increase in channel width because the area contraction or expansion ratio increases. The deceleration pressure drop is proportional to the change in quality across the test section.

11.8 Velocity in Microchannels

There are two distinct regions of flow in a microchannel: the entrance and regular flow region. When the fluid or gas enters the channel, the flow (velocity) profile changes from flat to a more rounded and eventually to the characteristic parabolic shape. Once this occurs, it is in the fully developed region of flow, as shown in Figure 11-8.

The parabolic profile is typical of laminar flow in channels, and is caused by the existence of the boundary layer. When the fluid first enters the channels, the velocity profile will not yet be parabolic. Instead, this profile will develop over a distance called the entrance length. The length of the entrance region for a circular duct is given by:

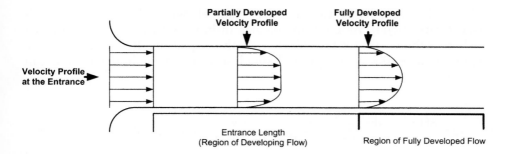

FIGURE 11-8. Developing velocity profiles from the entrance region to the fully developed region in a microchannel.

$$\frac{L_{FD}}{D} = \frac{0.6}{1+0.035*Re} = 0.056\,Re \qquad (11\text{-}64)$$

If the entrance is well rounded, the velocity profile is nearly uniform. Boundary layers form at the entrance as the fluid enters. The fluid acts according to the Continuity law, which says the fluid will slow down at the walls of the channel, while the fluid in the center of the wall will accelerate. There is an excess pressure drop across the entrance length due to the increased shear forces in the entrance boundary layers and the acceleration of the core.

In a circular pipe, the velocity distribution across the diameter of the pipe is given by:

$$v(r) = \frac{\Delta P}{4v}(R^2 - r^2) \qquad (11\text{-}65)$$

Microchannels with a rectangular shape are widely used in microfluidics. In the x direction, the two-dimensional velocity distribution, u(y, z), satisfies Poisson's equation:

$$\nabla^2 u(y, z) = -\frac{1}{\mu}\frac{dp}{dx} \qquad (11\text{-}66)$$

where μ is the dynamic viscosity, and dp/dx is the pressure gradient.

Since the velocity at the wall is zero, a Fourier series solution of the velocity field u(y, z) in a rectangular channel size of $-w/2 \le w/2$ and $-H/2 \le H/2$ can be written as:

$$u(y, z) = \frac{4w^2}{\mu\pi^3}\left(-\frac{dp}{dx}\right)\sum_{n=1}^{\infty}(-1)^{n-1}1 -$$

$$\frac{\cosh((2n-1)\pi z/w)}{\cosh((2n-1)\pi H/2w)}\frac{\cosh((2n-1)\pi y/w)}{(2n-1)^3} \qquad (11\text{-}67)$$

where the velocity varies in the channel with the width-to-depth aspect ratio (w/H). As cosh approaches infinity with n, it is a nontrivial task to achieve the exact solution of this equation. By integrating the equation over the area of the channel section, the volumetric rate can be written as:

$$Q = \frac{8Hw^3}{\mu\pi^4}\left(-\frac{dp}{dx}\right)\sum_{n=1}^{\infty}\left[\frac{1}{(2n-1)^4} - \frac{2w}{(2n-1)^5\pi H}\tanh((2n-1)\pi H/2w)\right] \quad (11\text{-}68)$$

where Q is the volumetric flow rate, and dp/dx is the pressure gradient along x. The pressure gradient can be related to the mean velocity, u, by:

$$\frac{dp}{dx} = -\frac{4k\mu u}{H^2} \quad (11\text{-}69)$$

where k is a constant related to the aspect ratio of a rectangular channel.

EXAMPLE 11-1: Plotting the Three-Dimensional Velocity Field

Using Equations 11-68 and 11-69 for a volumetric flow rate of 2e-6 m^3/s, create the three-dimensional velocity field for a fully developed flow in a rectangular channel. The width-to-depth aspect ratio is 4 (w/H = 4), and the viscosity is 1.002e-9.

Using MATLAB to solve:

```
%%%%%%%%%%%%%%%%%%%%%%%%%%%%%%%%%%%%%%%%%%%%%
%EXAMPLE 11-1: Plotting the 3-D velocity field
% UnitSystem SI
%%%%%%%%%%%%%%%%%%%%%%%%%%%%%%%%%%%%%%%%%%%%%
% Inputs

Q = 2e6;          % Flow rate (m^3/s)
muu = 1.002e-9;   % Viscosity
ymin = -200;
ymax = 200;
```

% Width of the channel

```
Zmin = -50;
zmax=50;
```

% Thickness of the channel

```
Npts = 40;
dy=(ymax-ymin)/npts;
dz=(zmax-zmin)/npts;
a=(ymax-ymin)/2;
b=(zmax-zmin)/2;
L = 5000;
Pi = 3.14159265359;
P1 = 0.0;
aspectratio = 4;    % Width to depth aspect ratio
ku = 14.2;
```

% Input of flow rate to calculate pressure gradient P

```
P = ku*muu*Q/(ymax-ymin)/(zmax-zmin)/(zmax-zmin)^2;

n = 1000;
ny = ((ymax-ymin)/dy);
nz = ((zmax-zmin)/dz);
for i = 1:ny+1
   for j = 1:nz+1
      u(i,j) = 0.0
   end
end
for i = 1:nz+1
   za(i) = zmin+(i-1)*dz;
end
for i = 1:ny+1
   ya(i) = ymin+(i-1)*dy;
end
for i = 1:ny+1
   for j = 1:nz+1
      for k = 1:800
u(i,j)  =   u(i,j)+(16*P*a^2)./(muu*pi^3).*((-1)^.(k-1).*(1-cosh((2.*k-1).*pi.*za(j)
   ./2./a)./...
cosh((2*k-1)*pi*b/2/a)).*cos((2*k-1)*pi*ya(i)/(2*a))./(2*k-1)^3);
      end
   end
end
```

```
% Plot the 3-D velocity flow field
figure1 = figure('Color',[1 1 1]);
hdlp = surf(ya,za,u);
colormap hsv
colorbar
xlabel('y = y/w','FontSize',12,'FontWeight','Bold');
ylabel('z = z/H','FontSize',12,'FontWeight','Bold');
zlabel('Normalized Velocity (u)','FontSize',12,'FontWeight','Bold');
set(hdlp,'LineWidth',1.5);
grid on;
I
```

Figure 11-9 shows the three-dimensional flow field in a rectangular channel with width to depth aspect ratio (w/H = 4) for example 11-1.

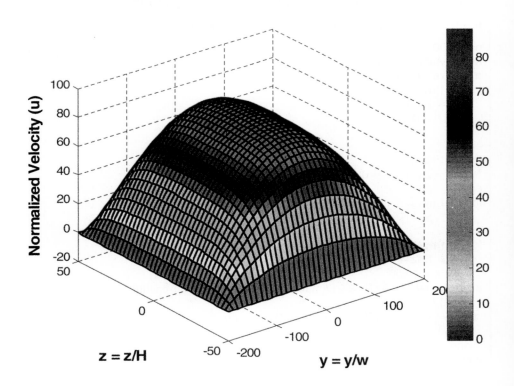

FIGURE 11-9. Three-dimensional velocity field in microchannels.

TABLE 11-3
Parameters for Example 11-2

	Inlet	Outlet
Quality	0.80	70
Vapor density (kg/m³)	100	95
Vapor viscosity (kg/ms)	1.4e-5	1.4e-5
Liquid density (kg/m³)	1000	1075
Liquid viscosity (kg/ms)	1.2e-4	1.24e-4

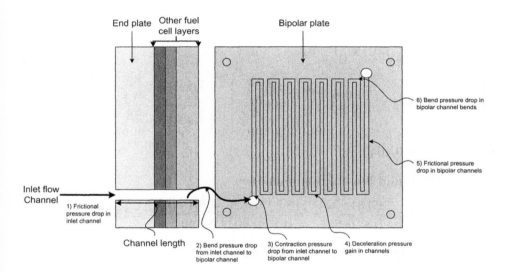

FIGURE 11-10. Illustration for Example 11-2.

EXAMPLE 11-2: Modeling the Two-Phase Pressure Drop in Microchannels

Create a two-phase pressure drop model for the microchannels using the equations introduced in this chapter. Figure 11-10 shows an illustration of the channels. The inlet channel diameter is 1.55 mm, with a mass flow rate of 2.18e-4 kg/s. The outlet value for G is 550. For the channels in the bipolar flow plate, the horizontal length is 0.0075 m, and the vertical length is 0.0015 m. There are 9 channels, and 8 "u" bends. Table 11-3 shows the other parameters required for calculating the pressure drop for this problem.

First calculate the area and G:

$$A_{inlet} = \pi \left(\frac{D_{inlet\_ID}}{2}\right)^2 = \pi \left(\frac{0.00155m}{2}\right)^2 = 1.8869e - 6m^2$$

$$G_{chan} = \frac{\dot{m}}{A_{inlet}} = \frac{2.18*10^{-4}\frac{kg}{s}}{1.8869e - 6} = 115.53\frac{kg}{m^2s}$$

The frictional pressure drop in the channel from the inlet to the bipolar channel entrance needs to be calculated next.

The void fraction is:

$$\alpha = \left[1 + \left(\frac{1-x}{x}\right)^{0.74}\left(\frac{\rho_v}{\rho_l}\right)^{0.65}\left(\frac{\mu_l}{\mu_v}\right)^{0.13}\right]^{-1}$$

$$= \left[1 + \left(\frac{1-0.80}{0.80}\right)^{0.74}\left(\frac{100\frac{kg}{m^3}}{1000\frac{kg}{m^3}}\right)^{0.65}\left(\frac{1.2\times10^{-4}\frac{kg}{m\cdot s}}{1.4\times10^{-5}\frac{kg}{m\cdot s}}\right)^{0.13}\right]^{-1}$$

$$= 0.9041$$

The liquid Reynold's number is:

$$Re_l = \frac{GD(1-x)}{(1+\sqrt{a})\mu_l} = \frac{115.53*1.89e-6*(1-0.80)}{(1+\sqrt{0.9041})*1.2e-4} = 152.99$$

The vapor Reynold's number is:

$$Re_v = \frac{GDx}{\mu_v\sqrt{a}} = \frac{115.53*1.89e-6*0.80}{1.4e-5\sqrt{0.9041}} = 1.08e-4$$

The friction factor for the laminar film is:

$$f_l = \frac{64}{Re_l} = \frac{64}{152.99} = 0.4183$$

The friction factor for vapor is:

$$f_v = 0.316 * Re_v^{-0.25} = 0.316*(1.08e-4)^{-0.25} = 0.0310$$

The pressure drops for the liquid and vapor phase are:

$$\left(\frac{dP}{dz}\right)_l = \frac{f_l G^2(1-x)^2}{2D\rho_l} = 72.05$$

$$\left(\frac{dP}{dz}\right)_v = \frac{f_v G^2 x^2}{2D\rho_v} = 854.94$$

The Martinelli parameter is calculated by:

$$X = \left(\frac{\left(\frac{dP}{dz}\right)_l}{\left(\frac{dP}{dz}\right)_v} \right)^{0.5} = 0.2903$$

The liquid superficial velocity is:

$$j_l = \frac{G(1-x)}{\rho_l(1-\alpha)} = 0.2409$$

The surface tension parameter is:

$$\psi = \frac{j_l \mu_l}{\sigma} = 0.0080$$

$$\frac{f_i}{f_l} = A \cdot X^a Re_l^b \psi^c = 0.06211$$

where the laminar region is <2300 and $A = 1.308 \times 10^{-3}$, $a = 0.4273$, $b = 0.9295$, and $c = -0.1211$.

$$\frac{\Delta P}{L} = \frac{1}{2} f_i \rho_v V_v^2 \cdot \frac{1}{D_i} = 22.015 Pa$$

The bend pressure drop from the flow channel to the bipolar channel entrance is:

$$\Delta P_{bend,in} = k_B \frac{G^2}{2\rho_l} \Psi_{H,x\_out} = 472.0 Pa$$

The contraction pressure drop from the flow channel to the bipolar channel entrance can be calculated by the following:

$$A_{bipolar} = \pi \left(\frac{D_{inlet\_ID}}{2} \right)^2 = 7.85e-9$$

$$\gamma_{con} = \frac{A_{inlet}}{A_{bipolar}} = 240.25$$

$$C_c = \frac{1}{0.639*\left(1 - \frac{1}{\gamma_{con}}\right)^{0.5} + 1} = 0.6106$$

Homogeneous flow model:

$$\Delta P_{con} = \frac{G_{chan}^2}{2\rho_l}\left[\left(\frac{1}{C_c}-1\right)^2 + 1 - \frac{1}{\gamma_{con}^2}\right]\Psi_H = 1.1065e3$$

The deceleration pressure gain is calculated using:

$$\Delta P_{\text{deceleration}} = \left[\frac{G^2 x^2}{\rho_v \alpha} + \frac{G^2 (1-x^2)}{\rho_l (1-\alpha)}\right]_{x=x_{in}} -$$

$$\left[\frac{G^2 x^2}{\rho_v \alpha} + \frac{G^2 (1-x^2)}{\rho_l (1-\alpha)}\right]_{x=x_{out}} = 280.51 \text{Pa}$$

The frictional pressure drop in the bipolar channels is calculated in the same manner as the frictional pressure drop from the inlet channel to the bipolar channel entrance. The pressure drops in the flow channel bends are also calculated using the same equation as for the bend from the inlet channel to the flow field channels. The last step involves summing all of the pressure drops to obtain the net pressure drop

$$\Delta P = \Delta P_{\text{chan,in}} + \Delta P_{\text{bend,in}} + \Delta P_{\text{con}} + \Delta P_{\text{deceleration}} + \Delta P_{\text{bipolar}} + \Delta P_{\text{bend,bipolar}}$$

Using MATLAB to solve:

%%

% EXAMPLE 11-2: Modeling the Two-Phase Pressure Drop in Microchannels

% UnitSystem SI

%%

% Pressure drop analysis for microchannels

```
D_chan = 0.00155;      % Channel diameter (m) (1.55 mm)
D_bipolar = 0.0001;    % Bipolar channel diameter (100 um)
m = 2.18e-4;           % Mass flow rate (kg/s)
x = 0.80;              % Quality
pv = 100;              % Vapor density (kg/m^3)
pl = 1000;             % Liquid density (kg/m^3)
mul = 1.2e-4;          % Liquid viscosity (kg/ms)
muv = 1.4e-5;          % Vapor viscosity (kg/ms)
sigma = 3.6e-3;        % (N/m)
A = 1.308e-3;
a = 0.4273;
b = 0.9295;
c = -0.1211;
L_chan = 0.01;         % Channel Length (10 mm)
kb = 0.6;
psi = 7.85;
```

% Calculate the area of the inlet channels

```
A_chan = pi*(D_chan/2)^2;
G_chan = m / A_chan;
```

% Frictional pressure drop in channel from inlet to the bipolar channel

% entrance

```
%%%%%%%%%%%%%%%%%%%%%%%%%%%%%%%%%%%%%%%
```

% Void fraction

```
alpha = (1 + (((1 − x) / x)^ 0.74)*((pv/pl)^0.65)*(mul/muv)^0.13)^(-1);
```

% Liquid Reynold's number

```
Re_l = (G_chan*D_chan*(1 − x))/ ((1+ sqrt(alpha))* mul);
```

% Vapor Reynold's number

```
Re_v = (G_chan*D_chan*x)/ (sqrt(alpha)*muv);
```

% Friction factor for laminar film

```
f_l = 64/Re_l;
```

% Vapor friction factor

```
f_v = 0.316*Re_v^(-0.25);
```

% Annular flow model

```
dPdz_l = (f_l*G_chan^2*(1 − x)^2)/(2*D_chan*pl); %Pa/m
dPdz_v = (f_v*G_chan^2*x^2)/(2*D_chan*pv); %Pa/m
```

% Martinelli parameter

```
Xm = (dPdz_l/dPdz_v)^0.5;
j_l = (G_chan*(1-x))/(pl*(1-alpha)); %m/s
phi = (j_l*mul)/sigma;
f_i = A*(Xm^a)*(Re_l^b)*(phi^c)*f_l; % For laminar region
deltaP_chan   =   0.5*f_i*G_chan^2/pv*(x^2)/(alpha^2.5)*(1/D_chan)*L_chan;
     % Pa
```

% Calculate the area of the bipolar channels

```
A_bipolar = pi*(D_bipolar/2)^2;
G_bipolar = m / (pi*D_chan*D_bipolar);
```

% Bend pressure drop from flow channel to bipolar channel entrance

%%%

% Homogenous flow model

deltaP_bend_in = kb * ((G_bipolar^2)/(2 * pl)) * psi % Pa

% Contraction pressure drop from the flow channel to bipolar channel entrance

%%%

gamma_con = A_chan/A_bipolar %
Cc = 1/(0.639 * ((1-(1/gamma_con))^0.5)+1)

% Homogenous flow model

deltaP_con_in =((G_bipolar^2)/(2 * pl)) * ((((1/Cc)-1)^2) + 1 -(1 / gamma_con^2)) * psi;
    % Pa

% Deceleration pressure gain in channels

%%%

```
x_in = 0.80;           % Quality in
pl_in = 1000;          % Liquid density in
mul_in=1.2e-4;         % Liquid viscosity in
pv_in = 100;           % Vapor density in
muv_in = 1.4e-5;       % Vapor viscosity in
pl_out = 1075;         % Liquid density out
mul_out = 1.24e-4;     % Liquid viscosity out
pv_out = 95;           % Vapor density out
muv_out = 1.4e-5;      % Vapor viscosity out
x_out = 0.70;          % Quality out
G = 550;
alphax_in    =    (1+    (((1-x_in)/x_in)^0.74) * ((pv_in/pl_in)^0.65) * ((mul_in/muv_
    in)^0.13))^(-1);
alphax_out= (1+ (((1-x_out)/x_out)^0.74) * ((pv_out/pl_out)^0.65) * ((mul_out/muv_
    out)^0.13))^(-1);
deltaP_decel = (((G^2) * (x_in^2)/(pv_in * alphax_in))+ ((G^2) * (1-x_in)^2)/(pl_in * (1-
    alphax_in)))- ... (((G^2) * (x_out^2)/(pv_out * alphax_out))+  ((G^2) * (1-x_out)^2)/
    (pl_out * (1-alphax_out)));
```

% Frictional pressure drop in bipolar channels

%%%

L_bipolar_hor = 0.0075*9; % horizontal length x number of channels
L_bipolar_vert = 0.0015*8; % % vertical length x number of u bends

% Void fraction

alpha_bipolar = (1 + (((1 − x) / x)^ 0.74) * ((pv/pl)^0.65) * (mul/muv)^0.13)^(-1);

% Liquid Reynold's number

```
Re_l = (G_bipolar * D_bipolar * (1 − x))/ ((1+ sqrt(alpha_bipolar)) * mul);
```

% Vapor Reynold's number

```
Re_v = (G_bipolar * D_bipolar * x)/ (sqrt(alpha_bipolar) * muv);
```

% Friction factor for laminar film

```
f_l = 64/Re_l;
```

% Vapor friction factor

```
f_v = 0.316 * Re_v^(-0.25);
```

% Annular flow model

```
dPdz_l = (f_l * G_bipolar^2 * (1 − x)^2)/(2 * D_bipolar * pl); %Pa/m
dPdz_v = (f_v * G_bipolar^2 * x^2)/(2 * D_bipolar * pv); %Pa/m
```

% Martinelli parameter

```
Xm = (dPdz_l/dPdz_v)^0.5;

j_l = (G_bipolar * (1-x))/(pl * (1-alpha_bipolar)); %m/s
phi = (j_l * mul)/sigma;

f_i = A * (Xm^a) * (Re_l^b) * (phi^c) * f_l; % For laminar region

deltaP_bipolar_hor     =     0.5 * f_i * G_bipolar^2/pv * (x^2)/(alpha_bipolar^2.5) *
    (1/D_bipolar) * L_bipolar_hor; % Pa
deltaP_bipolar_vert     =     0.5 * f_i * G_bipolar^2/pv * (x^2)/(alpha_bipolar^2.5) *
    (1/D_bipolar) * L_bipolar_vert; % Pa
```

% Bend pressure drop in bipolar channel bends

```
%%%%%%%%%%%%%%%%%%%%%%%%%%%%%%%%%%%%%%%
bends = 16; % "L" bends
```

% Homogenous flow model

```
deltaP_bend_bipolar = kb * ((G_bipolar^2)/(2 * pl)) * psi * bends; % Pa
```

% Net frictional pressure drop in channels

```
%%%%%%%%%%%%%%%%%%%%%%%%%%%%%%%%%%%%%%%

deltaP = deltaP_chan + deltaP_bend_in + deltaP_con_in + deltaP_decel + . . . deltaP_
    bipolar_hor +deltaP_bipolar_vert + deltaP_bend_bipolar; % Pa
```

% Convert to bar

```
deltaP = deltaP * 1e-5
```

Chapter Summary

When designing and modeling MEMS fuel cells, there are many properties that need to be considered that could be neglected in the macroscale fuel cell system models. Important micro fuel cell properties are surface effects, dead volumes, bubbles, and the consideration of both gas and liquid phases. In addition, the properties can differ greatly between 1 mm and 1 micron, therefore, the system must take into account the necessary parameters. Although micro fuel cells have been researched for several years, it appears that this science is still in its infancy based upon the current micro fuel cell designs in the literature. In order to progress in the area of micro fuel cells, mathematical modeling needs to be an integral part of the design process since most of the system variables cannot be measured. The lack of measurements is due to the small scale of the system, and the inability to measure internal variables while the system is operating.

Problems

- Design a micro fuel cell stack that has to operate at 50 °C with air and hydrogen pressures of 1 atm. The Pt/C loading is 1 mg/cm^2 and the cells use the Nafion 117 electrolyte. The total power should be 2 W.
- Create a new MEMS fuel cell design based upon the concepts introduced in this chapter. Describe the theory behind the design.
- Calculate the two-phase pressure drop for a bipolar plate with 25 channels and bends, with a length of 1 cm, width of 50 μm, and depth of 50 μm. The inlet flow rate is 0.5 mL/min.
- For Example 11-2, the bipolar plate is heated to 75 °C. What would be the new outlet parameters, and the associated pressure drop?

Endnotes

[1] Kandlikar, S.G., S. Garimella, D. Li, S. Colin, and M.R. King. 2005. *Heat Transfer and Fluid Flow in Minichannels and Microchannels*. Amsterdam: Elsevier Science.
[2] Ibid.
[3] Silva, V.S., J. Schirmer, R. Reissner, B. Ruffmann, H. Silva, A. Mendes, L.M. Madeira, and S.P. Nunes. Proton electrolyte membrane properties and direct methanol fuel cell performance. *J. Power Sources*. Vol. 140, 2005, pp. 41–49.
[4] Verma, A., and S. Basu. Direct use of alcohols and sodium borohydride as fuel in an alkaline fuel cell. *J. Power Sources*. Vol. 145, 2005, pp. 282–285.
[5] Liu, J.G., T.S. Zhao, Z.X. Liang, and R. Chen. Effect of membrane thickness on the performance and efficiency of passive direct methanol fuel cells. *J. Power Sources*. Vol. 153, 2006, pp. 61–67.
[6] U.S. Patent No. 7,029,781 B2. Microfuel Cell Having Anodic and Cathodic Microfluidic Channels and Related Methods. Lo Priore, S., M. Palmieri, and U. Mastromatteo. STMicroelectronics, Inc. Carollton, TX. April 18, 2006.

[7] Wee, J.-H. Performance of a unit cell equipped with a modified catalytic reformer in direct internal reforming molten carbonate fuel cell. *J. Power Sources.* Vol. 156, 2006, pp. 288–293.

[8] Bove, R., and P. Lumghi. Experimental comparison of MCFC performance using three different biogas types and methane. *J. Power Sources.* Vol. 145, 2005, pp. 588–593.

[9] Liu, Effect of membrane thickness on the performance and efficiency.

[10] Yan, Q., H. Toghiani, and J. Wu. Investigation of water transport through membrane in a PEM fuel cell by water balance experiments. *J. Power Sources.* Vol. 158, 2006, pp. 316–325.

[11] Fabian, T., J.D. Posner, R. O'Hayre, S.-W. Cha, J.K. Eaton, F.B. Prinz, and J.G. Santiago. The role of ambient conditions on the performance of a planar, air-breathing hydrogen PEM fuel cell. *J. Power Sources.* Vol. 161, 2006, pp. 168–182.

[12] Kim, J.-Y., O.J. Kwon, S.-M. Hwang, M.S. Kang, and J.J. Kim. Development of a miniaturized polymer electrolyte membrane fuel cell with silicon separators. *J. Power Sources.* Vol. 161, 2006, pp. 432–436.

[13] Sousa, R., Jr., and E. Gonzalez. Mathematical modeling of polymer electrolyte fuel cells. *J. Power Sources.* Vol. 147, 2005, pp. 32–45.

[14] Ibid.

[15] Silva, Proton electrolyte membrane properties.

[16] Yamada, M., and I. Honma. Biomembranes for fuel cell electrodes employing anhydrous proton conducting uracil composites. *Biosensors Bioelectronics.* Vol. 21, 2006, pp. 2064–2069.

[17] Rowshanzamir, S., and M. Kazemeini. A new immobilized-alkali H2/O2 fuel cell. *J. Power Sources.* Vol. 88, 2000, pp. 262–268.

[18] Verma, Direct use of alcohols and sodium borohydride.

[19] Lim, C., and C.Y. Wang. Development of high power electrodes for a liquid feed direct methanol fuel cell. *J. Power Sources.* Vol. 113, 2003, pp. 145–150.

[20] Liu, Effect of membrane thickness on the performance and efficiency.

[21] Lo Priore, U.S. Patent No. 7,029,781 B2. Microfuel Cell.

[22] Kandlikar, *Heat Transfer and Fluid Flow in Minichannels and Microchannels.*

[23] Aravamudhan, S., A. Rahman, and S. Bhansali. Porous silicon-based orientation independent, self-priming micro-direct ethanol fuel cell. *Sensors Actuators A.* Vol. 123–124, 2005, pp. 497–504.

[24] Spiegel, C.S. Designing and Building Fuel Cells. 2007. New York: McGraw-Hill.

[25] Springer et al. Polymer electrolyte fuel cell model. *J. Electrochem. Soc.* Vol. 138, No. 8, 1991, pp. 2334–2342.

[26] Spiegel, *Designing and Building Fuel Cells.*

[27] Aravamudhan, *Porous silicon-based orientation independent, self-priming micro-direct ethanol fuel cell.*

[28] Ibid.

[29] Spiegel, *Designing and Building Fuel Cells.*

[30] Nguyen, N.-T., and S.H. Chan. Micromachined polymer electrolyte membrane and direct methanol fuel cells—A review. *J. Micromech. Microeng.* Vol. 16, 2006, pp. R1–R12.

[31] U.S. Patent Application Publication No. US 2006/0003217 A1. Planar MEMS-braneless Microchannel Fuel Cell. Cohen, J.L., D.J. Volpe, D.A. Westly, A. Pechenik, and H.D. Abruna. January 5, 2006.

[32] Choban, E.R., L.J. Markoski, A. Wieckowski, and P.J.A. Kenis. Microfluidic fuel cell based on laminar flow. *J. Power Sources.* Vol. 128, 2004, pp. 54–60.
[33] Li, X. 2006. *Principles of Fuel Cells.* New York: Taylor & Francis Group.
[34] Spiegel, *Designing and Building Fuel Cells.*
[35] Springer. Polymer electrolyte fuel cell model.
[36] Spiegel, *Designing and Building Fuel Cells.*
[37] Kandlikar, *Heat Transfer and Fluid Flow in Minichannels and Microchannels.*
[38] Ibid.

Bibliography

Barbir, F. *PEM Fuel Cells: Theory and Practice.* 2005. Burlington, MA: Elsevier Academic Press.

Blum, A., T. Duvdevani, M. Philosoph, N. Rudoy, and E. Peled. Water neutral micro direct methanol fuel cell (DMFC) for portable applications. *J. Power Sources.* Vol. 117, 2003, pp. 22–25.

Bosco, A. *General Motors Fuel Cell Research.* Hy-Wire. Available at: http://www.ansoft.com/workshops/altpoweree/Andy_Bosco_GM.pdf. Accessed November 4, 2006.

Cha, S.W., R. O'Hayre, Y. Saito, and F.B. Prinz. The influence of size scale on the performance of fuel cells. *Solid State Ionics.* Vol. 175, 2004. pp. 789–795.

Chang, P.A.C., J. St-Pierre, J. Stumper, and B. Wetton. Flow distribution in proton exchange membrane fuel cell stacks. *J. Power Sources.* Vol. 162, 2006, pp. 340–355.

Chen, R., and T.S. Zhao. Mathematical modeling of a passive feed DMFC with heat transfer effect. *J. Power Sources.* Vol. 152, 2005, pp. 122–130.

Chung, B.W., C.N. Chervin, J.J. Haslam, A.-Q. Pham, and R.S. Glass. Development and characterization of a high performance thin-film planar SOFC stack. *J. Electrochem. Soc.* Vol. 152, No. 2, 2005, pp. A265–A269.

Coutanceau, C., L. Demarconnay, C. Lamy, and J.M. Leger. Development of electrocatalysts for solid alkaline fuel cell (SAFC). *J. Power Sources.* Vol. 156, 2006, pp. 14–19.

Dyer, C.K. Fuel cells for portable applications. *J. Power Sources.* Vol. 106, 2002, pp. 31–34.

Gulzow, E., M. Schulze, and U. Gerke. Bipolar concept for alkaline fuel cells. *J. Power Sources.* Vol. 156, 2006, pp. 1–7.

Heinzel, A., C. Hebling, M. Muller, M. Zedda, and C. Muller. Fuel cells for low power applications. *J. Power Sources.* Vol. 105, 2002, pp. 250–255.

Koh, J.-H., H.-K. Seo, C.G. Lee, Y.-S. Yoo, and H.C. Lim. Pressure and flow distribution in internal gas manifolds of a fuel cell stack. *J. Power Sources.* Vol. 115, 2003, pp. 54–65.

Lee, S.-J., C.-D. Hsu, and C.-H. Huang. Analyses of the fuel cell stack assembly pressure. *J. Power Sources.* Vol. 145, 2005, pp. 353–361.

Lin, B. 1999. Conceptual design and modeling of a fuel cell scooter for urban Asia. Princeton University, master's thesis.

Liu, S., Q. Pu, L. Gao, C. Korzeniewski, and C. Matzke. From nanochannel induced proton conduction enhancement to a nanochannel based fuel cell. *Nano Lett.* Vol. 5, No. 7, 2005, pp. 1389–1393.

Lu, G.Q., and C.Y. Wang. Development of micro direct methanol fuel cells for high power applications. *J. Power Sources*. Vol. 144, 2005, pp. 141–145.

Mench, M.M., C.-Y. Wang, and S.T. Tynell. *An Introduction to Fuel Cells and Related Transport Phenomena*. Department of Mechanical and Nuclear Engineering, Pennsylvania State University. Draft. Available at: http://mtrl1.mne.psu.edu/Document/jtpoverview.pdf. Accessed March 4, 2007.

Mench, M.M., Z.H. Wang, K. Bhatia, and C.Y. Wang. 2001. *Design of a Micro-direct Methanol Fuel Cell*. Electrochemical Engine Center, Department of Mechanical and Nuclear Engineering, Pennsylvania State University.

Mitrovski, S.M., L.C.C. Elliott, and R.G. Nuzzo. Microfluidic devices for energy conversion: Planar integration and performance of a passive, fully immersed H2-O2 fuel cell. *Langmuir*. Vol. 20, 2004, pp. 6974–6976.

Modroukas, D., V. Modi, and L.G. Frechette. Micromachined silicon structures for free-convection PEM fuel cells. *J. Micromech. Microeng.* Vol. 15, 2005, pp. S193–S201.

Morita, H., M. Komoda, Y. Mugikura, Y. Izaki, T. Watanabe, Y. Masuda, and T. Matsuyama. Performance analysis of molten carbonate fuel cell using a Li/Na electrolyte. *J. Power Sources*. Vol. 112, 2002, pp. 509–518.

Nakagawa, N., and Y. Xiu. Performance of a direct methanol fuel cell operated. *J. Power Sources*. Vol. 118, 2003, pp. 248–255.

Rowe, A., and X. Li. Mathematical modeling of proton exchange MEMSbrane fuel cells. *J. Power Sources*. Vol. 102, 2001, pp. 82–96.

O'Hayre, R., S.-W. Cha, W. Colella, and F.B. Prinz. 2006. *Fuel Cell Fundamentals*. New York: John Wiley & Sons.

Soler, J., T. Gonzalez, M.J. Escudero, T. Rodrigo, and L. Daza. Endurance test on a single cell of a novel cathode material for MCFC. *J. Power Sources*. Vol. 106, 2002, pp. 189–195.

Song, R.-H., and D.R. Shin. Influence of CO concentration and reactant gas pressure on cell performance in PAFC. *Int. J. Hydrogen Energy*. Vol. 26, 2001, pp. 1259–1262.

Subhash, S.C. July 2003. *High Temperature Solid Oxide Fuel Cells Fundamentals, Design and Applications*. Rio de Janeiro, Brazil: Pan American Advanced Studies Institute, Pacific Northwest National Laboratory, U.S. Department of Energy.

U.S. Patent No. 4,490,444. High Temperature Solid Oxide Fuel Cell Configurations and Interconnections. A.O. Isenberg, Forest Hills, PA. Westinghouse Electric Corporation. December 25, 1984.

U.S. Patent No. 7,125,625 B2. Electrochemical Cell and Bipolar Assembly for an Electrochemical Cell. Cisar, A.J., C.C. Andrews, C.J. Greenwald, O.J. Murphy, C. Boyer, R.C. Yalamanchili, and C.E. Salinas. Lynntech, Inc. College Station, TX. October 24, 2006.

Wells, B., and H. Voss. May 17, 2006. *DMFC Power Supply for All-Day True Wireless Mobile Computing*. Polyfuel, Inc. Available at: http://www.hydrogen.energy.gov/pdfs/review06/fcp_39_wells.pdf. Accessed November 4, 2006.

Wong, C.W., T.S. Zhao, Q. Ye, and J.G. Liu. Experimental investigations of the anode flow field of a micro direct methanol fuel cell. *J. Power Sources*. Vol. 155, 2006, pp. 291–296.

Wozniak, K., D. Johansson, M. Bring, A. Sanz-Velasco, and P. Enoksson. A micro direct methanol fuel cell demonstrator. *J. Micromech. Microeng.* Vol. 14, 2004, pp. S59–S63.

Yoon, K.H., J.H. Jang, and Y.S. Cho. Impedance characteristics of a phosphoric acid fuel cell. *J. Mater. Sci. Lett.* Vol. 17, 1998, pp. 1755–1758.

Yoon, K.H., J.Y. Choi, J.H. Jang, Y.S. Cho, and K.H. Jo. Electrode/matrix interfacial characteristics in a phosphoric acid fuel cell. *J. Appl. Electrochem.* Vol. 30, 2000, pp. 121–124.

You, L., and H. Liu. A two-phase flow and transport model for PEM fuel cells. *J. Power Sources.* Vol. 155, 2006, pp. 219–230.

Zhu, B. Proton and oxygen ion-mixed conducting ceramic composites and fuel cells. *Solid State Ionics.* Vol. 145, 2001, pp. 371–380.

CHAPTER 12

Modeling Fuel Cell Stacks

12.1 Introduction

As seen in the previous chapters, many parameters must be considered when designing and modeling fuel cells. The calculations involved with designing fuel cell stacks are very basic, but sometimes are unknown to newcomers in the field. The most commonly used stack configuration is the bipolar configuration, which is very similar to how batteries are designed. There are also many alternative stack configurations, however, the materials, designs, and methods of fabricating the components are similar. When considering the design of a fuel cell stack, usually several limitations should be considered. Some of these limitations may include the following:

- Size, weight, and volume at the desired power
- Cost
- Water management
- Fuel and oxidant distribution

Figure 12-1 illustrates a PEM fuel cell stack.

This chapter explains the basics of modeling stack design, stack clamping, and adequate fuel distribution.

12.2 Fuel Cell Stack Sizing

The sizing of a fuel cell stack is very simple; there are two independent variables that must be considered—voltage and current. The known requirements are the maximum power, voltage, and/or current. Recall that power output is a product of stack voltage and current:

$$W_{FC} = V_{st} \cdot I \qquad (12\text{-}1)$$

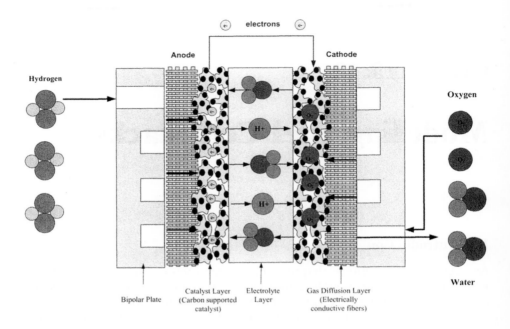

FIGURE 12-1. Schematic of a PEM fuel cell stack.

Other initial considerations that are helpful when designing a fuel cell stack are the current and power density. These are often unavailable initially, and can be calculated from the desired power output, stack voltage, efficiency, and volume and weight limitations. The current is a product of the current density and the cell active area:

$$I = i * A_{cell} \qquad (12\text{-}2)$$

As mentioned previously, the cell potential and the current density are related by the polarization curve:

$$V_{cell} = f(i) \qquad (12\text{-}3)$$

Figure 12-2 shows example polarization curves for single PEM fuel cells from the literature. Most fuel cell developers use a nominal voltage of 0.6 to 0.7 V at nominal power. Fuel cell systems can easily be designed at nominal voltages of 0.8 V per cell or higher if the correct design, materi-

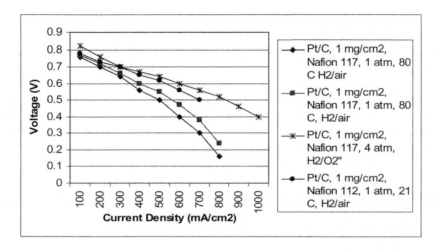

FIGURE 12-2. Polarization curves for PEM fuel cell single cells[1,2,3].

als, operating conditions, balance-of-plant, and electronics are selected. Balance-of-plant components, are discussed in Chapter 13.

12.3 Number of Cells

The number of cells in the stack is often determined by the maximum voltage requirement and the desired operating voltage. The total stack potential is a sum of the stack voltages or the product of the average cell potential and number of cells in the stack[4]:

$$V_{st} = \sum_{i=1}^{N_{cell}} V_i = \bar{V}_{cell} * N_{cell}$$ (12-4)

The cell area must be designed to obtain the required current for the stack. When this is multiplied by the total stack voltage, the maximum power requirement for the stack will be obtained. The average voltage and corresponding current density selected can have a large impact upon stack size and efficiency. The fuel cell stack efficiency can be approximated with the following equation:

$$\eta_{stack} = \frac{V_{cell}}{1.482}$$ (12-5)

EXAMPLE 12-1: Designing the Fuel Cell Stack

Design a fuel cell with a voltage, current, cell area, and number of cells to power a scooter with a power requirement of 5.9 kW.

The number of cells depends upon the required operating voltage. The electric scooter industry in Taiwan is standardizing on 48-V electric motors, so the number of cells is chosen to have the stack operate in the vicinity of 48 V at the most common power demand. Note that in a fuel cell, as the total power output changes, the voltage varies as well[5].

Using MATLAB to solve:

%%

% EXAMPLE 12-1: Designing the Fuel Cell Stack

% UnitSystem SI

%%

% Inputs

```
Power = 5900;     % Required stack power (W)
Voltage = 48;     % Stack Voltage (V)
V_cell = 0.6;     % Cell voltage
i = 0.5;          % Current Density (A/cm^2)
```

% Calculate the required stack current

```
I = Power / Voltage; % The current is the power divided by the voltage
```

% Assume that the fuel cell voltage is 0.6 V per cell

```
N_cells = Voltage/V_cell;
```

% Assume the current density is 0.5 A/cm^2, therefore, the current needed per cell is

```
i_cell = I/N_cells;
```

% The area required per cell is

```
A_cell = i_cell/i; % cm^2
```

12.4 Stack Configuration

In the traditional bipolar stack design, the fuel cell stack has many cells in series, and the cathode of one cell is connected to the anode of the next cell. The MEAs, gaskets, bipolar plates, and end plates are the layers of the fuel cell. The stack is held together by bolts, rods, or another pressure device to clamp the cells together. When contemplating a fuel cell design, the following should be considered[6]:

- Fuel and oxidant should be uniformly distributed through each cell, and across its surface area.
- The temperature must be uniform throughout the stack.
- If designing a fuel cell with a polymer electrolyte, the membrane must not dry out or become flooded with water.
- The resistive losses should be kept to a minimum.
- The stack must be properly sealed to ensure no gas leakage.
- The stack must be sturdy and able to withstand the necessary environments it will be used in.

The most common fuel cell configuration is shown in Figure 12-3 and has been shown throughout the book. Each cell (MEA) is separated by a plate with flow fields on both sides to distribute the fuel and oxidant. The fuel cell stack end plates have only a single-sided flow field. The majority of fuel cell stacks, regardless of fuel cell type, size, and fuels used, is of this configuration.

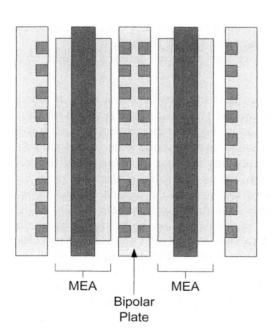

FIGURE 12-3. Typical fuel cell stack configuration (a two-cell stack).

12.5 Distribution of Fuel and Oxidants to the Cells

Fuel cell performance is dependent upon the flow rate of the reactants. Uneven flow distribution can result in uneven performance between cells. Reactant gases need to be supplied to all cells in the same stack through common manifolds. Some stacks rely on external manifolds, while others use an internal manifolding system. One advantage of external manifolding is its simplicity, which allows a low pressure drop in the manifold, and permits good flow distribution between cells. A disadvantage is that the gas flows may flow in crossflow, which can cause uneven temperature distribution over the electrodes and gas leakage. Internal manifolding distributes gases through channels in the fuel cell itself. An advantage of internal manifolding is more flexibility in the direction of flow of the gases. One of the most common methods is ducts formed by the holes in the separator plates that are aligned once the stack is assembled. Internal manifolding allows a great deal of flexibility in the stack design. The main disadvantage is that the bipolar plate design may get complex, depending on the fuel flow channel distribution design. The manifolds that feed gases to the cells and collect gases have to be properly sized. The pressure drop through the manifolds should be an order of magnitude lower than the pressure drop through each cell in order to ensure uniform flow distribution. When analyzing the flow for the cells[7]:

1. The flow into each junction should equal the flow out of it.
2. The flow in each segment has a pressure drop that is a function of the flow rate and length through it.
3. The sum of the pressure drops around a closed loop must be zero.

Some of the factors that need to be considered when designing manifold stacks include manifold structure, size, number of manifolds, overall gas flow pattern, gas channel depth, and the active area for electrode reactions. The manifold holes can vary in shape from rectangular to circular. The area of the holes is important because it determines the velocity and type of flow. The flow pattern is typically a U-shape (reverse flow), where the outlet gas flows in the opposite direction to the inlet gas, or a Z-shape (parallel flow), where the directions of the inlet and outlet gas flows are the same as shown in Figures 12-4 and 12-5. The pressure change in manifolds is much lower than that in the gas channels on the electrodes in order to ensure a uniform flow distribution among cells piled in a stack[8,9].

The pressure drop, flow rates, velocity, and mole fractions can be calculated in a similar manner to the equations introduced in Chapters 5, 10, and 11. The pressure drop can be calculated from the Bernoulli equation as follows:

$$\Delta P(i) = -\rho \frac{[u(i)]^2 - [u(i-1)]^2}{2} + f\rho \frac{L_s}{D_H} \frac{[u(i)]^2}{2} + K_f \rho \frac{[u(i-1)]^2}{2} \qquad (12\text{-}6)$$

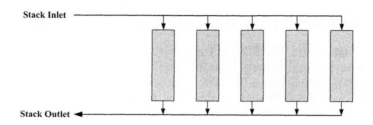

FIGURE 12-4. A U-type manifold.

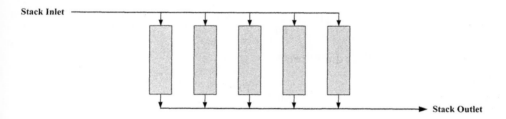

FIGURE 12-5. A Z-type manifold.

where ρ is the density of the gas (kg/m³), v is the velocity (m/s), f is the friction coefficient, L_s is the length of the segment (m), D_H is the hydraulic diameter of the manifold segment (m), and K_f is the local pressure loss coefficient.

For laminar flow (Re < 2300), the friction coefficient f for a circular conduit (as mentioned previously) is:

$$f = \frac{64}{Re} \tag{12-7}$$

The walls of the fuel cell manifolds are considered "rough" when the stack has bipolar plates that are clamped together. The friction coefficient for turbulent flow is a function of wall roughness. The friction coefficient is

$$f = \frac{1}{\left(1.14 - 2\log\dfrac{\varepsilon}{D}\right)^2} \tag{12-8}$$

where $\dfrac{\varepsilon}{D}$ is the relative roughness, which can be as high as 0.1.

EXAMPLE 12-2: Transient Pressure Drop Model

Create a transient MATLAB program that will calculate the pressure drop of six fuel cell layers. Use the fuel cell parameters introduced in Example 10-3 in Chapter 10. The code created in this example can act as a start for a program that calculates the flow rates, velocities, mass flows, and pressure drops through the layers in a fuel cell stack. Plot the flow rates after a 20 and 120 second simulation time.

The pressure drops into and out of each fuel cell layer need to be calculated. The pressure drop for a circular channel is:

$$\Delta P = f\frac{L}{D_H}\rho\frac{\bar{v}^2}{2} + \Sigma K_L \rho\frac{\bar{v}^2}{2}$$

where the hydraulic diameter is:

$$D_H = \frac{4 \times A_c}{P_{cs}}$$

and Reynold's number is:

$$Re = \frac{\rho v_m D_{ch}}{\mu}$$

The friction factor is:

$$f = \frac{64}{Re}$$

Using MATLAB to solve:

```
%%%%%%%%%%%%%%%%%%%%%%%%%%%%%%%%%%%%%%%%%%
% EXAMPLE 12-2: Transient Pressure Drop Model
% UnitSystem SI
%%%%%%%%%%%%%%%%%%%%%%%%%%%%%%%%%%%%%%%%%%
% Inputs
function [t,P] = flow
% FUELCELL Fuel Cell Stack pressure and velocity model
% Best viewed with a monospaced font with 4 char tabs.
%%%%%%%%%%%%%%%%%%%%%%%%%%%%%%%%%%%%%%%%%%
```

% Constants

const.N = 6;	% Number of layers
const.tfinal = 50;	% Simulation time (s)
const.R = 8.314472;	% Ideal gas constant (J/K-mol)
const.v_H2_in = 1.7e-8;	% Volumetric flow rate of wet hydrogen (m^3/s)
const.v_air_in = 1.e-8;	% Volumetric flow rate of air (m^3/s)
const.P_tot = 2;	% Total pressure (bar)
const.Tf_in = 353.2;	% Initial fluid temperature (K)
const.Tf_air = 273.5;	% Initial air temperature (K)
const.mu_H2 = 8.6e-6;	% Viscosity of wet hydrogen (Pa-s)
const.mu_air = 8.6e-6;	% Viscosity of air (Pa-s)
const.be = 0;	% # of bends in channel
const.rho = 0.08988;	% Density of hydrogen (kg/m^3)
%const.damp = 0.6;	% ODE solver damping factor (to avoid ringing of the solution)

% Convert volumetric flow rate to molar flow rate using ideal gas law

const.n_air_in = const.v_air_in * (const.P_tot./const.Tf_air) * (1/0.0831) * 1000;
    % mol/s

% Convert volumetric flow rate to molar flow rate using ideal gas law

const.n_H2_in = const.v_H2_in * (const.P_tot./const.Tf_in) * (1/0.0831) * 1000;
    % mol/s

% Layers

% 1 – Left end plate
% 2 – Gasket
% 3 – Contact (Copper)
% 4 – Contact
% 5 – Gasket
% 6 – End plate

% Parameters defined at the layer boundaries

% 1 2 3 4 5 6 7
% Gas temperature
param.T_f = [353.2, 353.2 353.2 353.2, 353.2, 353.2, 353.2];

% Parmeters defined at the layer centers

% 1 2 3 4 5 6
% Area (m^2)
param.A = [0.0367, 0.0367, 0.0367, 0.0367, 0.0367, 0.0367];

% Channel Area (m^2)

param.Ach = [1.5708e-006, 1.5708e-006, 1.5708e-006, 1.5708e-006, 1.5708e-006, 1.5708e-006];

% Thickness (m)

```
param.thick = [0.025, 0.025, 0.002, 0.002, 0.025, 0.025];
```

% Number of slices within layer

```
param.M = [1, 1, 1, 1, 1, 1];
```

% Channel radius

```
param.r = [0.000625, 0.000625, 0.000625, 0.000625, 0.000625, 0.000625];
```

% Channel width (rectangular channels)

```
param.wc = [0.000625, 0.000625, 0.000625, 0.000625, 0.000625, 0.000625];
```

% Channel depth (rectangular channels)

```
param.dc = [0.000625, 0.000625, 0.000625, 0.000625, 0.000625, 0.000625];
```

% Channel length

```
param.L = [0.000625, 0.000625, 0.000625, 0.000625, 0.000625, 0.000625];
```

%%%%%%%%%%%%%%%%%%%%%%%%%%%%%%%%%%%%%%

% Grid definition

```
% x – interslice coordinates
% n – molar flow rates at slice boundary
```

%%%%%%%%%%%%%%%%%%%%%%%%%%%%%%%%%%%%%%

```
% Grid up mass flows. Assume each layer abuts the
% next one. The mass flow rate is at the boundary of each slice. x is at the
% edge of each slice (like a stair plot).
x = 0;
layer = [];
for i = 1:const.N,
x = [x, x(end) + (1:param.M(i)) * param.thick(i)/param.M(i)]; %Boundary points
    layer = [layer, i * ones(1,param.M(i))];
end
```

% Slice thicknesses

```
dx = diff(x); %gives approximate derivatives between x's
```

% Initial pressure (defined at layer boundaries)

```
P = zeros(size(x)); % Pressure
u_m = zeros(size(x)); % Velocity
```

% Intial pressure equals the outside pressure

```
P(1) = const.P_tot; % mol/s
P(end) = const.P_tot; % mol/s
```

% Mean velocity in a channel with area A and molar flow rate of MOL. P is the pressure and T is the temperature

```
% Use ideal gas law to calculate initial velocity of hydrogen coming in
u_m(1) = (const.n_H2_in ./ (const.P_tot ./ const.Tf_in * (1/0.0831) * 1000))./ param.
   A(1); % m^3/s
u_m(end) = (const.n_air_in ./ (const.P_tot ./ const.Tf_air * (1/0.0831) * 1000))./
   param.A(end); % m^3/s

f = @(t,P) pressure(t,P,u_m,x,layer,param,const,dx');

options = odeset('OutputFcn',@(t,P,opt) pressureplot(t,P,opt,x));
[t,P] = ode45(f, [linspace(0,const.tfinal,100)], P, options);

end % of function

%

function dPdt = pressure(t,P,u_m,x,layer,param,const,dx)
% Pressure calculations for fuel cell
```

% Make a convenient place to set a breakpoint

```
if (t > 30)
  s = 1;
end
```

% Preallocate output

```
dPdt = zeros(size(P));

% Fluid flows from left to right for layers 1-3 and
% right to left for layers 4-6
inlet = [find(layer < 4) find(layer > 3)+1];
outlet = [find(layer < 4)+1 find(layer > 3)];

% Treat P as a row vector
P = P(:)';

% Calculate outlet pressure drop in flow channel from each layer

% Calculate velocity (m/s) % Calculation is on the flows into and out of each
% block
u_m(outlet)=6*u_m(inlet)*(x(outlet)/param.dc(layer)-(x(outlet)/param.dc(layer))^2);
   % velocity (m/s)

%Hydraulic diameter % Calculation is dependent upon the number of layers
Dh_outlet = (4*param.Ach(layer))./(2*pi*param.r(layer)); %circular flow channel
   (m)
```

```
%Reynold's number at the channel exit
Re_outlet = (const.rho .* u_m(outlet) .* Dh_outlet)./ const.mu_H2;

%Friction Factor
f_outlet = 64./Re_outlet; %for circular channels

%Pressure Drop
Kl_outlet = const.be * 30 * f_outlet(layer);

P_outlet = (f_outlet.* param.thick(layer)./ Dh_outlet).* const.rho.* ((u_m(outlet).^2)./
2)+(Kl_outlet.* const.rho.* ((u_m(outlet).^2)./ 2)); %bar

%%%%%%%%%%%%%%%%%%%%%%%%%%%%%%%%%%%%%%%%
%

% Combine into rate of change of P

%

% Do this in a loop since more than one layer outlet may contribute to
% a given element of dPdt.
for i = 1:length(outlet)
    dPdt(outlet(i)) = dPdt(outlet(i)) + P_outlet(i) – P(outlet(i));
end

% Use damping factor to help numerical convergence
%dPdt = const.damp * dPdt;

end % of function

%

function status = pressureplot(t,P,opt,x)
if isempty(opt)
    plot(t,P), title(['t = ',num2str(t)]), hold on
    status = 0;
    drawnow
end

end % of function
```

12.6 Stack Clamping

The stacking design and cell assembly parameters significantly affect the performance of fuel cells. Adequate contact pressure is needed to hold together the fuel cell stack components to prevent leaking of the reactants, and minimize the contact resistance between layers. The required clamping force is equal to the force required to compress the fuel cell layers adequately while not impeding flow. The assembly pressure affects the characteristics of the contact interfaces between components. If inadequate or nonuniform assembly pressure is used, there will be stack-sealing problems, such as fuel leakage, internal combustion, and unacceptable contact resistance. Too much pressure may impede flow through the GDL, or damage the MEA, resulting in a broken porous structure and a blockage of the gas diffusion passage. In both cases, the clamping pressure can decrease the cell performance. Every stack has a unique assembly pressure due to differences in fuel cell materials and stack design. Due to thin dimensions and the low mechanical strength of the electrodes and electrolyte layer versus the gaskets, bipolar plates, and end plates, the most important goal in the stack design and assembly is to achieve a proper and uniform pressure distribution.

12.6.1 Clamping Using Bolts

The most common method of clamping the stack is by using bolts. When considering the optimal clamping pressure on the properties of the fuel cell stack, sometimes an overlooked factor is the torque required for the bolts, and the factors that contribute to the ideal torque. The optimal torque is not merely due to the ideal clamping pressure on the fuel cell layers, but it is also affected by the shape and material of the bolt and nut, the bolt seating and threading, the stack layers, thickness, and number of layers. Materials bolted together withstand moment loads by clamping the surfaces together, where the edge of the part acts as a fulcrum, and the bolt acts as a force to resist the moment created by an external force or moment. Figure 12-6 shows forces exerted by the clamped materials (fuel cell layers) on a clamping bolt and nut.

Tightening the bolts stretches the bolts and compresses the stack materials. If an external force is applied to the stack, the optimal torque usually means that the stack stays compressed. This ensures proper stiffness of the stack. Figure 12-7 shows how the region under a bolt head acts like a spring.

12.6.2 Force Required on the Stack for Optimal Compression of the GDL

The contact resistance and GDL permeability are governed by the material properties of the contacting GDL and bipolar plate layers. The contact resistance between the catalyst and membrane layers is low because they are fused together, and the contact resistance between the bipolar plates and other layers is low because the materials are typically nonporous with similar material properties (high density, with similar Poisson's ratios and Young's moduli). The GDL and the bipolar plate layers have several characteristics that make the contact resistance and permeability larger than

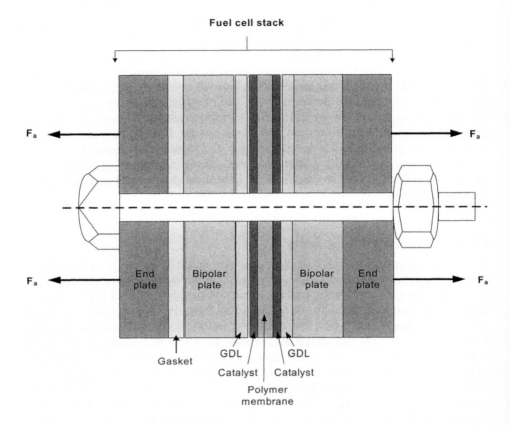

FIGURE 12-6. The forces exerted by the clamped materials (fuel cell layers) on the bolt and nut.

between the other layers: (1) the Poisson's ratios and Young's moduli have large differences (a hard material with a soft material); (2) the GDL layer is porous, and the permeability has been reduced due to the reduction in pore volume or porosity; and (3) part of the GDL layer blocks the flow channels that are in the bipolar plate creating less permeability through the GDL as the compression increases.

The amount of compression of the GDL and bipolar plate material layers can be determined using a Herzian compression equation. The calculations assume that the surfaces in contact are not perfectly smooth, that the elastic limits of the materials are not exceeded, that the materials are homogeneous, and that there are no frictional forces within the contact area. The compression formula for two spheres in contact is:

$$\alpha_{comp} = \frac{(3\pi)^{2/3}}{2} \cdot F^{2/3} \cdot (V_1 + V_2)^{2/3} \cdot \left(\frac{1}{D_1} + \frac{1}{D_2}\right)^{1/3} \qquad (12\text{-}9)$$

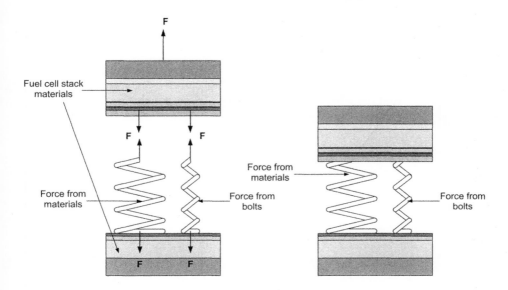

FIGURE 12-7. The forces exerted by the clamped materials and bolt.

where α_{comp} is the total elastic compression at the point of contact of two bodies, measured along the line of applied force, F is the total applied force, D is the diameter of the active area of the material (width of MEA), and

$$V = \frac{(1 - v^2)}{\pi E} \tag{12-10}$$

where v is Poisson's ratio, and E is the Young's modulus.

As shown in Figure 12-8, both the in-plane and through-plane conductivities increase as the compressed thickness of the GDL was decreased. The conductivities have a linear dependence on the GDL compressed thickness. This may be due to the reduced porosity of the GDL, which leads to shorter distances between conductive carbon fibers and better contact between the fibers.

12.6.3 The Stiffness of Bolted Layers

In order to accurately determine the ideal clamping pressure (tightening torque) for a fuel cell stack, the stiffness of the materials between the bolts has to be estimated. The stiffness of the materials includes the compressive stiffness of the materials under the bolt head in series with the stiffness of the physical interface, which increases with pressure, and the stiffness of

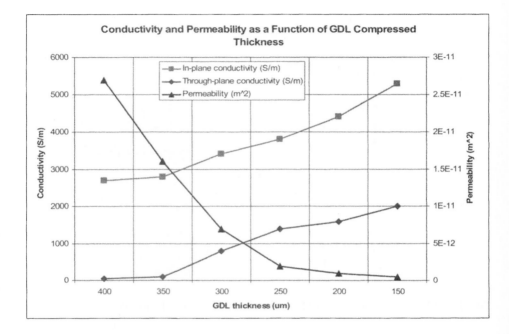

FIGURE 12-8. Conductivity and permeability as a function of GDL compressed thickness.

the threaded material. Some of the dimensions used in the bolt and layer stiffness calculations are shown in Figure 12-9.

In order to determine the stiffness of the cone-like section under the bolt head, the first step is to calculate the stiffness of each layer of the fuel cell stack:

$$k_{layer} = \frac{4 * h_{layer}}{\pi E \left(d_{bolthead} * \left(d_{bolthead} + 2\frac{h_{stack}}{2} * \cos\left(\frac{\alpha\pi}{180}\right)\right) - d_{bore}^2 \right)} \quad (12\text{-}11)$$

$$k_{layer} \approx \frac{4 * h_{layer}}{\pi E_{mod} \left(\left(De_{seat} + \left(\frac{h_{stack} * 0.2}{2}\right)\right)^2 - d_{bore}^2 \right)} \quad (12\text{-}12)$$

where k_{layer} is the stiffness of the fuel cell layer (such as the end plate or bipolar plate), h_{layer} is the thickness of that particular layer, E_{mod} is the modulus of elasticity in tension (MPa) of the material, $d_{bolthead}$ is the diameter of the bolt head, h_{stack} is the stack thickness (total thickness of all

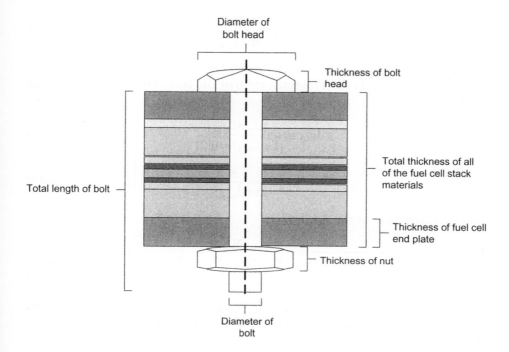

FIGURE 12-9. Dimensions used in the bolt and layer stiffness calculations.

materials), α is the effective cone angle, De_{seat} is the outer diameter of the seating face, and d_{bore} is the clearance hole diameter.

The stiffness of the bolt, head, shaft, and nut is calculated in a similar fashion. The tensile stiffness of the bolt shaft is:

$$k_{boltshaft} \approx \frac{\pi \left(\dfrac{d_{bolt\_dia}}{2} \right)^2 E_{bolt}}{L_{bolt}} \qquad (12\text{-}13)$$

$$k_{boltshaft} \approx \frac{4 * L_{bolt}}{\pi * d_{bolt\_dia}^2} \qquad (12\text{-}14)$$

where d_{bolt_dia} is the bolt diameter $\left(d_{bolt\_dia} = \dfrac{d_{pitch} + d_{rt}}{2} \right)$, E_{bolt} is the Young's modulus of the bolt, and L_{bolt} is the bolt length. The shear stiffness of the bolt head is:

$$k_{bolthead} = \frac{h_{bolthead} \pi E_{bolt}}{(1 + v_{bolt}) * \ln(2)} \qquad (12\text{-}15)$$

$$k_{bolthead} \approx \frac{1.5}{\pi * d_{bolt\_dia}} \qquad (12\text{-}16)$$

where $h_{bolthead}$ is the thickness of the bolt head, E_{bolt} is the Young's modulus of the bolt, and v_{bolt} is the Poisson's ratio of the bolt. The shear stiffness in the nut is:

$$k_{nut} = \frac{h_{nut}\pi E_{bolt}}{(1 + v_{bolt}) * \ln(2)} \qquad (12\text{-}17)$$

$$k_{nut} \approx \frac{1.8}{\pi * d_{pitch}} \qquad (12\text{-}18)$$

The total stiffness of the stack is:

$$k_{stack} = \frac{N}{k_{endplate} + k_{gasket} + k_{contact} + k_{ffplate} + k_{GDL} + k_{mem} + k_{GDL} + k_{ffplate} + k_{contact} + k_{gasket} + k_{endplate}} \qquad (12\text{-}19)$$

where N is the number of bolts in the stack. The stiffness of the bolt shaft in tension, and the head and nut (if a nut is used) in shear, all act in series, so their stiffness combines to give the total stiffness of the bolt:

$$k_{bolt} = \frac{E_{bolt}}{k_{boltshaft} + k_{bolthead} + k_{nut}} \qquad (12\text{-}20)$$

As the stack thickness increases, the length of the bolt to pass through the stack thickness also increases, so the bolt stiffness decreases in a linear fashion. On the other hand, the diameter of the strain cone increases, which offsets much of the height increase, and the stack stiffness decreases far more slowly than that of the bolt.

The ratio of flange to bolt stiffness is:

$$k_{s-b} = \frac{k_{stack}}{k_{bolt}} \qquad (12\text{-}21)$$

The total stiffness can be expressed by:

$$k_{tot} = k_{bolt} + k_{stack} \qquad (12\text{-}22)$$

12.6.4 Calculating the Tightening Torque

The stiffness of the group of surcharged parts of the stack is:

$$c_1 = \frac{1}{k_{bolt}} + \frac{(1-n)}{k_{stack}} \qquad (12\text{-}23)$$

where n is the coefficient of implementation of the operational force (0.5). The resulting stiffness of the group of relieved parts of the stack is:

$$c_2 = \frac{k_{stack}}{n} \qquad (12\text{-}24)$$

The part of the operational force relieving the clamped parts is:

$$F_2 = \frac{F * c_2}{(c_1 + c_2)} \qquad (12\text{-}25)$$

where F is the force required for the ideal compression of the GDL by 75 microns (from figure 12-8). The bolt seating coefficient is calculated by:

$$m_{seat1} = m_c * \frac{(De_{seat} + Di_{seat})}{2} \qquad (12\text{-}26)$$

where De_{seat} is the outer diameter of the seating face, Di_{seat} is the inner diameter of the seating face, and m_c is the friction coefficient in the seating face of head (nut) of the bolt.

The assembly force of the stack can be calculated by:

$$F_0 = q_a * F * F_2 + F_{0T} + 0.05 \qquad (12\text{-}27)$$

where q_a is the desired coefficient of tightness, and F_{0T} is the change of force required due to the heating of the connection. F_{0T} was assumed to be zero for all of the calculations since the stacks used for validating the model were all air-breathing fuel cell stacks tested at room temperature. The bolt seating is calculated by:

$$M_{seat} = m_{seat1} * F_0 \qquad (12\text{-}28)$$

The tightening torque is then:

$$M = F_0 * d_{pitch} * M_{seat} * \frac{thr_{pitch} * \pi * d_{pitch} * m_i}{\pi * d_{pitch} - thr_{pitch} * m_i} \qquad (12\text{-}29)$$

where F_0 is the assembly force of the stack, d_{pitch} is the pitch diameter, thr_{pitch} is the thread pitch, and m_i is the friction coefficient in thread (0.15).

12.6.5 Relating Torque to the Total Clamping Pressure

The average interface contact pressure, P_{avg}, can be calculated by dividing the total clamp force (product of the number of bolts, and the individual bolt clamp force) with the interface contact area, A_{int}[5,6]:

$$P_{avg} = \frac{N * F_0}{A_{int}} \qquad (12\text{-}30)$$

The average contact pressure is a linear function of bolt torque.

TABLE 12-1
Material Properties Used for Material Stiffness and Compression Calculations for Example 12-3

Fuel Cell Layer/Material	Thickness (mm)	Modulus of Elasticity in Tension (MPa)	Young's Modulus (N/mm²)	Poisson's Ratio
Polycarbonate end plate	10	2896	2200	0.37
Gasket: black conductive rubber	1	2	100	0.48
SS flow field plate	0.5	206,000	200,000	0.31
Carbon cloth	0.4	2	300	0.4
Nafion	0.05	2	236	0.487
Carbon cloth	0.4	2	300	0.4
SS flow field plate	0.5	206,000	200,000	0.31
Gasket: black conductive rubber	1	2	100	0.48
Polycarbonate end plate	10	2896	2200	0.37

TABLE 12-2
Bolt Properties Used for Bolt Stiffness and Torque Calculations for Example 12-3

Property	Stack Bolts
No. of bolts	4
Material	SS 316
Hex key size	5/32"
Bolt diameter (mm)	4.826
Bolt thread root diameter (mm)	3.451
Thread pitch	1.058
Pitch diameter (mm)	4.139
Bolt head diameter (mm)	8
Thickness of bolt head (mm)	5
Bolt length (between bolt head & nut) (mm)	25
Outer diameter of annulus seating face (mm)	7.925
Inner diameter of annulus seating face (mm)	5.232
Nut thickness (mm)	3
Bolt clearance hole (mm)	5.232

EXAMPLE 12-3: Calculate Optimal Tightening Torque

Using the parameters given in Tables 12-1 and 12-2, calculate the optimal tightening torque for the stack using equations 12-9–12-30.
Using MATLAB to solve:

```
%%%%%%%%%%%%%%%%%%%%%%%%%%%%%%%%%%%%%%%%%
% EXAMPLE 12-3: Calculate the Optimal Tightening Torque
% UnitSystem SI
%%%%%%%%%%%%%%%%%%%%%%%%%%%%%%%%%%%%%%%%%
% Inputs

%Design of optimal bolt connection through stack. Force and pressure applied to
   the stack
%The bolt is connected with a through bolt, and the loading is in the bolt axis. The
   course of loading is static.
clear;
N = 4; % number of bolts

% Material properties of the bolt
%%%%%%%%%%%%%%%%%%%%%%%%%%%%%%%%%%%%%%%%%
% Material of bolt

E_bolt = 200000;        % Young's Modulus of the bolt (N/mm)
v_bolt = 0.31;          % Poisson's ratio of the bolt
Em_bolt= 210000;        % Modulus of elasticity in tension (MPa) or 30000 ksi

% Bolt and Thread Parameters

dia_bolt = 4.826;       % Thickness
dia_rt = 3.451;         % bolt thread root diameter (mm)
thread_pitch = 1.058;   % Thread Pitch
pitch_dia = 4.139;      % Pitch diameter
dbolthead = 8;          % Bolt head diameter (mm)
hnut = 3;               % Nut thickness (0 if threaded into flange)
hbolthead = 5;          % Thickness of bolt head (mm)
Lbolt = 35;             % Bolt length
L_bolt = 25.4;          % Bolt length between the bolt head and nut

% Geometry of the bolt connection – Calculated for annulus
seating face

seatDe = 7.925;         % Outer diameter of the seating face (mm)
seatDi = 5.232;         % Inner diameter of the seating face (mm)
```

% bolt hole

```
dbore = 5.232;          % Bolt clearance hole (mm)
alpha = 45;             % Cone angle
qa = 0.5;               % Desired coefficient of tightness
n = 0.5;                % Coefficient of implementation of the operational force
mi = 0.150;             % Friction coefficient in thread
mc = 0.150;             % Friction coefficient in seating face of head (nut) of the
                          bolt
```

% Material properties of end plate (polycarbonate)

```
%%%%%%%%%%%%%%%%%%%%%%%%%%%%%%%%%%%%%%%%%

Em_end = 2896;          % Modulus of elasticity in tension (MPa)
thick_end = 10;         % Thickness (mm)
```

%Material properties of rubber gasket

```
%%%%%%%%%%%%%%%%%%%%%%%%%%%%%%%%%%%%%%%%%

Em_gask = 2;            % Modulus of elasticity in tension (MPa)
thick_gask = 1;         % Thickness (mm)
```

% Material properties of stainless steel

```
%%%%%%%%%%%%%%%%%%%%%%%%%%%%%%%%%%%%%%%%%

Em_nia = 206000;        % Modulus of elasticity in tension (MPa) or 30000 ksi
thick_nia = 0.5;        % Thickness (mm)
```

% Material properties of carbon cloth

```
%%%%%%%%%%%%%%%%%%%%%%%%%%%%%%%%%%%%%%%%%

Em_cc = 2;              % Modulus of elasticity in tension (MPa) or 0.3 ksi
thick_cc = 0.4;         % Thickness (mm)
```

% Material properties of Nafion

```
%%%%%%%%%%%%%%%%%%%%%%%%%%%%%%%%%%%%%%%%%

Em_naf = 2;             % Modulus of elasticity in tension (MPa) or 0.3 ksi
thick_naf = 0.05;       % Thickness (mm)
```

% Stiffness calculations

```
%%%%%%%%%%%%%%%%%%%%%%%%%%%%%%%%%%%%%%%%%
```

% Total thickness of parts

```
tot_thick = (thick_end + thick_gask + thick_nia + thick_cc + thick_naf + thick_cc +
    thick_nia + thick_gask + thick_end);
De = seatDe + (tot_thick*0.2)/2;
```

% Bolt stiffness

```
boltdia = (pitch_dia + dia_rt)/2;
nut =1.8/(pi*pitch_dia);
head = 1.5/(pi*boltdia);
kboltshaft = (4*L_bolt)/(pi*boltdia^2);
cb = Em_bolt/(kboltshaft + head + nut);
```

% Calculate stiffness of each part

```
kcomp_end=(4*thick_end)/(pi*(De^2-dbore^2)*Em_end);
kcomp_gask =(4*thick_gask)/(pi*(De^2-dbore^2)*Em_gask);
kcomp_nia = (4*thick_nia)/(pi*(De^2-dbore^2)*Em_nia);
kcomp_cc = (4*thick_cc)/(pi*(De^2-dbore^2)*Em_cc); %multiply by the area
kcomp_naf = (4*thick_naf)/(pi*(De^2-dbore^2)*Em_naf);
```

% Total stiffness of the clamped parts

```
cm1 = 1/(kcomp_end + kcomp_gask + kcomp_nia + kcomp_cc + kcomp_naf +
    kcomp_cc + kcomp_nia + kcomp_gask + kcomp_end);
cm = N/(kcomp_end + kcomp_gask + kcomp_nia + kcomp_cc + kcomp_naf +
    kcomp_cc + kcomp_nia + kcomp_gask + kcomp_end);
kjoint = cb + cm1;    % total layers stiffness
```

% Force calculations

```
%%%%%%%%%%%%%%%%%%%%%%%%%%%%%%%%%%%%%%%%
d_gdl = 40;           % diameter (mm)
d_bpp = 40;           % diameter (mm)
rho_GDL = 1.1e4;      % Electrical resistivity through plane (800 u-ohm-m) uohm
                        mm
rho2 = 300;           % Electrical resistivity through plane of bipolar plate (190 u-
                        ohm-m) uohm mm
void = 0.8;           % GDL porosity
E_gdl = 3;            % Young's modulus of carbon cloth (3 GPa)
E_bpp = 200;          % Young's modulus of stainless steel
v_gdl = 0.4;          % Poisson's ratio of Carbon Cloth
v_bpp = 0.31;         % Poisson's ratio of stainless steel
V_gdl =(1-(v_gdl^2))/(pi*E_gdl);
V_bpp =(1-(v_bpp^2))/(pi*E_bpp);
% Amount of compression in microns
alpha  =  (((3*pi)^(2/3))/2*(F_bolt_max^(2/3))*((V_gdl  +  V_bpp)^(2/3))*((1/d_
    gdl)+(1/d_bpp))^(1/3))*10
```

% Electrical constriction resistance of the single contact

```
rho1 = rho_GDL*(1 – void)^1.5;  % transverse electrical resistivity of GDL
r = d_gdl/2;
Resist = (rho1 + rho2)./(4 .*r) % Resistivity
```

% Maximum force

F_bolt_max = 310;　　% compression is 75.92 um GDL, Resistivity = 16.05

%%

% Tightening torque calculation

%%

c1 = 1/((1/cb) + ((1-n)/cm));　　% Resulting stiffness of the group of surcharged parts of the joint

c2 = cm/n;　　% Resulting stiffnes of the group of relieved parts of the joint

F2 = F_bolt_max * c2/(c1 + c2);　　% Part of operational force relieving clamped parts

mseat1 = mc * (seatDe + seatDi)/2;

F0T = 0;　　% Change of prestressing due to the heating of the connection

F0L = 0;　　% Loss of prestressing due to the deformation of the connection

F0 = qa * F_bolt_max + F2 + F0T − F0L + 0.5 % Assembly prestressing of the joint (N)−based upon axial load only

Mseat = (mseat1 * F0)/(2 * 1000);

% Tightening torque

M = F0 * pitch_dia * (thread_pitch * pi * pitch_dia * mi)/(pi * pitch_dia − thread_pitch * mi)/(2 * 1000)+ Mseat

% N-m

Mm = 141.61 * M % oz-in

% Average interface contact pressure using bolt clamp force

F = F0/1000; % Force in kN
w = 8/100; % width in m
l = 8/100; %length in m
A_int = w * l; %m^2
P_avg_MPa = (N * F /A_int)/1000 % MPa
P_avg_bar = P_avg_MPa * 10

Figure 12-10 shows the actual polarization curves of the PEM fuel cell stack described in Example 12-3 under five different clamping pressures. The current is dynamically stable for four of the five clamping pressures. The lowest clamping pressure of 0.20 Nm (28 oz-in) displayed the worst I-V performance, due to mass-transfer limitations and high contact resistance. The polarization curves continuously increase until a torque of 0.25 Nm (36 oz-in) is reached. As the torque continues to increase to 0.31 Nm (44 oz-in), the polarization curves again begin to decrease.

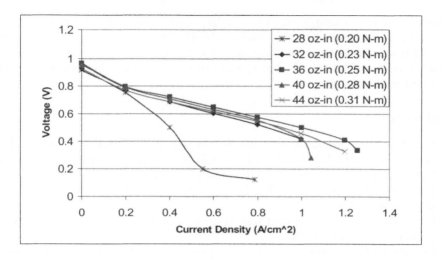

FIGURE 12-10. Polarization curves with tightening torques of 0.20 to 0.31 Nm (28 oz-in to 44 oz-in).

TABLE 12-3
Calculated Force, Tightening Torque, and Contact Pressure

Total force on the stack	310.8 N
Optimal tightening torque	36.35 oz-in (0.257 N-m)
Average interface contact pressure	0.194 MPa (1.94 bar)

The material and bolt properties from Tables 12-1 and 12-2 were entered into the numerical model in Example 12-3, and the optimal force, pressure, and tightening torque were calculated. The results are shown in Table 12-3.

The values in Table 12-3 show that the calculated optimal tightening torque matches the tightening torque associated with the best fuel cell I-V curve in Figure 12-10.

Chapter Summary

Many parameters must be considered when designing a fuel cell. Some of the most basic design considerations include power required, size, weight, volume, cost, transient response, and operating conditions. From these initial requirements, the more detailed design requirements (such as the number of cells, material and component selections, flow field design, etc.) can be chosen. The most commonly used stack configuration is the bipolar configuration, which has been described in previous chapters. Common manifold types for even reactant flow through the cell and alternate cell interconnections for electron flow through the stack are also presented.

Although it is not emphasized in the literature, the clamping pressure is a critical parameter for optimal fuel cell performance. A model was created to estimate the optimal torque, and to emphasize how the mechanical characteristics of the stack, bolts, and pressure have a substantial affect on fuel cell performance.

Problems

- Design a fuel cell stack that has to operate at 80 °C with air and hydrogen pressures of 1 atm. The Pt/C loading is 1 mg/cm^2 and the cells use the Nafion 117 electrolyte. The total power should be 250 W.
- Design a PEMFC stack that has to operate at 0 °C. The total power should be 250 W.
- Design a PEMFC stack that has to operate at 60 °C and 3 atm. The total power should be 100 W.
- Design a PEMFC stack that has to operate at 25 °C and 1 atm. The total power should be 50 W.

Endnotes

[1] Yan, Q., H. Toghiani, and J. Wu. Investigation of water transport through membrane in a PEM fuel cell by water balance experiments. *J. Power Sources.* Vol. 158, 2006, pp. 316–325.
[2] Fabian, T., J.D. Posner, R. O'Hayre, S.-W. Cha, J.K. Eaton, F.B. Prinz, and J.G. Santiago. The role of ambient conditions on the performance of a planar, air-breathing hydrogen PEM fuel cell. *J. Power Sources.* Vol. 161, 2006, pp. 168–182.
[3] Spiegel, C.S. Designing and Building Fuel Cells. 2007. New York: McGraw-Hill.
[4] Ibid.
[5] Lin, B. 1999. Conceptual design and modeling of a fuel cell scooter for urban Asia. Princeton University, master's thesis.
[6] Lee, *Analyses of the fuel cell stack assembly pressure.*
[7] Ibid.
[8] Lee, S.-J., C.-D. Hsu, and C.-H. Huang. Analyses of the fuel cell stack assembly pressure. *J. Power Sources.* Vol. 145, 2005, pp. 353–361.
[9] Lee, *Analyses of the fuel cell stack assembly pressure.*
[10] Li, X. 2006. *Principles of Fuel Cells.* New York: Taylor & Francis Group.
[11] Song, R.-H., and D.R. Shin. Influence of CO concentration and reactant gas pressure on cell performance in PAFC. *Int. J. Hydrogen Energy.* Vol. 26, 2001, pp. 1259–1262.

Bibliography

Aravamudhan, S., Abdur Rub Abdur Rahman, and S. Bhansali. Porous silicon-based orientation independent, self-priming micro-direct ethanol fuel cell. *Sensors Actuators A.* Vol. 123–124, 2005, pp. 497–504.

Barbir, F. 2005. *PEM Fuel Cells: Theory and Practice.* Burlington, MA: Elsevier Academic Press.

Blum, A., T. Duvdevani, M. Philosoph, N. Rudoy, and E. Peled. Water neutral micro direct methanol fuel cell (DMFC) for portable applications. *J. Power Sources.* Vol. 117, 2003, pp. 22–25.

Bosco, A. *General Motors Fuel Cell Research.* Hy-Wire. Available at: http://www .ansoft.com/workshops/altpoweree/Andy_Bosco_GM.pdf. Accessed November 4, 2006.

Bove, R., and P. Lumghi. Experimental comparison of MCFC performance using three different biogas types and methane. *J. Power Sources.* Vol. 145, 2005, pp. 588–593.

Cha, S.W., R. O'Hayre, Y. Saito, and F.B. Prinz. The influence of size scale on the performance of fuel cells. *Solid State Ionics.* Vol. 175, 2004. pp. 789–795.

Chang, P.A.C., J. St-Pierre, J. Stumper, and B. Wetton. Flow distribution in proton exchange membrane fuel cell stacks. *J. Power Sources.* Vol. 162, 2006, pp. 340–355.

Chen, R., and T.S. Zhao. Mathematical modeling of a passive feed DMFC with heat transfer effect. *J. Power Sources.* Vol. 152, 2005, pp. 122–130.

Choban, E.R., L.J. Markoski, A. Wieckowski, and P.J.A. Kenis. Microfluidic fuel cell based on laminar flow. *J. Power Sources.* Vol. 128, 2004, pp. 54–60.

Chung, B.W., C.N. Chervin, J.J. Haslam, A.-Q. Pham, and R.S. Glass. Development and characterization of a high performance thin-film planar SOFC stack. *J. Electrochem. Soc.* Vol. 152, No. 2, 2005, pp. A265–A269.

Coutanceau, C., L. Demarconnay, C. Lamy, and J.M. Leger. Development of electrocatalysts for solid alkaline fuel cell (SAFC). *J. Power Sources.* Vol. 156, 2006, pp. 14–19.

Dyer, C.K. Fuel cells for portable applications. *J. Power Sources.* Vol. 106, 2002, pp. 31–34.

Gulzow, E., M. Schulze, and U. Gerke. Bipolar concept for alkaline fuel cells. *J. Power Sources.* Vol. 156, 2006, pp. 1–7.

Heinzel, A., C. Hebling, M. Muller, M. Zedda, and C. Muller. Fuel cells for low power applications. *J. Power Sources.* Vol. 105, 2002, pp. 250–255.

Kim, J.-Y., O.J. Kwon, S.-M. Hwang, M.S. Kang, and J.J. Kim. Development of a miniaturized polymer electrolyte membrane fuel cell with silicon separators. *J. Power Sources.* Vol. 161, 2006, pp. 432–436.

Koh, J.-H., H.-K. Seo, C.G. Lee, Y.-S. Yoo, and H.C. Lim. Pressure and flow distribution in internal Gs manifolds of a fuel cell stack. *J. Power Sources.* Vol. 115, 2003, pp. 54–65.

Lim, C., and C.Y. Wang. Development of high power electrodes for a liquid feed direct methanol fuel cell. *J. Power Sources.* Vol. 113, 2003, pp. 145–150.

Liu, J.G., T.S. Zhao, Z.X. Liang, and R. Chen. Effect of membrane thickness on the performance and efficiency of passive direct methanol fuel cells. *J. Power Sources.* Vol. 153, 2006, pp. 61–67.

Liu, S., Q. Pu, L. Gao, C. Korzeniewski, and C. Matzke. From nanochannel induced proton conduction enhancement to a nanochannel based fuel cell. *Nano Lett.* Vol. 5, No. 7, 2005, pp. 1389–1393.

Lu, G.Q., and C.Y. Wang. Development of micro direct methanol fuel cells for high power applications. *J. Power Sources.* Vol. 144, 2005, pp. 141–145.

Mench, M.M., C.-Y. Wang, and S.T. Tynell. *An Introduction to Fuel Cells and Related Transport Phenomena.* Department of Mechanical and Nuclear Engineering, Pennsylvania State University, PA, USA. Draft. Available at: http://mtrl1.mne.psu.edu/Document/jtpoverview.pdf. Accessed March 4, 2007.

Mench, M.M., Z.H. Wang, K. Bhatia, and C.Y. Wang. 2001. *Design of a Micro-direct Methanol Fuel Cell.* Electrochemical Engine Center, Department of Mechanical and Nuclear Engineering, Pennsylvania State University, PA, USA.

Mitrovski, S.M., L.C.C. Elliott, and R.G. Nuzzo. Microfluidic devices for energy conversion: Planar integration and performance of a passive, fully immersed H2-O2 fuel cell. *Langmuir.* Vol. 20, 2004, pp. 6974–6976.

Modroukas, D., V. Modi, and L.G. Frechette. Micromachined silicon structures for free-convection PEM fuel cells. *J. Micromech. Microeng.* Vol. 15, 2005, pp. S193–S201.

Morita, H., M. Komoda, Y. Mugikura, Y. Izaki, T. Watanabe, Y. Masuda, and T. Matsuyama. Performance analysis of molten carbonate fuel cell using a Li/Na electrolyte. *J. Power Sources.* Vol. 112, 2002, pp. 509–518.

Nakagawa, N., and Y. Xiu. Performance of a direct methanol fuel cell operated. *J. Power Sources.* Vol. 118, 2003, pp. 248–255.

Nguyen, N.-T., and S.H. Chan. Micromachined polymer electrolyte membrane and direct methanol fuel cells—A review. *J. Micromech. Microeng.* Vol. 16, 2006, pp. R1–R12.

O'Hayre, R., S.-W. Cha, W. Colella, and F.B. Prinz. 2006. *Fuel Cell Fundamentals.* New York: John Wiley & Sons.

Rowe, A., and X. Li. Mathematical modeling of proton exchange membrane fuel cells. *J. Power Sources.* Vol. 102, 2001, pp. 82–96.

Rowshanzamir, S., and M. Kazemeini. A new immobilized-alkali H2/O2 fuel cell. *J. Power Sources.* Vol. 88, 2000, pp. 262–268.

Silva, V.S., J. Schirmer, R. Reissner, B. Ruffmann, H. Silva, A. Mendes, L.M. Madeira, and S.P. Nunes. Proton electrolyte membrane properties and direct methanol fuel cell performance. *J. Power Sources.* Vol. 140, 2005, pp. 41–49.

Soler, J., T. Gonzalez, M.J. Escudero, T. Rodrigo, and L. Daza. Endurance test on a single cell of a novel cathode material for MCFC. *J. Power Sources.* Vol. 106, 2002, pp. 189–195.

Sousa, R., Jr., and E. Gonzalez. Mathematical modeling of polymer electrolyte fuel cells. *J. Power Sources.* Vol. 147, 2005, pp. 32–45.

Springer et al. Polymer electrolyte fuel cell model. *J. Electrochem. Soc.* Vol. 138, No. 8, 1991, pp. 2334–2342.

Subhash, S.C. July 2003. *High Temperature Solid Oxide Fuel Cells Fundamentals, Design and Applications.* Rio de Janeiro, Brazil: Pan American Advanced Studies Institute, Pacific Northwest National Laboratory, U.S. Department of Energy.

U.S. Patent Application Publication No. US 2006/0003217 A1. Planar Membraneless Microchannel Fuel Cell. Cohen, J.L., D.J. Volpe, D.A. Westly, A. Pechenik, and H.D. Abruna. January 5, 2006.

U.S. Patent No. 4490444. High Temperature Solid Oxide Fuel Cell Configurations and Interconnections. A.O. Isenberg, Forest Hills, PA. Westinghouse Electric Corporation. December 25, 1984.

U.S. Patent No. 7,029,781 B2. Microfuel Cell Having Anodic and Cathodic Microfluidic Channels and Related Methods. Lo Priore, S., M. Palmieri, and U. Mastromatteo. STMicroelectronics, Inc. Carollton, TX. April 18, 2006.

U.S. Patent No. 7,125,625 B2. Electrochemical Cell and Bipolar Assembly for an Electrochemical Cell. Cisar, A.J., C.C. Andrews, C.J. Greenwald, O.J. Murphy, C. Boyer, R.C. Yalamanchili, and C.E. Salinas. Lynntech, Inc. College Station, TX. October 24, 2006.

Verma, A., and S. Basu. Direct use of alcohols and sodium borohydride as fuel in an alkaline fuel cell. *J. Power Sources.* Vol. 145, 2005, pp. 282–285.

Wee, J.-H. Performance of a unit cell equipped with a modified catalytic reformer in direct internal reforming molten carbonate fuel cell. *J. Power Sources.* Vol. 156, 2006, pp. 288–293.

Wells, B., and H. Voss. May 17, 2006. *DMFC Power Supply for All-Day True Wireless Mobile Computing.* Polyfuel, Inc. Available at: http://www.hydrogen .energy.gov/pdfs/review06/fcp_39_wells.pdf. Accessed November 4, 2006.

Wong, C.W., T.S. Zhao, Q. Ye, and J.G. Liu. Experimental investigations of the anode flowfield of a micro direct methanol fuel cell. *J. Power Sources.* Vol. 155, 2006, pp. 291–296.

Wozniak, K., D. Johansson, M. Bring, A. Sanz-Velasco, and P. Enoksson. A micro direct methanol fuel cell demonstrator. *J. Micromech. Microeng.* Vol. 14, 2004, pp. S59–S63.

Yamada, M., and I. Honma. Biomembranes for fuel cell electrodes employing anhydrous proton conducting uracil composites. *Biosensors Bioelectronics.* Vol. 21, 2006, pp. 2064–2069.

Yoon, K.H., J.Y. Choi, J.H. Jang, Y.S. Cho, and K.H. Jo. Electrode/matrix interfacial characteristics in a phosphoric acid fuel cell. *J. Appl. Electrochem.* Vol. 30, 2000, pp. 121–124.

Yoon, K.H., J.H. Jang, and Y.S. Cho. Impedance characteristics of a phosphoric acid fuel cell. *J. Mater. Sci. Lett.* Vol. 17, 1998, pp. 1755–1758.

You, L., and H. Liu. A two-phase flow and transport model for PEM fuel cells. *J. Power Sources.* Vol. 155, 2006, pp. 219–230.

Zhu, B. Proton and oxygen ion-mixed conducting ceramic composites and fuel cells. *Solid State Ionics.* Vol. 145, 2001, pp. 371–380.

CHAPTER 13

Fuel Cell System Design

13.1 Introduction

In order to obtain optimum performance from a fuel cell stack, the hydrogen and oxidant flow, water removal, and the voltage output should be optimized using external plant components. Fuel cell system designs range from very simple to very complex depending upon the fuel cell application and the system efficiency desired. A fuel cell system can be very efficient with a few plant components, as shown in Figure 13-1. Usually, the larger the fuel cell stack, the more complex the fuel cell plant subsystem. Modeling the fuel cell system efficiency and output allows the system designer to be more efficient when creating new systems.

Most of the fuel cell system components shown in Figure 13-1 are used to distribute air and hydrogen flows into and out of the fuel cell stack. The part of the fuel cell system that is responsible for air flow includes a particulate filter for cleaning the system, humidification module, and a pressure transducer. There is also a pump to ensure an adequate supply of air into the fuel cell stack. The hydrogen flows into the fuel cell stack using a pressurized tank. A mass flow controller should be installed in this system to monitor the flow rate. The water and hydrogen coming out of the system are cleaned and the pressure is monitored before it exits the system[1].

As the fuel cell system increases in size, it becomes more complex as the temperature, pressure, water, and heat become more problematic, and need to be monitored more closely. In addition, if a carbon-based fuel is converted to hydrogen for electrical power and heat, fuel processing units and gas cleanup units may be necessary. Other additional components often found in fuel cell plant include: heat exchangers, pumps, fans, blowers, compressors, electrical power inverters, converters and conditioners, water handling devices, and control systems.

Only a few sensors and pressure transducers are included in Figure 13-1. A fully developed control system will consist of thermocouples,

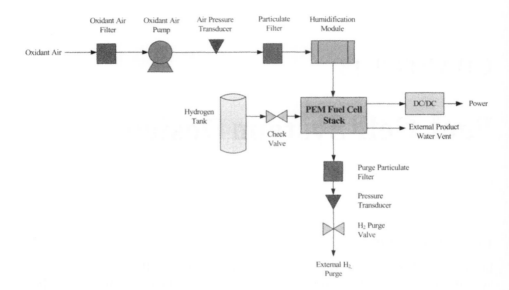

FIGURE 13-1. Simple PEM fuel cell system[2].

pressure transducers, methanol/hydrogen sensors, and mass flow controllers, which will measure and control data using a data acquisition program. As described in Chapters 7 through 12, the fuel cell catalyst, membranes, and flow field plates are very important areas for fuel cell design, modeling, and improvement, but stack optimization is equally important. This chapter focuses on modeling these important subsystems.

13.2 Fuel Subsystem

As seen in Figure 13-1, the fuel subsystem is very important because the reactants may need to undergo several processes before they are ultimately delivered to the fuel cell with the required conditions. Plant components such as blowers, compressors, pumps, and humidification systems have to be used to deliver the gases to the fuel cell with the proper temperature, humidity, flow rate, and pressure. Other plant components, such as turbines, are also useful because they can harness energy from the heated exhaust gases leaving the fuel cell. This chapter describes these plant components and gives some of the relevant equations needed for producing quick models for the fuel cell plant subsystem.

13.2.1 Humidification Systems

In PEM fuel cells, a hydrogen humidification system may be required to prevent the fuel cell PEM from dehydrating under the load. As discussed previously, water management is a challenge in the PEM fuel cell because there is ohmic heating under high current flow, which will dry out the polymer membrane and slow ionic transport. Some fuel cell stacks may not require any humidification due to water generation at the cathode. In larger fuel cell systems, either the air or the hydrogen or both the air and hydrogen must be humidified at the fuel inlets. The gases can be humidified by bubbling the gases through water, water or steam injection, flash evaporation, or through a water/heat exchanger device. Examples of these humidification methods are shown in Figure 13-2.

When the total pressure is constant, the humidity depends upon the partial pressure of vapor in the mixture. For a vapor–gas system where the vapor is component A and the fixed phase is component B[3,4]:

$$\Phi = \frac{M_A p_A}{M_B(p_{tot} - p_A)} \tag{13-1}$$

For an air–water system:

$$\Phi = \frac{M_{H2O} p_{H2O}}{M_{air}(p_{tot} - p_{H2O})} = \frac{18 * p_{H2O}}{29 * (p_{tot} - p_{H2O})}$$

The mole fraction of the vapor is:

$$x_{vap} = \frac{\dfrac{\Phi}{M_{H2O}}}{\dfrac{1}{M_{air}} + \dfrac{\Phi}{M_{H2O}}} \tag{13-2}$$

The saturation humidity is where the gas has vapor in equilibrium with the liquid at gas temperature:

$$\Phi_s = \frac{M_{H2O} p_{H2O}^0}{M_{air}(p_{tot} - p_{H2O}^0)} = \frac{18 * p_{H2O}^0}{29 * (p_{tot} - p_{H2O}^0)} \tag{13-3}$$

The percent humidity is the ratio of the actual humidity to the saturation humidity:

$$\Phi_{water} = 100 * \frac{\Phi}{\Phi_s} = \Phi_R \frac{p_{tot} - p_{H2O}^0}{p_{tot} - p_{H2O}} \tag{13-4}$$

The heat that is required to increase the temperature of one pound of gas and the vapor it contains by 1 °F for the air–water system is:

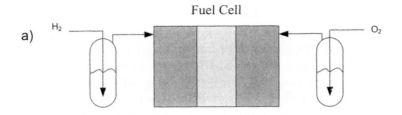

Dewpoint Humidification

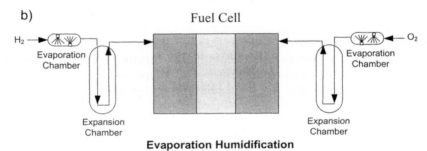

Evaporation Humidification

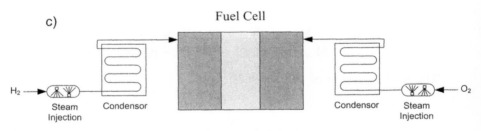

Steam Injection Humidification

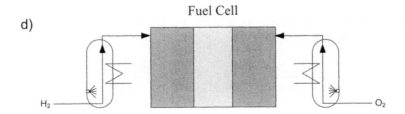

Flash Evaporation Humidification

FIGURE 13-2. Conventional humidification methods: (a) dewpoint humidification, (b) evaporation humidification, (c) steam injection humidification, and (d) flash evaporation humidification[5].

$$c_s = (c_p)_{H2O} + \Phi(c_p)_{air} \approx 0.24 + 0.45 * \Phi \tag{13-5}$$

The wet bulb temperature is the dynamic equilibrium temperature attained by the liquid surface when the rate of heat transfer to the surface by convection equals the rate of heat required for evaporation away from the surface[6]. The partial pressure and the vapor pressure are usually small relative to the total pressure; therefore, the wet bulb equation can be expressed in terms of humidity conditions[7]:

$$\Phi_s - \Phi = h_c \frac{T - T_w}{\lambda k_1} \tag{13-6}$$

For the air–water system, this equation becomes:

$$\Phi_s - \Phi = 18h_c \frac{T - T_w}{29\lambda k_1}$$

The adiabatic saturation temperature is reached when a stream of air is mixed with water at a temperature, T_s, in an adiabatic system. This can be expressed as:

$$\Phi_s - \Phi = c_s \frac{T - T_s}{\lambda} \tag{13-7}$$

A humidity (psychrometric) chart provides a way to determine the properties of a gas–vapor mixture. Figure 13-3 shows an example of a psychrometric chart for a mixture of air and water. Any point on the chart represents a specific mixture of air and water. Points above and below the saturation lines represent a mixture of saturated air as a function of air temperature. The curved lines between the saturation line and the temperature axis represent mixtures of air and water at specific percentage humidities.

Commercial humidifiers usually use heating coils and a warm water spray to bring the gas to the desired temperature and humidity. Figure 13-4 shows the process that can occur in a humidifying device. Point A represents the entering air with the dry bulb temperature, T_1 and humidity H_1. A dry bulb temperature of T_2 and humidity of H_2 is desired (point B). The method of reaching point B is by going from point A to point * by water spray, and then heating to reach point B. In addition, the length of the pipe required for sufficient mixing after the steam injection needs to be calculated to ensure a uniformly humidified gas is entering the fuel cell.

13.2.2 Fans and Blowers

A commonly used method of providing air to a fuel cell is through the use of fans or blowers. The fan or blower is driven by an electric motor, which

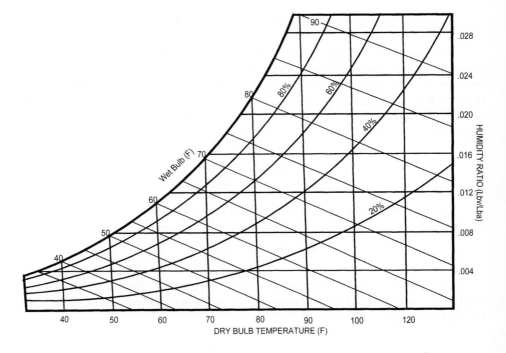

FIGURE 13-3. Example psychrometric chart[8,9].

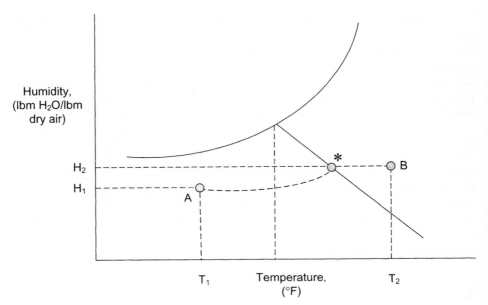

FIGURE 13-4. Example humidification process[10].

requires power from the fuel cell or other source to run. One of the most commonly used fans is the axial fan, which is effective in moving air over parts, but not effective across large pressure differentials. The back pressure of this fan type is very low at 0.5 cm of water[11]. These fans are well suited for many hydrogen–air PEM fuel cell designs. The actual fan power is given by the following equation[12]:

$$W_{act} = \frac{W_{ideal}}{\eta_s} \tag{13-8}$$

The ideal power can be calculated by:

$$W_{ideal} = m c_{p,avg}(T_2 - T_1) + \frac{v_2^2}{2} - \frac{v_1^2}{2} \tag{13-9}$$

where $c_{p,avg}$ is the specific heat at the average temperature of the inlet and outlet. The ideal exit temperature can be calculated from Equation (13-10):

$$\frac{T_2}{T_1} = \left(\frac{P_2}{P_1}\right)^{(\gamma-1)/\gamma} \tag{13-10}$$

where T_2 is the isentropic temperature, and γ is the ratio of the specific heat capacities of the gas, C_p/C_v. The actual exit temperature is then:

$$T_2 = T_1 + \frac{1}{c_{p,avg}}\left\{\frac{W_{act}}{m} - \left(\frac{v_2^2}{2} - \frac{v_1^2}{2}\right)\right\} \tag{13-11}$$

The actual speed and power required can be found from the manufacturer's table once the inlet volume rate and pressure boost are specified. Fan data can sometimes be represented in terms of dimensionless parameters. These are defined as:

Discharge Coefficient:

$$C_Q = \frac{\dot{V}}{ND^3} \tag{13-12}$$

Pressure Coefficient:

$$C_H = \frac{\Delta P}{\rho N^2 D^2} \tag{13-13}$$

Isentropic Efficiency:

$$\eta_{fan} = \frac{W_{ideal}}{W_{actual}} \tag{13-14}$$

Specific Speed:

$$N_s = \frac{N\dot{V}^{1/2}\rho^{3/4}}{(\Delta P)^{3/4}} \tag{13-15}$$

where $\dot{V}$ is the volumetric flow rate, ρ is the density of the fluid, D is the wheel blade diameter, N is the fan speed, ΔP is the fan pressure boost, and W is the fan power. The ideal fan power is:

$$W_{ideal} = m\left[c_{p,avg}T_1\left(\left(\frac{P_2}{P_1}\right)^{(\gamma-1)/\gamma} - 1\right) + \left(\frac{v_2^2}{2} - \frac{v_1^2}{2}\right)\right] \tag{13-16}$$

Often the temperature rise of the fluid as it passes through the fan is often neglected, and the following equation can be used to calculate the fan power:

$$W_{act} = V\left[\Delta P + \frac{\rho v_2^2 - v_1^2}{2}\right] \tag{13-17}$$

Since the total pressure can be expressed as $P_0 = P + \frac{\rho v_2^2}{2}$, then:

$$W_{act} = V * \Delta P_0 \tag{13-18}$$

Larger pressure differences can be obtained by using centrifugal fans. Centrifugal fans have air or gases entering in the axial direction, and discharge air or gases in the radial direction[13]. These are used for circulating cooling air through small- to medium-sized fuel cells. The pressure created by these fans is from 3 to 10 cm of water[14,15].

Blowers are also used in atmospheric systems to draw air into the fuel cell. The blower is typically powered by a battery for startup, and then some of the power output of the fuel cell is used to keep the blower running (like other plant components). The blower power required is

$$W = (\Delta PV)/\eta_{blower} \tag{13-19}$$

where η_{blower} is the blower efficiency.

EXAMPLE 13-1: Calculate the Fan Efficiency

A fan is used to move air into a fuel cell system at 20 °C and 100 kPa. The fan moves 0.05 m³/s of air with a pressure boost of 0.6 kPa. The actual power is 50 W, and the outlet velocity is 1 m/s. Determine the efficiency of the fan.

With the pressure boost, $P_2 = 100.6$ kPa. The ideal work must first be calculated to determine the efficiency:

$$w_{ideal} = c_p(T_2 - T_1)$$

where T_2 comes from our isentropic ratio, and γ is 1.38:

$$\frac{T_2}{T_1} = \left(\frac{P_2}{P_1}\right)^{(\gamma-1)/\gamma} = (293.15) * \left(\frac{100.6}{100}\right)^{(1.38-1)/1.38} = 293.63\,K$$

$$w_{ideal} = (1.005) * (293.63 - 293.15) = 0.4858\ kJ/kg$$

Then:

$$\eta_s = \frac{W_{ideal}}{W_{act}} = \frac{W_{ideal}}{\dot{w}_{act}/\dot{m}}$$

where

$$m = \dot{v}/V = \dot{v}/(RT/P) = (0.05 * 100)/(0.286 * 293.15) = 0.0596\ kg/s$$

Therefore,

$$\eta_s = \frac{W_{ideal}}{W_{act}} = \frac{W_{ideal}}{\dot{w}_{act}/\dot{m}} = \frac{0.4858\,kJ/kg}{(0.05\,kW/0.0596\,kg/s)} = 0.579$$

Using MATLAB to solve:

```
%%%%%%%%%%%%%%%%%%%%%%%%%%%%%%%%%%%%%%%%%
% EXAMPLE 13-1: Calculate the Fan Efficiency
% UnitSystem SI
%%%%%%%%%%%%%%%%%%%%%%%%%%%%%%%%%%%%%%%%%
% Inputs

T = 20;          % Operating temperature (degrees C)
P = 100;         % Operating pressure (kPa)
PBoost = 0.6;    % PBoost: the boost in the pressure
```

```
fanPower = 0.05;        % fanPower(cm3/s)
actualPower = 50;       % actualPower: the actual power generated in Watts
outletVelocity = 1;     % outletVelocity (m/s)
cp = 1.005;             % Specific heat
T = T + 273.15;         % Convert temperature to K
gamma = 1.38;
R = 0.286;
```

% Calculate the exit temperature from isentropic ratio

```
T2 = T.*((P + PBoost)./P).^((gamma- 1)./gamma);
```

% Ideal work (kJ/kg)

```
W_ideal = cp.*(T2 – T)
```

% Mass flow rate (kg/s)

```
m = fanPower ./ (R.*T ./ P);
```

% Fan Efficiency

```
etha = W_ideal ./ ((actualPower ./ 1000) ./ m)
```

13.2.3 Compressors

Compressors are used to compress air, which allows a greater concentration of oxygen per volume per time, and therefore, increases the fuel cell efficiency. This enables the dropoff in voltage due to mass transport to be delayed until higher current densities. If the pressure is higher, a lower volumetric flow rate can be used for the same molar flow rate, and humidification requires less water for saturation (per mole of air). The compression can be isothermal or adiabatic. Isothermal compression allows temperature equilibration with the environment, and adiabatic uses compression without any heat exchange with the environment[16].

The most common type of compressor is the centrifugal compressor. It uses kinetic energy to create a pressure increase. The centrifugal compressor can be operated with high efficiencies through a high range of flow rates by changing both the flow rate and the pressure. This compressor type is commonly found on engine turbocharging systems. Figure 13-5 shows an example of a motor-driven turbocompressor for PEM fuel cells[17].

The efficiency of the compressor is important for the overall efficiency of the fuel cell system. The efficiency is found by using the ratio of actual work done to raise the pressure from P_1 to P_2:

$$\frac{T_2}{T_1} = \left(\frac{P_2}{P_1}\right)^{(\gamma-1)/\gamma}$$

(13-20)

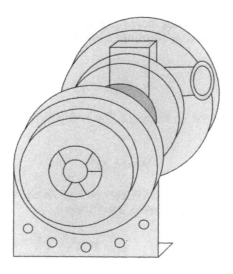

FIGURE 13-5. Example of a motor-driven turbocompressor for PEM fuel cells[18].

where T_2 is the isentropic temperature, and γ is the ratio of the specific heat capacities of the gas, C_p/C_v. In order to use Equation (13-21), the heat flow from the compressor and the kinetic energy of the gas through the compressor should be considered negligible. The gas is also considered to be ideal, therefore, it can be assumed that the specific heat is constant at a constant pressure.

The actual work done by the system is:

$$W = c_p(T_2 - T_1)m \qquad (13\text{-}21)$$

where m is the mass of the gas compressed (air flow rate, g/s), T_1 and T_2 are the inlet and exit temperatures, respectively, and c_p is the specific heat at constant pressure (J/gK). The efficiency is the ratio of these two quantities of work:

$$\eta = \text{isentropic work/real work} = \frac{c_p(T_2 - T_1)m}{c_p(T_2 - T_1)m} \quad \text{and} \quad \eta_c = \frac{(T_2 - T_1)}{(T_2 - T_1)} \qquad (13\text{-}22)$$

The change in temperature at the end of compression can be found from the following equation:

$$\Delta T = T_2 - T_1 = \frac{T_1}{\eta_c}\left(\left(\frac{P_2}{P_1}\right)^{\frac{\gamma-1}{\gamma}} - 1\right) \qquad (13\text{-}23)$$

where y is the ratio of specific heats (for diatomic gases, k = 1.4). The total efficiency is the compressor efficiency multiplied by the mechanical efficiency of the shaft:

$$\eta_T = \eta_m \times \eta_c \tag{13-24}$$

The power required to increase the temperature of the gas is defined as:

$$W = c_p \Delta T \dot{m} \tag{13-25}$$

Taking into account the inefficiencies with the compression process, and substituting this into equation 13-25, then:

$$P_{compressor} = c_p \frac{T_1}{\eta_c} \left(\left(\frac{P_2}{P_1} \right)^{\frac{\gamma-1}{\gamma}} - 1 \right) \dot{m} \tag{13-26}$$

Many compressors are manufactured commercially, and when designing the fuel cell system, the important factors to consider are the temperature, pressure, type of gas handled, reliability, efficiency, and corrosion-free materials.

EXAMPLE 13-2: Designing an Air Compressor

Air at 100 kPa and 298 K enters a compressor at 2 kg/s and receives a pressure boost of 100 kPa. Determine the power required, adiabatic efficiency, and exit temperature for an (a) ideal compressor and (b) adiabatic compressor with 75% efficiency.

Using the ideal gas law:

$$v_1 = \frac{RT_1}{P_1} = \frac{0.287 * 298}{100} = 0.8553$$

The volumetric flow rate can be calculated by:

$$V_1 = v_1 * m = 0.8553 * 2 = 1.710 \frac{m^3}{s}$$

(a) The exit pressure can be estimated by:

$$P_2 = P_1 + \Delta P = 100 + 100 = 200 \text{ kPa}$$

The ideal calculation of T2 can be estimated by:

$$\frac{T_2}{T_1} = \left(\frac{P_2}{P_1}\right)^{(\gamma-1)/\gamma} = (298) * \left(\frac{200}{100}\right)^{(1.38-1)/1.38} = 360.68\,K$$

The ideal work can be calculated by:

$$W = c_{p,avg}(T2 - T1)m = 1.005 * (360.68 - 298) * (2) = 125.98\ kW$$

The efficiency of an ideal compressor is 100%.

(b) For an adiabatic compressor with an efficiency of 75%:

$$W_{act} = \frac{W_{ideal}}{\eta_s} = \frac{125.98}{0.75} = 167.98\,kW$$

The actual exit temperature can be calculated by:

$$T_2 = T_1 + \frac{W_{act}}{m * c_{p,avg}} = 298 + \frac{167.98}{2 * 1.005} = 381.57\ K$$

Using MATLAB to solve:

```
%%%%%%%%%%%%%%%%%%%%%%%%%%%%%%%%%%%%%%%%%%
% EXAMPLE 13-2: Designing an Air Compressor
% UnitSystem SI
%%%%%%%%%%%%%%%%%%%%%%%%%%%%%%%%%%%%%%%%%%
% Inputs
T = 298;          % Operating temperature (K)
P = 100;          % Operating pressure (kPa)
PBoost = 100;     % PBoost: the boost in the pressure
eta = 0.75;       % Efficiency
m_air = 2;        % Mass flow rate (kg/s)
gamma = 1.38;
cp = 1.005;       % Specific heat
R = 0.286;        % Ideal gas constant
% Calculate volume using the ideal gas law
v1 = R*T/P;
% Calculate volumetric flow rate (m³/s)
V1 = v1*m_air;
```

% Ideal exit temperature

T_exit_ideal = T*((P + PBoost)/P)^((gamma- 1)/gamma)

% Ideal work

W_ideal = cp*(T_exit_ideal – T)*m_air

% Actual work

W_actual = W_ideal / eta

% Actual exit temperature

T_exit_actual = T + W_actual/ (cp*m_air)

EXAMPLE 13-3: Designing an Air Compressor

A fuel cell stack with an output power of 50 kW operates with a pressure of 2 bar. Air is fed to the stack using a screw compressor at 1.0 bar and 22 °C with a stoichiometry of 1.5. The average cell voltage is 0.7 V. The rotor speed factor is 300 rev/minK1/2, and the efficiency is 0.6. Find the required rotational speed of the air compressor, the temperature of the air as it leaves the compressor, and the power of the electric motor needed to drive the compressor.

The mass flow rate of air should be found first:

$$m = 3.57 \times 10^{-7} \times \lambda \times \frac{P_e}{V_c} \, kg/s,$$

where λ is the stoichiometry

$$m = \frac{3.57 \times 10^{-7} \times 1.5 \times 50,000}{0.70} = 0.03825 \, kg/s$$

This should then be converted to the mass flow factor:

$$m_{ff} = \frac{0.03825 \times \sqrt{295.15}}{1.0} = \frac{0.657 \, kg}{s\sqrt{K} \, bar}$$

The mass flow factor helps to find the rotor speed factor and efficiency from many standard compressor performance charts. In this example, the rotor speed factor was given as 300 rev/minK1/2, and the efficiency is 0.6. The rotor speed calculation is

$$300 \times \sqrt{295.15} = 5153.98 \, rpm$$

The temperature rise is calculated:

$$\Delta T = \frac{295.15}{0.6}\left(\left(\frac{2.0}{1.0}\right)^{0.286} - 1\right) = 107.86\,K$$

Because the inlet air temperature is 22 °C, the exit temperature will be 129.86 °C. This indicates that the PEM fuel cell would need cooling. The compressor power is

$$W_{compressor} = 1004 \times \frac{295.15}{0.6}\left(\left(\frac{2.0}{1.0}\right)^{0.286} - 1\right)*0.03825 = 4142.15\,kW$$

This power ignores the mechanical losses in the bearings and drive shafts. Many PEM fuel cells also require that the inlet air be humidified, which will alter the specific heat capacity and the ratio of specific heat capacities, and thus alter the performance of the compressor. Sometimes the air will also be humidified after the compression when the air is hotter.

Using MATLAB to solve:

```
%%%%%%%%%%%%%%%%%%%%%%%%%%%%%%%%%%%%%%%%%%
% EXAMPLE 13-3: Designing an Air Compressor
% UnitSystem SI
%%%%%%%%%%%%%%%%%%%%%%%%%%%%%%%%%%%%%%%%%%
% Inputs

T_input = 22;              % Input Temperature (C)
P_input = 1;               % Input Pressure (kPa)
P_op = 2;                  % Operating Pressure (kPa)
S = 1.5;                   % Air stoichiometry
rotorSpeedFactor = 300;    % Rotor speed factor in rev/(min * sqrt(K))
eta = 0.6;                 % Efficiency
P_output = 50;             % Output_power (kW)
V = 0.7;                   % Average cell voltage (V)
T_input = T_input + 273.15;  % Convert temperature to K
R = 0.286;                 % Ideal gas constant

% Calculate the mass flow rate of air (kg/s)

m = 3.57 * 10^-4 * S * P_output / V;

% Mass flow factor

mff = m * sqrt(T_input) / P_input;
```

```
% Rotor speed calculation

rotorSpeed = rotorSpeedFactor * sqrt(T_input);

% Temperature rise

T_rise = (T_input / eta) * ((P_op/P_input)^R – 1)

% Outlet temperature

T_outlet = T_input + T_rise – 273.15;

% Compressor power

powerNeeded = 1004 * T_rise * m;
```

13.2.4 Turbines

In pressurized fuel cell systems, the outlet gas is typically warm and pressurized (though lower than the inlet pressure). This hot gas from fuel cells can be turned into mechanical work through the use of turbines. This energy can be used to generate work that may offset the work needed to compress the air. An example of a turbocompressor system was shown in Figure 13-5. The efficiency of the turbine determines whether it should be incorporated into the fuel cell system[19,20].

Like compressors and fans, the efficiency of the turbine can be determined using the following equation:

$$\frac{T_2}{T_1} = \left(\frac{P_2}{P_1}\right)^{(\gamma-1)/\gamma} \tag{13-27}$$

By substituting the proper equations, the efficiency becomes:

$$\eta_{turbine} = \frac{T_1 - T_2}{T_1 - T_2} = \frac{T_1 - T_2}{T_1}\left(1 - \left(\frac{P_2}{P_1}\right)^{\frac{\gamma-1}{\gamma}} - 1\right) \tag{13-28}$$

where η_c is the ratio between the actual work and the ideal isentropic work between P_1 and P_2. The temperature at the end of expansion is:

$$\Delta T = T_2 - T_1 = \eta_c T_1\left(\left(\frac{P_2}{P_1}\right)^{\frac{\gamma-1}{\gamma}} - 1\right) \tag{13-29}$$

The power of the turbine can be found using the same equation as the compressor:

$$W = c_p \Delta T \dot{m} \tag{13-30}$$

$$P_{turbine} = c_p \eta_c T_1 \left(\left(\frac{P_2}{P_1} \right)^{\frac{\gamma-1}{\gamma}} - 1 \right) \dot{m} \tag{13-31}$$

Due to the inefficiency of the compression and expansion process, a turbine may only recover a portion of work that is required of other system components. If the temperature of the exhaust is high, the turbine may generate all of the required power needed for other fuel cell subsystems.

EXAMPLE 13-4: Calculate Available Power

For the fuel cell analyzed in Example 13-2, the exit temperature is 100°C (383.15 K) and the pressure is 1.8 bar. The efficiency and rotor speed of the turbine are 0.55 and 4000 rev/(minK$^{1/2}$) respectively. Use cp = 1100 J/kgK and $\gamma = 1.38$. What power will be available from the exit gases?

The cathode exit gas mass would have been increased by the water present in the fuel cell, but since the mass change will be very small, we will consider it negligible for this problem, so the value of 0.03825 kg/s will be used. First, calculate the mass flow factor:

$$m_{ff} = \frac{0.03825 \times \sqrt{383.15}}{1.8} = \frac{0.416 \text{ kg}}{s\sqrt{K} \text{ bar}}$$

The rotor speed is

$$Sp_{rotor} = 5000 \times \sqrt{383.15} = 97,871.09 \text{ RPM}$$

Since the efficiency is 0.55, the available power is

$$P_{turbine} = 1100 \times 0.55 \times 383.15 \left(\left(\frac{1.0}{1.8} \right)^{0.275} - 1 \right) * 0.03825 \approx -1323.37 \text{ kW}$$

This is the amount of power given out, which is a useful addition to the 50 kW generated by the fuel cell, but it is not nearly enough to drive the compressor.

Using MATLAB to solve:

```
%%%%%%%%%%%%%%%%%%%%%%%%%%%%%%%%%%%%%%%%
% EXAMPLE 13-4:  Calculate Available Power
% UnitSystem SI
%%%%%%%%%%%%%%%%%%%%%%%%%%%%%%%%%%%%%%%%
T = 110;                        % Operating temperature (C)
P_input = 1;                    % Input Pressure (kPa)
P_op = 1.8;                     % Operating Pressure (kPa)
S = 1.5;                        % Air stoichiometry
rotorSpeedFactor = 5000;        % Rotor speed factor in rev/(min * sqrt(K))
eta = 0.55;                     % Efficiency
P_output = 50;                  % Output_power (kW)
V = 0.7;                        % Average cell voltage (V)
T = T + 273.15;                 % Convert temperature to K
cp = 1100;                      % Specific heat
gamma = 1.38;
```

% Calculate the mass flow rate of air (kg/s)

```
m = 3.57 * 10^-4 * S * P_output / V;
```

% Mass flow factor

```
mff = m * sqrt(T) / P_input;
```

% Calculate rotor speed

```
rotorSpeed = rotorSpeedFactor * sqrt(T);
```

% Calculate the temperature rise

```
T2 = T * ((P_op/P_input)^((gamma- 1)/gamma) - 1);
```

% Compressor power

```
powerGenerated = cp * eta * T2 * m;
```

13.2.5 Fuel Cell Pumps

Pumps, like blowers, compressors, and fans, are among the most important components in the fuel cell plant system. These components are required to move fuels, gases, and condensate through the system and are important factors in the fuel cell system efficiency. Small- to medium-sized PEM fuel cells for portable applications have a back pressure of about 10 kPa or 1 m of water[21]. This is too high for most axial or centrifugal fans, as discussed earlier[22].

Choosing the correct pump for the fuel cell application is important. As in fans, blowers, and compressors, factors to consider are efficiency, reliability, corrosion-free materials, and the ability to work with the required temperatures, pressures, and flow rates for the specific fuel cell system. The appropriate matching of a high-efficiency pump with the appropriate motor speed/torque curve may allow for a more efficient fuel cell stack and system. The equations that describe pump performance characteristics are the same as the fan performance characteristics (Equations (13-13–13-16)).

EXAMPLE 13-5: Pump Design

A pump needs to be selected to move 10 kg/s of water with a pressure boost of 100 kPa. Calculate the work required, exit temperature and, if applicable, the pump speed and diameter for (a) the ideal pump, (b) an axial flow pump at maximum efficiency, and (c) a centrifugal pump at maximum efficiency. The maximum efficiency for an axial flow pump has the following parameters:

$$(C_Q)_{max} = 0.6398 \quad (C_H)_{max} = 1.4922 \quad \eta_{max} = 0.8488$$

The maximum efficiency for a centrifugal flow pump has the following parameters:

$$(C_Q)_{max} = 0.11509 \quad (C_H)_{max} = 5.3317 \quad \eta_{max} = 0.93508$$

(a) The ideal work can be calculated by:

$$W_{ideal} = \dot{m}v\Delta P = (10)(1.002 * 10^{-3}) * 100 = 1.002 \text{ kW}$$

The exit temperature is equal to the inlet temperature.

(b) The actual work can be calculated by:

$$W_{act} = \frac{W_{ideal}}{\eta_{max}} = \frac{1.002}{0.8488} = 1.18 \text{ kW}$$

The temperature change is then given by:

$$\Delta T = \frac{W_{act} - W_{ideal}}{mc_p} = \frac{1.18 - 1.002}{(10)(4.181)} = 0.00427 \text{ K}$$

At maximum efficiency, the following equations apply for the pump diameter and speed:

$$D = \left[\frac{(C_H)_{max} m^2}{\rho \Delta P (C_Q)_{max}^2} \right]^{1/4} \quad \text{and} \quad N = \frac{m}{\rho D^3 (C_Q)_{max}}$$

$$D = \left[\frac{(C_H)_{max} m^2}{\rho \Delta P (C_Q)_{max}^2} \right]^{1/4} = \left[\frac{1.4922 * (10)^2}{(998) * (100) * 0.6398^2} \right]^{0.25} = 0.246 \text{ m}$$

$$N = \frac{m}{\rho D^3 (C_Q)_{max}} = \frac{10}{998 * 0.246^3 * 0.6398} = 1.0546 \text{ rps} = 63.28 \text{ rpm}$$

(c) The actual work can be calculated by:

$$W_{act} = \frac{W_{ideal}}{\eta_{max}} = \frac{1.002}{0.93508} = 1.072 \text{ kW}$$

The temperature change is then given by:

$$\Delta T = \frac{W_{act} - W_{ideal}}{m c_p} = \frac{1.072 - 1.002}{(10)(4.181)} = 0.00166 \text{ K}$$

At maximum efficiency:

$$D = \left[\frac{(C_H)_{max} m^2}{\rho \Delta P (C_Q)_{max}^2} \right]^{1/4} = \left[\frac{5.3317 * (10)^2}{(998) * (100) * 0.11509^2} \right]^{0.25} = 0.403 \text{ m}$$

$$N = \frac{m}{\rho D^3 (C_Q)_{max}} = \frac{10}{998 * 0.403^3 * 0.11509} = 1.330 \text{ rps} = 79.81 \text{ rpm}$$

Using MATLAB to solve:

```
%%%%%%%%%%%%%%%%%%%%%%%%%%%%%%%%%%%%%%%%%
% EXAMPLE 13-5: Pump Design
% UnitSystem SI
%%%%%%%%%%%%%%%%%%%%%%%%%%%%%%%%%%%%%%%%%
% Inputs

m_H2O = 10;        % Mass flow rate (kg/s)
PBoost = 100;      % PBoost: the boost in pressure
cp = 4.181;        % Specific heat
rho = 998;         % Density
```

```
eta_axial = 0.8488;      % Axial pump efficiency
actual_CQ_max = 0.6398;
actual_CH_max = 1.4922;
eta_cent = 0.93508;      % Centrifugal pump efficiency
cent_CQ_max = 0.11509;
cent_CH_max = 5.3317;
```

% Part a: Calculate ideal work

```
W_ideal = m_H2O * 0.001002 * PBoost;
```

```
%%%%%%%%%%%%%%%%%%%%%%%%%%%%%%%%%%%%%
```

% Part b: Calculate actual work of axial flow pump @ max efficiency

```
W_actual = W_ideal / eta_actual;
```

% Temperature change

```
delT_actual = (W_actual - W_ideal) / (m_H2O * cp);
```

% Pump diameter

```
d_axial = ((actual_CH_max * m_H2O^2)/(rho * PBoost * actual_CQ_max^2))^0.25;
```

% Pump speed

```
pumpSpeed_axial = m_H2O / (rho * (d_axial^3) * actual_CQ_max) * 60;
```

```
%%%%%%%%%%%%%%%%%%%%%%%%%%%%%%%%%%%%%
```

% Part c: Calculate actual work of centrifugal flow pump @ max efficiency

```
W_cent = W_ideal / eta_cent;
```

% Temperature change

```
delT_cent = (W_cent - W_ideal) /(m_H2O * cp);
```

% Pump diameter

```
d_cent = ((cent_CH_max * m_H2O^2)/(rho * PBoost * cent_CQ_max^2))^0.25;
```

% Pump speed

```
pumpSpeed_cent = m_H2O / (rho * d_cent^3 * cent_CQ_max) * 60;
```

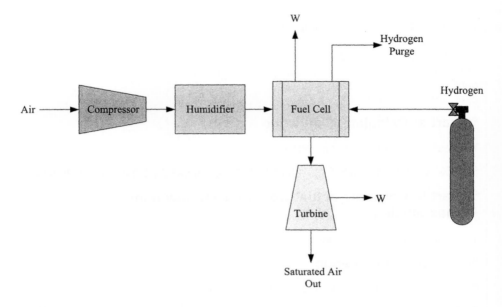

FIGURE 13-6. Fuel cell system for Example 13-6.

EXAMPLE 13-6: System Design

A 1-kW PEM fuel cell operates at a temperature of 80°C and 3 atm with a cell voltage of 0.65 V, as shown in Figure 13-6. The air stoichiometry is 2.5. The compressor and turbine efficiency is 0.6. The temperature and pressure of the air entering the compressor are 25°C and 1 atm, respectively. (a) Find the amount of power required for the compressor, and the associated temperature change. (b) Is a humidifier needed for this fuel cell system? (c) Would putting a turbine into the fuel cell system be useful? If so, calculate the associated temperature change with the turbine. (d) Calculate the amount of heat generated by the stack.

(a) With the molar mass of air being 28.97×10^{-3} kg/mol and the mole fraction of air that is oxygen is 0.21, the inlet air flow rate can be estimated by:

$$m_{air,in} = \frac{S_{O2} \times M_{air} \times P}{4 \times x_{O2} \times V_{cell} \times F}$$

$$m_{air,in} = \frac{2.5 \times 28.97 \times 10^{-3} \times 1000}{4 \times 0.21 \times 0.65 \times 96,485} = 0.001375 \, \text{kg/s}$$

Using air at 25 °C, $\gamma = 1.4$, the temperature rise associated with the compression of air can be calculated by:

$$\Delta T = T_2 - T_1 = \frac{T_1}{\eta_c}\left(\left(\frac{P_2}{P_1}\right)^{\frac{\gamma-1}{\gamma}} - 1\right)$$

$$\Delta T = \frac{298}{0.6}\left(\left(\frac{303,975.03}{101,325.01}\right)^{\frac{1.4-1}{1.4}} - 1\right) = 183.35 \text{ K}$$

With cp = 1004 J/kg*K, the power needed to compress the air is:

$$P_{compressor} = c_p \frac{T_1}{\eta_c}\left(\left(\frac{P_2}{P_1}\right)^{\frac{\gamma-1}{\gamma}} - 1\right)\dot{m}$$

$$P_{compressor} = 1004 * \frac{298}{0.6}\left(\left(\frac{303,975.03}{101,325.01}\right)^{\frac{1.4-1}{1.4}} - 1\right) * 0.001375 \text{ kg/s} = 253.12 \text{ W}$$

There is a large temperature rise, which will need to be compensated for by cooling the air before it enters the cell.

(b) The exit air flow rate can be estimated by:

$$m_{air,out} = \frac{S_{O2} \times M_{air} \times P}{4 \times x_{O2} \times V_{cell} \times F} - \frac{32 \times 10^{-3} \times P}{4 \times V_{cell} \times F}$$

$$m_{air,out} = \frac{2.5 \times 28.97 \times 10^{-3} \times 1000}{4 \times 0.21 \times 0.65 \times 96,485} - \frac{32 \times 10^{-3} \times 1000}{4 \times 0.65 \times 96,485} = 0.001246 \text{ kg/s}$$

If the exit air is at 100% humidity and the saturated vapor pressure of water at 80 °C is 47.39 kPa, then the pressure of the dry air is the total pressure of the exit air minus the saturated vapor pressure. If we estimate the exit pressure to be less than the entry pressure due to pressure drop through the flow fields, we estimate a 20 kPa pressure drop, then 303,975.03 − 20,000 = 283,975.03 Pa, and the pressure of dry air is 283,975.03 Pa − 47,390 Pa = 236,585.03 Pa.

With the molecular mass of water is 18 and the molecular mass of air is 28.97, then the humidity ratio is:

$$\omega = \frac{m_w}{m_a} = \frac{18 \times P_W}{28.97 \times P_a} = \frac{18 \times 47,390}{28.97 \times 236,585.03} = 0.1245$$

Therefore, the mass flow rate of water leaving the cell is:

$$m_w = \omega m_a = 0.1245 \times 0.001246 \text{ kg/s} = 0.000155 \text{ kg/s}$$

The rate of water production:

$$m_{H2O} = \frac{M_{air} \times P}{2 \times V_{cell} \times F} = \frac{28.97 \times 10^{-3} \times 1000}{2 \times 0.65 \times 96,485} = 0.000231 \text{kg/s}$$

The rate that water should enter the cell can be estimated by $0.000155 - 0.000231 = -0.000076$, which implies that no water needs to enter the cell. Therefore, a humidifier is not needed. The total exit flow rate is the dry air flow rate plus the water flow rate, which is $0.001246 \text{ kg/s} + 0.000155 \text{ kg/s} = 0.001401 \text{ kg/s}$.

(c) If cp = 1100 J/kg∗K, and $\gamma = 1.33$, and the turbine exit pressure is still above atmospheric pressure ($\approx 283,975.03 - 100,000$ Pa = 183,975.03), then the turbine power can be calculated by:

$$P_{turbine} = c_p \eta_c T_1 \left(\left(\frac{P_2}{P_1} \right)^{\frac{\gamma-1}{\gamma}} - 1 \right) \dot{m}$$

$$P_{turbine} = 1100 * 0.6 * 353.15 \left(\left(\frac{183,975.03}{283,975.03} \right)^{\frac{1.33-1}{1.33}} - 1 \right) * 0.001401 = -33.33 \text{ W}$$

About 33 W would make a useful contribution to the 253 W needed. The temperature change through the turbine can be calculated by:

$$\Delta T = \eta_c T_1 \left(\left(\frac{P_2}{P_1} \right)^{\frac{\gamma-1}{\gamma}} - 1 \right)$$

$$\Delta T = 0.6 * 353.15 \left(\left(\frac{183,975.03}{283,975.03} \right)^{\frac{1.33-1}{1.33}} - 1 \right) = -21.62 \text{ K}$$

This would bring the exit gas temperature down to 58.38 K.

(d) The heating rate is:

$$q = P * \left(\frac{1.25}{V_{cell}} - 1 \right) = 1000 * \left(\frac{1.25}{0.65} - 1 \right) = 923.077 \text{ W}$$

Using MATLAB to solve:

```
%%%%%%%%%%%%%%%%%%%%%%%%%%%%%%%%%%%%%%%%%
% EXAMPLE 13-6:  System Design
% UnitSystem SI
%%%%%%%%%%%%%%%%%%%%%%%%%%%%%%%%%%%%%%%%%
T_in = 25;              % Inlet temperature (C)
T_op = 80;              % Operating temperature (C)
P_op = 3000;            % Operating Pressure (kPa)
P_input = 1000;         % Inlet Pressure (kPa)
S = 2.5;                % Air stoichiometry
V = 0.65;               % Average cell voltage (V)
eta = 0.5;              % Efficiency
P =1000;                % Power
F = 96485;              % Faraday's law
M_air = 28.97e-3;       % Molecular weight of air
x_O2 = 0.21;            % Oxygen mole fraction
cp = 1004;              % Specific heat
P_W = 47390;            % Pressure drop of water (Pa)
P_A = 236585.03;        % Pressure drop of air (Pa)
gamma = 1.4;
T_in = T_in + 273.15;   % Convert temperature to K
T_op = T_op + 273.15;   % Convert temperature to K
```

% Inlet air flow rate

```
m_air = S*M_air*P_input / (4*x_O2*V*F);
```

% Temperature rise due to the compression of air

```
delta_temp = T_in / eta*((P_op/P_input)^((gamma- 1)/gamma) - 1);
```

% Power needed to compress the air

```
P_compressor = cp*delta_temp*m_air;
```

% Exit air flow rate

```
m_air_out = m_air - 32e-3*P_input / (4*V*F);
```

% Humidity ratio

```
w = 18*P_W / (28.97*P_A);
```

% The mass flow rate of water leaving the cell

```
m_w = w*m_air_out;
```

% Rate of water production

```
m_H2O = M_air*P_input / (2*V*F);
```

% Remaining water

```
remainingWater = m_w – m_H2O;
if (remainingWater > 0)
'humidifier required'
else
'no humidifier required'
end
cp = 1100;          % Specific heat
gamma = 1.33;
P1 = 283975.03;   % Pressure drop
P2 = 183975.03;   % Pressure drop
```

% Temperature rise due to the compression of air

```
delta_temp2 = T_op*eta*((P2/P1)^((gamma- 1)/gamma) – 1);
```

% Turbine power

```
P_turbin = cp*delta_temp2*(m_w + m_air_out);
```

% Heating rate

```
heating_rate = P_input*(1.25/V – 1);
```

Chapter Summary

A fuel cell system can be very efficient with a simple plant and electrical subsystem—or a very complex one. Typically, the larger the fuel cell stack, the more complex the fuel cell plant subsystem will be. The number of ways to design and optimize the fuel cell plant and electrical subsystems are endless. The plant components reviewed in this chapter include humidifiers, fans, blowers, compressors, turbines, and pumps. A series of quick models (like those presented in this chapter) can aid the fuel cell stack designer to make the best design decisions.

Problems

- A 100 kW fuel cell operates at 0.8 V per cell at 60°C at ambient pressure with an oxygen stoichiometry of 2.0. Liquid water is separated from the cathode exhaust. Calculate the amount of water that needs to be stored for 3 days of operation.
- Calculate the temperature in a compressor that provides air for a 50 kW fuel cell system. Assume that air is dry at 25°C and 1 atm, and the delivery pressure is 2 atm. The fuel cell generates 0.70 V per cell and operates with an oxygen stoichiometric ratio of 2. The compressor efficiency is 0.65.

- For the fuel cell described in the second problem, calculate the amount of power that can be recovered if the exhaust air is run through the turbine. The exhaust gas has 100% humidity with a pressure drop of 0.3 atm. The turbine efficiency is 65%.
- For the fuel cell system described in the second problem, calculate the amount of water (g/s) needed to fully saturate the air at the fuel cell entrance at 70°C.
- For the fuel cell system described in the second problem, calculate the amount of heat (W) needed for the air humidification process.

Endnotes

[1] Spiegel, C.S. 2007. *Designing and Building Fuel Cells*. New York: McGraw-Hill.
[2] Ibid.
[3] Mohan, N., T. Undeland, and W. Robbins. 1995. *Power Electronics: Converters, Applications and Design*. 2nd ed. New York: John Wiley & Sons, Inc.
[4] Spiegel, *Designing and Building Fuel Cells*.
[5] Ibid.
[6] Mohan, *Power Electronics: Converters, Applications and Design*, 2nd ed.
[7] Ibid.
[8] Robinson, R.N. 1996. *Chemical Engineering Reference Manual for the PE Exam*. 5th ed. Belmont, CA: Professional Publications, Inc.
[9] Spiegel, *Designing and Building Fuel Cells*.
[10] Ibid.
[11] Incropera, F.P., and D.P. Dewitt. 2002. *Introduction to Heat Transfer*. 4th ed. New York: John Wiley & Sons, Inc.
[12] Spiegel, *Designing and Building Fuel Cells*.
[13] Li, X. 2006. *Principles of Fuel Cells*. New York: Taylor & Francis Group.
[14] Incropera, *Introduction to Heat Transfer*, 4th ed.
[15] Spiegel, *Designing and Building Fuel Cells*.
[16] Ibid.
[17] Ibid.
[18] Ibid.
[19] Incropera, *Introduction to Heat Transfer*, 4th ed.
[20] Spiegel, *Designing and Building Fuel Cells*.
[21] Larminie, J., and A. Dicks. 2000. *Fuel Cell Systems Explained*. Chichester, England. John Wiley & Sons, Ltd.
[22] Spiegel, *Designing and Building Fuel Cells*.

Bibliography

Barbir, F. 2005. *PEM Fuel Cells: Theory and Practice*. Burlington, MA: Elsevier Academic Press.
Brunstein, V. November 1, 2002. *DC-DC Converters Condition Fuel Cell Outputs*. New Milford, NJ: Advanced Power Associates Corp.

Chen, F.C., Z. Gao, R.O. Loutfy, and M. Hecht. Analysis of optimal heat transfer in a PEM fuel cell cooling plate. *Fuel Cells.* Vol. 3, No. 4, 2003.

Dinsdale, J. January 2004. *New Energy Conservation Technologies.* Ceramic Fuel Cells Limited, ESAA Residential School, Queensland University of Technology and University of Queensland.

EG&G Technical Services. 2000. *The Fuel Cell Handbook.* 5th ed. Washington, DC: U.S. Department of Energy.

GeoCool Lab. Department of Mechanical Engineering, University of Alabama. Available at: http://bama.ua.edu/~geocool/PsychWB.doc. Accessed March 24, 2007.

Kimball, J.W. Power converters for micro fuel cells. *Power Electronics Technol.* October, 2004.

Li, X. 2006. *Principles of Fuel Cells.* New York: Taylor & Francis Group.

Lin, B. 1999. Conceptual design and modeling of a fuel cell scooter for urban Asia. Princeton University, master's thesis.

Izenson, M.G., and R.W. Hill. Water and thermal balance in PEM fuel cells. *J. Fuel Cell Science and Technology.* Vol. 1, 2004, pp. 10–17.

Makuch, G. February 1, 2004. *Micro Fuel Cells Strive for Commercialization.* Bothel, WA: Neah Power Systems.

Mench, M.M., Z.H. Wang, K. Bhatia, and C.Y. Wang. 2001. *Design of a Micro-direct Methanol Fuel Cell.* Electrochemical Engine Center, Department of Mechanical and Nuclear Engineering, Pennsylvania State University.

Mikkola, M. Experimental studies on polymer electrolyte membrane fuel cell stacks. Helsinki University of Technology, Department of Engineering Physics and Mathematics, master's thesis, 2001.

National Fuel Cell Research Center. Power Electronics for Fuel Cells Workshop, August 8–9, 2003.

Ozpineci, B., L.M. Tolbert, and Z. Du. *Multiple Input Converters for Fuel Cells.* Presented at the *Industry Applications Conference 39th IAS Annual Meeting, 2004. Conference Record of the 2004 IEEE.* Vol. 2, No. 3–7, October 2004, pp. 791–797.

Perry, R.H., and D.W. Green. 1997. *Perry's Chemical Engineers' Handbook.* 7th ed. New York: McGraw-Hill.

Rayment, C., and S. Sherwin. May 2, 2003. *Introduction to Fuel Cell Technology.* Department of Aerospace and Mechanical Engineering, University of Notre Dame.

Smithsonian Institution, 2001. *Fuel Cell Origins.* Available at: http://fuelcells.si.edu.

Tse, L., and D. Bong. *A Brief History of Fuel Cells.* Available at: http://www.visionengineer.com.

Zhang, Y., M. Ouyang, Q. Lu, J. Luo, and X. Li. A model predicting performance of proton exchange membrane fuel cell stack thermal system. *Appl. Thermal Eng.* Vol. 24, 2004, pp. 501–513.

CHAPTER 14

Model Validation

14.1 Introduction

Model validation is the most important step in the model-building process. However, it is often neglected. Validating a mathematical model usually consists of quoting the R^2 statistic from the fit. Unfortunately, a high R^2 value does not mean that the data actually fit well. If the model does not fit the data well, this negates the purpose of building the model in the first place.

There are many statistical tools that can be used for model validation. This chapter covers the basic concepts for model validation, which include residuals, normal distribution of random errors, missing terms in the functional part of the model, and unnecessary terms in the model. The most useful is graphical residual analysis. There are many types of plots of residuals that allow the model accuracy to be evaluated. There are also several methods that are important to confirm the adequacy of graphical techniques. To help interpret a borderline residual plot, a lack-of-fit test for assessing the correctness of the functional part of the model can be used. The number of plots that can be used for model validation is limited when the number of parameters being estimated is relatively close to the size of the data set. This occurs when there are designed experiments. In this case, residual plots are often difficult to interpret because of the number of unknown parameters.

14.2 Residuals

The residuals from a fitted model are the differences of the responses at each combination of variables, and the predicted response using the regression function. The definition of the residual for the ith observation in the data set can be written as:

$$e_{ij} = y_{ij} - \bar{y}_{ij} \qquad (14\text{-}1)$$

with y_{ij} denoting the ith response in the data set and $\bar{y}_{ij}$ represents the list of explanatory variables, each set at the corresponding values found in the ith observation in the data set. If a model is adequate, the residuals should have no obvious patterns or systematic structure. The primary method of determining whether the residuals have any particular pattern is by studying the scatterplots. Scatterplots of the residuals are used to check the assumption of constant standard deviation of random errors.

14.2.1 Drifts in the Measurement Process

Drifts in the measurement process can be checked by creating a "run order" or "run sequence" plot of the residuals. These are scatterplots where each residual is plotted versus an index that indicates the order (in time) in which the data were collected. This is useful when the data have been collected in a randomized run order, or an order that is not increasing or decreasing in any of the predictor variables. If the data are increasing or decreasing with the predictor variables, then the drift in process may not be separated from the functional relationship between the predictors and the response. This is why randomization is encouraged in the design of experiments[1,2].

14.2.2 Independent Random Errors

A lag plot of residuals helps to assess whether the random errors are independent from one to the next. If the errors are independent, the estimate of the error in the standard deviation will be biased, which leads to improper inferences about the process. The lag plot works by plotting each residual value versus the value of the successive residual. Due to the way that the residuals are paired, there will be one less point than most other types of residual plots.

There will be no pattern or structure in the lag plot if the errors are independent. The points will appear randomly scattered across the plot, and if there is a significant dependence between errors, there will be some sort of deterministic pattern that is evident.

Example 14-1 shows the types of scatterplots used for determining consistent standard deviation.

EXAMPLE 14-1: Plotting Residuals

A parametric model has been developed to predict the performance of a PEM fuel cell over a range of operating currents and temperatures. The parametric equation predicts activation overvoltage from a linear regression analysis. Table 14-1 shows the run order, experimental temperature, experimental current, calculated experimental activation overpotential, and predicted activation overpotential from the model given in Amphlett et al.[3]. (a) Calculate and plot the residuals versus the experimental factors using scatterplots, (b) create the run order plot, and (c) create the lag plot.

Using MATLAB to solve:

TABLE 14-1
Experimental and Calculated Results for Example 14-1

Run Order	Temperature (K)	Current (Amps)	Experimental Activation Overvoltage (V)	Predicted Activation Overvoltage (V)
1	358	2.72	−0.2717	−0.2647
2	328	6.66	−0.4017	−0.4014
3	343	6.66	−0.3522	−0.35
4	358	6.66	−0.3038	−0.3025
5	343	6.66	−0.3341	−0.3283
6	343	6.66	−0.3756	−0.3775
7	328	6.66	−0.3727	−0.3747
8	343	2.72	−0.322	−0.3188
9	343	6.66	−0.3492	−0.35
10	343	6.66	−0.3472	−0.35
11	328	2.72	−0.3141	−0.3193
12	328	6.66	−0.352	−0.3541
13	343	6.66	−0.3482	−0.35
14	358	6.66	−0.3473	−0.3537
15	358	6.66	−0.325	−0.3252
16	358	6.66	−0.3218	−0.3252
17	343	16.33	−0.4075	−0.4083
18	343	16.33	−0.3902	−0.3868
19	343	6.66	−0.3492	−0.35
20	343	6.66	−0.3788	−0.3775
21	358	16.33	−0.3834	−0.386
22	343	2.72	−0.2969	−0.292
23	343	6.66	−0.3249	−0.3283
24	343	2.72	−0.2868	−0.292
25	343	16.33	−0.4453	−0.4379
26	343	6.66	−0.3502	−0.35
27	328	6.66	−0.3793	−0.3747
28	343	16.33	−0.4062	−0.4083

```
%%%%%%%%%%%%%%%%%%%%%%%%%%%%%%%%%%%%%%%%%%%%
% EXAMPLE 14-1: Plotting Residuals
% UnitSystem SI
%%%%%%%%%%%%%%%%%%%%%%%%%%%%%%%%%%%%%%%%%%%%
% Inputs

% Number of experiments
run_number = [1:28];
```

% Temperature of each run

```
T = [358 328 343 358 343 343 328 343 343 343 328 328 343 358 358 358
    343 . . . 343 343 343 358 343 343 343 343 343 328 343];
```

% Current density of each run

```
i = [2.72 6.66 6.66 6.66 6.66 6.66 6.66 2.72 6.66 6.66 2.72 6.66 6.66 . . . 6.66 6.66
    6.66 16.33 16.33 6.66 6.66 16.33 2.72 6.66 2.72 16.33 6.66 . . . 6.66 16.33];
```

% Activation polarization (experimental)

```
V_act_ex = [–0.2717 –0.4017 –0.3522 –0.3038 –0.3341 –0.3756 –0.3727
    –0.322 . . . –0.3492 –0.3472 –0.3141 –0.352 –0.3482 –0.3473 –0.325 –0.3218
    –0.4075 . . . –0.3902 –0.3492 –0.3788 –0.3834 –0.2969 –0.3249 –0.2868
    –0.4453 –0.3502 . . . –0.3793 –0.4062];
```

% Activation polarization (predicted)

```
V_act_sim = [–0.2647 –0.4014 –0.35 –0.3025 –0.3283 –0.3775 –0.3747
    –0.3188 . . . –0.35 –0.35 –0.3193 –0.3541 –0.35 –0.3537 –0.3252 –0.3252
    –0.4083 . . . –0.3868 –0.35 –0.3775 –0.386 –0.292 –0.3283 –0.292 –0.4379
    –0.35 . . . –0.3747 –0.4083];
```

% Obtain residuals

```
e = V_act_ex - V_act_sim;
```

% Plot residuals versus temperature

```
figure1 = figure('Color',[1 1 1]);
hdlp=scatter(T,e);
xlabel('Temperature (K)','FontSize',12,'FontWeight','Bold');
ylabel('Residuals','FontSize',12,'FontWeight','Bold');
set(hdlp,'LineWidth',1.5);
grid on;
```

% Plot residuals versus current density

```
figure2 = figure('Color',[1 1 1]);
hdlp=scatter(i,e);
xlabel('Current Density (A/cm2)','FontSize',12,'FontWeight','Bold');
ylabel('Residuals','FontSize',12,'FontWeight','Bold');
set(hdlp,'LineWidth',1.5);
grid on;
```

% Plot residuals versus run order

```
figure3 = figure('Color',[1 1 1]);
hdlp=scatter(run_number,e);
xlabel('Run Order','FontSize',12,'FontWeight','Bold');
ylabel('Residuals','FontSize',12,'FontWeight','Bold');
set(hdlp,'LineWidth',1.5);
grid on;
```

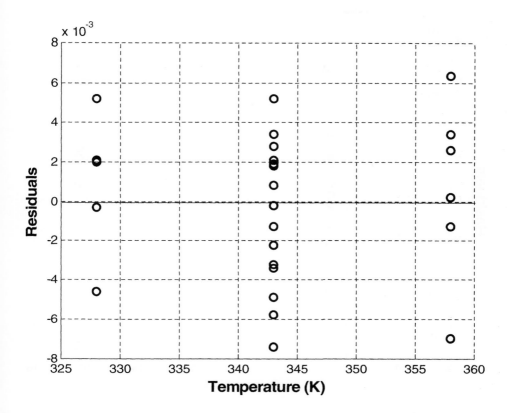

FIGURE 14-1. Scatterplot of temperature versus residuals.

```
% Plot residuals versus residuals (Lag plot)
figure4 = figure('Color',[1 1 1]);
hdlp=scatter(e(2:28),e(1:27));
xlabel('Residuals','FontSize',12,'FontWeight','Bold');
ylabel('Residuals','FontSize',12,'FontWeight','Bold');
set(hdlp,'LineWidth',1.5);
grid on;
```

Figures 14-1 through 14-4 show the residuals versus the experimental factors, the run order plot, and the lag plot for Example 14-2.

For Figures 14-1 and 14-2, the range of the residuals in these figures looks essentially constant across the levels of the predictor variables, which are temperature, and current. There are points randomly scattered above and below $y = 0$ line. This suggests that the standard deviation of the

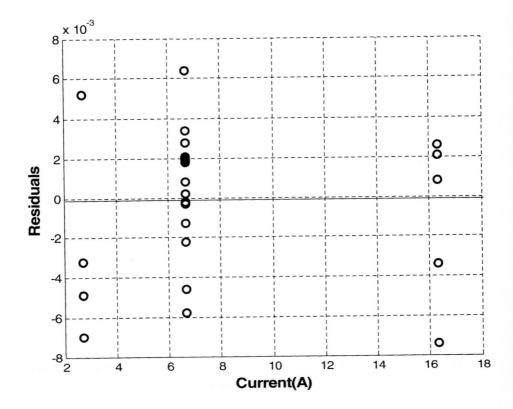

FIGURE 14-2. Scatterplot of current versus residuals.

random error is the same for the responses observed at each temperature and current. These scatter plots indicate that the parametric model is most likely a good fit to the experimental data.

14.3 Normal Distribution of Normal Random Errors

14.3.1 Histograms and Normal Probability Plots

In order to determine the normal distribution of normal random errors, there are two plots that can be used: the histogram and the normal probability plot. These plots can be used to verify if the random error in the process has been obtained from a normal distribution. When making a decision about the model and process, the normality assumption is needed to obtain more information about the error rates. If the random errors are not from a normal distribution, incorrect decisions will be made more or less often than the stated confidence levels indicate.

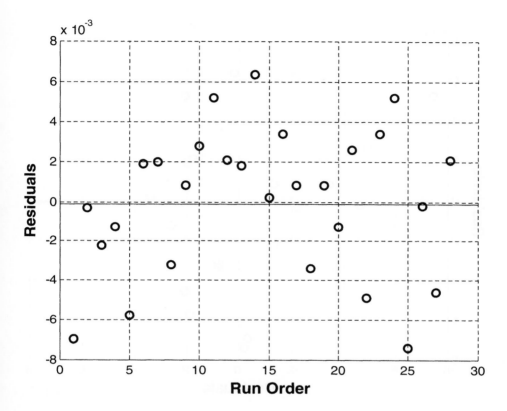

FIGURE 14-3. Run order plot for Example 14-1.

The normal probability plot is a plot of the sorted values of the residuals versus the associated theoretical values from the standard normal distribution. Unlike most residual scatterplots, a random scatter of points does not validate that the assumption being checked was met. If the random errors are normally distributed, the plotted points will lie close to a straight, diagonal line. If points deviate significantly from the line, the random errors are probably not randomly distributed, and the data have some outliers in them.

If the normal probability plot suggests that the errors come from a non-normal distribution, then a histogram can be used to determine what the distribution looks like. If the histogram is bell-shaped, the conclusions of the normal probability plots are correct. It is important to note that information about the distribution of random errors from the process can only be obtained if the functional part of the model is correctly specified, the standard deviation is constant, there is no drift in the process, and the random errors are not dependent on the run[4,5].

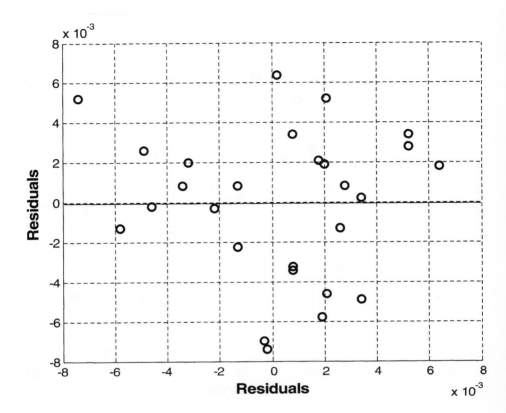

FIGURE 14-4. Lag plot for Example 14-1.

EXAMPLE 14-2: Creating a Histogram and Normal Probability Plot

Using the residuals from Example 14-1, create a normal probability plot and histogram to determine if the data are normally distributed. Another method of presenting the error data is through the use of a box plot. Use a box plot to view the residuals from Example 14-1.

Using MATLAB to solve:

```
%%%%%%%%%%%%%%%%%%%%%%%%%%%%%%%%%%%%%%%%%%%
% EXAMPLE 14-2: Creating a Histogram and Probability Plot
% UnitSystem SI
%%%%%%%%%%%%%%%%%%%%%%%%%%%%%%%%%%%%%%%%%%%
% Plot normal probability plot
figure1 = figure('Color',[1 1 1]);
hdlp=normplot(e);
xlabel('Quantities from Standard Normal Distribution','FontSize',12,'FontWeight',
    'Bold');
ylabel('Residuals','FontSize',12,'FontWeight','Bold');
set(hdlp,'LineWidth',1.5);
grid on;

% Plot Histogram
n = length(e);
b = -0.01:0.002:0.01;
figure2 = figure('Color',[1 1 1]);
hdlp = hist(e,b);
maxn=max(hdlp);
cs = cumsum(hdlp * maxn/n);
bar(b,hdlp,0.95,'g')
axis([-0.01,0.01,0,maxn])
box off
hold on
plot(b,cs,'k-s')
xlabel('Residuals','FontSize',12,'FontWeight','Bold');
ylabel('Count','FontSize',12,'FontWeight','Bold');
plot([-0.01 0.01],[maxn maxn],'k',[0.01 0.01],[0 maxn],'k')
j=0:0.1:1;
lenj=length(j);
text(repmat(0.011,lenj,1), maxn. * j',num2str(j',2))
plot([repmat(0.01,1,lenj);repmat(0.01,1,lenj)],[maxn * j;maxn * j],'k')

% Box plot
figure3 = figure('Color',[1 1 1]);
boxplot(e,'Notch','on');
```

Figures 14-5 through 14-7 show the normal probability plot, histogram, and box plot for Example 14-2.

14.4 Missing Terms in the Functional Part of the Model

Residual plots are the most valuable tool for assessing whether variables are missing in the functional part of the model. However, if the results are

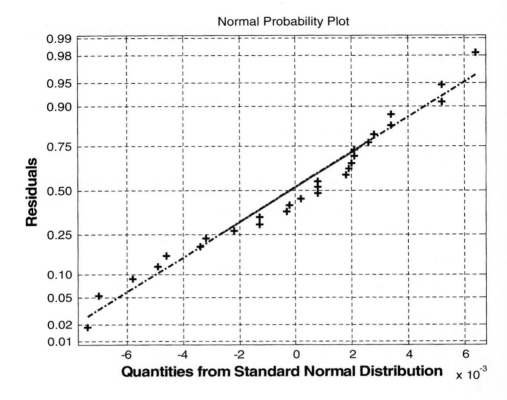

FIGURE 14-5. Normal probability plot for Example 14-2.

nebulous, it may be helpful to use statistical tests for the hypothesis of the model. One may wonder if it may be more useful to jump directly to the statistical tests (since they are more quantitative), however, residual plots provide the best overall feedback of the model fit. These quantitative tests are termed "lack-of-fit" tests, and there are many illustrated in any statistics textbook.

The most commonly used strategy is to compare the amount of variation in the residuals with an estimate of the random variation in the model is to use an additional data set. If the random variation is similar, then it can be assumed that no terms are missing from the model. If the random variation from the model is larger than the random variation from the independent data set, then terms may be missing or misspecified in the functional part of the model.

Comparing the variation between experimental and model data sets is very useful, however, there are many instances where a replicate measurement is not available. If this is the case, the lack-of-fit statistics can be

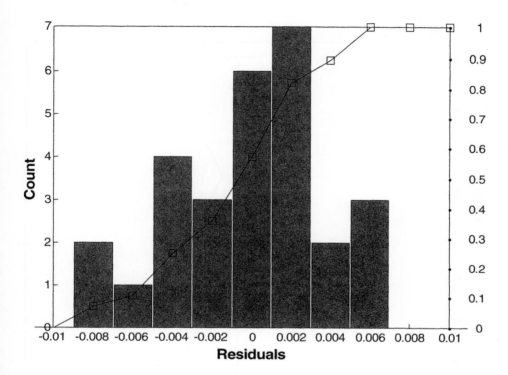

FIGURE 14-6. Histogram for Example 14-2.

calculated by partitioning the residual standard deviation into two independent estimators of the random variation in the process.

One estimator depends upon the model, and the means of the replicated sets of data (σ_m), and the other estimator is a standard deviation of the variation observed in each set of replicated measurements (σ_r). The squares of these two estimators are often called "mean square for lack-of-fit." The model estimator can be calculated by[6]:

$$\sigma_m = \sqrt{\frac{1}{(n_u - p)} \sum_{i=1}^{n_u} n_i (y_{ij} - \bar{y}_{ij})^2} \qquad (14\text{-}2)$$

where p is the number of unknown parameters in the model, n is the sample size of the data set used to fit the model, n_u is the number of combinations of predictor variable levels, and n_i is the number of replicated observations at the ith combination of predictor variable levels.

If the model is a good fit, the value of the function would be a good estimate of the mean value of response for every combination of predictor variable values. If the function provides good estimates of the mean response

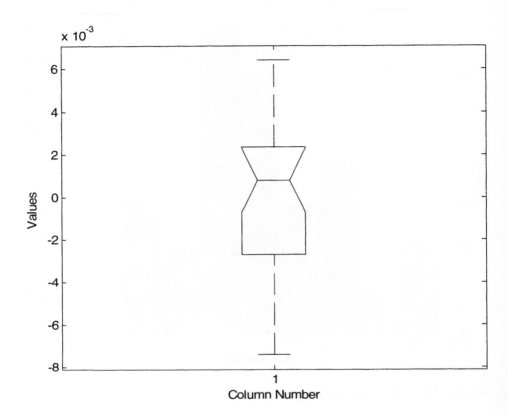

FIGURE 14-7. Box plot for Example 14-2.

at the ith combination, then σ_m should be close in value to σ_r and should also be a good estimate of σ. If the model is missing any important terms, or any of the terms are correctly specified, then the function will provide a poor estimate of the mean response for some combination of predictors, and σ_m will probably be greater than σ_r.

The model-dependent estimator can be calculated using[7]:

$$\sigma_r = \sqrt{\frac{1}{(n - n_u)} \sum_{i=1}^{n_u} \sum_{i=1}^{n_i} (y_{ij} - \bar{y}_{ij})^2} \qquad (14\text{-}3)$$

Since σ_r depends only on the data and not on the functional part of the model, this indicates that σ_r will be a good estimator of σ, regardless of whether the model is a complete description of the process. Typically, if $\sigma_m > \sigma_r$, then one or more parts of the model may be missing or improperly specified. Due to random error in the model, sometimes σ_m will be greater than σ_r even when the model is accurate. To ensure that the model hypothesis is not rejected by accident, it is necessary to understand how much greater σ_r can possibly be. This will ensure that the hypothesis is only rejected when σ_m is greater than σ_r. A ratio that can be used when the model fits the data is[8]:

$$L = \frac{\sigma_m^2}{\sigma_r^2} \tag{14-4}$$

The probability of rejecting the hypothesis is controlled by the probability distribution that describes the behavior of the statistic, L. One method of defining the cut-off value is using the value of L when it is greater than the upper-tail cut-off value from the F distribution. This allows a quantitative method of determining when σ_m is greater than σ_r.

The probability specified by the cut-off value from the F distribution is called the "significance level" of the test. The most commonly used significance value is $\alpha = .05$, which means that the hypothesis of an adequate model will only be rejected in 5% of tests for which the model really is adequate. The cut-off values can be calculated using the F distribution described in most statistics textbooks.

14.5 Unnecessary Terms in the Model

Sometimes models fit the data very well, but there are additional unnecessary terms. These models are said to "overfit" the data. Since the parameters for any unnecessary terms in the model usually have values near zero, it may seem harmless to leave them in the model. However, if there are many extra terms in the model, there could be occurrences where the error from the model may be larger than necessary, and this may affect conclusions drawn from the data.

Overfitting often occurs when developing purely empirical models for experimental data, with little understanding of the total and random variation in the data. This happens when regression methods fit the data set instead of using functions to describe the structure in the data. There are models that are sometimes are made to fit very complex patterns, but these may actually be finding structure in the noise if the model is analyzed carefully.

To determine if a model has too many terms, statistical tests can also be used. The tests for overfitting of data is one area in which statistical tests are more effective than residual plots. In this case, individual tests for each parameter in the model are used rather than a single test. The test statistics for testing whether or not each parameter is zero is typically based on T distribution. Each parameter estimated in the model is measured in terms of how many standard deviations it is from its hypothesized value of zero. If the parameter's estimated value is close enough to the hypothesized value, then any additional deviation can be attributed to random error, and the hypothesis that the parameter's true value is not zero is accepted. However, if the parameter's estimated value is so far away from the hypothesized value that the deviation cannot be plausibly explained by random error, the hypothesis that the true value of the parameter is zero is rejected.

The test statistic for each of these tests is simply the estimated parameter value divided by its estimated standard deviation:

$$T = \frac{\beta_i}{\sigma_{\beta_i}} \tag{14-5}$$

Equation 14-5 provides a measure of the distance between the estimated and hypothesized values of the parameter in standard deviations. Since the random errors are normally distributed, and the value of the parameter is zero, the test statistic has a Student's T distribution with $n - p$ degrees of freedom. Therefore, the cut-off values from the T distribution can be used to determine the amount of variable that is due to random error. These tests should each be used with cut-off values with a significance level of $\alpha/2$ since these tests are generally used to simultaneously test whether or not a parameter value is greater than or less than zero. This will ensure that the hypothesis of each parameter equals zero will be rejected by chance with probability α.

EXAMPLE 14-3: Creating ANOVA Tables and Confidence Intervals

Using the residual data given in Example 14-1, create the ANOVA table for the activation voltage, temperature, and pressure. Find the confidence interval for error mean at a 95% confidence level.

Using MATLAB to solve:

```
%%%%%%%%%%%%%%%%%%%%%%%%%%%%%%%%%%%%%%%%%%
% EXAMPLE 14-3: Creating ANOVA Tables and
Confidence Intervals
% UnitSystem SI
%%%%%%%%%%%%%%%%%%%%%%%%%%%%%%%%%%%%%%%%%%
% ANOVA Tables
% ANOVA with the temperature and current as factors
p = anovan(V_act_ex, {T i})
```

Analysis of Variance

Source	Sum Sq.	d.f.	Mean Sq.	F	Prob>F
X1	0.00696	2	0.00348	11.45	0.0004
X2	0.03096	2	0.01548	50.93	0
Error	0.00699	23	0.0003		
Total	0.04324	27			

FIGURE 14-8. ANOVA for temperature and current for Example 14-3.

% Confidence Interval

```
meen = mean(e);
L = length(e);
q = std(e)*tinv(0.975,L-1)/sqrt(L);
disp(['Sample mean =' num2str(meen)])
disp('Confidence interval for error mean at 95% confidence level –')
disp([' ' num2str(meen-q) '<=Error mean<=' num2str(meen+q)])
```

Figure 14-8 illustrates the ANOVA table with the temperature and current as factors.

The confidence limits printed in the MATLAB workspace from Example 14-3 are as follows:

Sample mean = –3.5714e-006
Confidence interval for sample mean at 95% confidence level –
–0.0014334 <= Sample mean <= 0.0014262

In Figure 14-8, the model F-values of 50.93 and 11.45 indicate that the model terms are significant. There is a 0.0% and 0.04% chance that a model F-Value this large can be due to noise. When the values of Prob > F are less than 0.05, this typically indicates that the model terms are significant. If the values are greater than 0.100, this indicates that the model terms are not significant. If there are many insignificant model terms, model reduction may improve the model.

Chapter Summary

Fuel cell validation is the most important step in the model-building process. However, little attention is usually given to this step. A fast method for analyzing the validity of a model is to look at plots of residuals versus the experimental factors, run plots, and lag plots. These plots give a good feel for how accurately a model fits the experimental data, and how dependable it is. Selecting various statistical techniques, or using a combination of them, will tell the user if there are any unnecessary portions of the model, or will help determine the amount of noise. Some of the techniques that are useful in comparing experimental and calculated results are histograms, normal probability plots, T and F distributions, analysis of variance, and confidence levels. If residual scatterplots are used with one or more of the tests listed above, there will be substantial evidence that a model is a good fit to the experimental data.

Problems

- Perform a regression analysis for the data in Example 14-1.
- Determine the T and F distribution tests for the data in Example 14-1.
- How well do you think that the calculated data in Example 14-1 fit the experimental data?
- Perform a lack-of-fit test for the data in Example 14-1.

Endnotes

[1]   *NIST/SEMATECH e-Handbook of Statistical Methods*, http://www.itl.nist .gov/div898/handbook/. Date created 06/01/2003. Last updated 07/18/2006.
[2]   Montgomery, D.C. *Design and Analysis of Experiments*. 5th ed. 2001. New York: John Wiley & Sons.
[3]   Amphlett, J.C., R.M. Baumert, R.F. Mann, B.A. Peppley, P.R. Roberge, and T.J. Harris. Performance modeling of the Ballard Marck IV solid polymer electrolyte fuel cell. J. Electrochem. Soc. Vol. 142, No. 1, January 1995.
[4]   *NIST/SEMATECH e-Handbook of Statistical Methods*.
[5]   Montgomery, *Design and Analysis of Experiments*.
[6]   *NIST/SEMATECH e-Handbook of Statistical Methods*.
[7]   Ibid.
[8]   Ibid.

Bibliography

Barbir, F. 2005. *PEM Fuel Cells: Theory and Practice*. Burlington, MA: Elsevier Academic Press.
Lu, G.Q., and C.Y. Wang. Development of micro direct methanol fuel cells for high power applications. *J. Power Sources*. Vol. 144, 2005, pp. 141–145.
O'Hayre, R., S.-W. Cha, W. Colella, and F.B. Prinz. 2006. *Fuel Cell Fundamentals*. New York: John Wiley & Sons.
Pekula, N., K. Heller, P.A. Chuang, A. Turhan, M.M. Mench, J.S. Brenzier, and K. Unlu. Study of water distribution and transport in a polymer electrolyte fuel cell using neutron imaging. *Nucl. Instrum. Methods Phys. Res. A*. Vol. 542, 2005, pp. 134–141.
Raposa, G. Performing AC impedance spectroscopy measurements on fuel cells. *Fuel Cell Magazine*. February/March 2003.
Smith, M., K. Cooper, D. Johnson, and L. Scribner. Comparison of fuel cell electrolyte resistance measurement techniques. *Fuel Cell Magazine*. April/May 2005.
Turhan, A., K. Heller, J.S. Brenizer, and M.M Mench. Quantification of liquid water accumulation and distribution in a polymer electrolyte fuel cell using neutron imaging. *J. Power Sources*. Vol. 160, 2006, pp. 1195–1203.
The U.S. Fuel Cell Council's Joint Hydrogen Quality Task Force. November 29, 2004. *Primer on Fuel Cell Component Testing: Primer for Generating Test Plans*. Document No. USFCC 04-003. Available at: www.usfcc.com.

APPENDIX A

Physical Constants

The following table lists constants and conversions that may be helpful for fuel cell design calculations. These constants are used in many equations throughout the book.

Avogadro's Number	N_A	6.02×10^{23} Atoms/mol
Universal gas constant	R	0.08205 L atm/mol K
		8.314 J/mol K
		0.08314 bar m^3/mol K
		8.314 kPa m^3/mol K
Planck's constant	h	6.626×10^{-34} J s
		4.136×10^{-15} eV·s
Boltzmann's constant	k	1.38×10^{-23} J/K
		8.61×10^{-5} eV/K
Electron mass	m	9.11×10^{-31} kg
Electron charge	q	1.60×10^{-19} C
Faraday's constant	F	96,485.34 C/mol

APPENDIX B

The following table lists enthalpy of formation, the Gibbs function of formation, and absolute entropy for selected substances at 25 °C and 1 atm.

Substance	Formula	H_f (J/mol)	G_f (J/mol)	S (J/mol*K)
Acetylene	$C_2H_2(g)$	+226,730	+209,170	200.85
Ammonia	$NH_3(g)$	−46,190	−16,590	192.33
Benzene	C_6H_6 (g)	+82,930	+129,660	269.20
Carbon	C(s)	0	0	5.74
Carbon dioxide	$CO_2(g)$	−393,522	−394,360	213.80
Carbon monoxide	CO(g)	−110,530	−137,150	197.65
Ethane	$C_2H_6(g)$	−84,680	−32,890	229.49
Ethyl alcohol	$C_2H_5OH(g)$	−235,310	−168,570	282.59
Ethyl alcohol	$C_2H_5OH(l)$	−277,690	−174,891	160.70
Ethylene	C_2H_4 (g)	+52,280	+68,120	219.83
Hydrogen	H_2 (g)	0	0	130.68
Hydrogen	H(g)	+217,999	+203,290	114.72
Hydrogen peroxide	$H_2O_2(g)$	−136,310	−105,600	232.63
Hydroxyl	OH(g)	+38,987	+34,280	183.70
Methane	$CH_4(g)$	−74,850	−50,790	186.16
Methyl alcohol	$CH_3OH(g)$	−200,670	−162,000	239.70
Methyl alcohol	$CH_3OH(l)$	−238,660	−166,360	126.80
n-Butane	$C_4H_{10}(g)$	−126,150	−15,710	310.12
n-Dodecane	$C_{12}H_2(g)$	−291,010	+50,150	622.83
Nitrogen	$N_2(g)$	0	0	191.61
Nitrogen	N(g)	+472,680	+455,510	153.30
n-Octane	$C_8H_{18}(g)$	−208,450	+16,530	466.73
n-Octane	C_8H_{18} (l)	−249,950	+6,610	360.79
Oxygen	$O_2(g)$	0	0	205.14
Oxygen	O(g)	+249,170	+231,770	161.06
Propane	$C_3H_8(g)$	−130,850	−23,490	269.91
Propylene	C_3H_6 (g)	+20,410	+62,720	266.94
Water	$H_2O(g)$	−241,826	228,590	188.83
Water	$H_2O(l)$	−285,826	237,180	69.92

APPENDIX C

The following table lists the ideal gas-specific heat at constant pressure for selected substances at 300 K.

Gas	Formula	Molecular Weight W kg/kmol	Gas Constant kJ/(kg K)	Specific Heat kJ/(kg K)
Air	—	28.97	0.2870	1.005
Argon	Ar	39.948	0.2081	0.5203
Butane	C_4H_{10}	58.124	0.1433	1.7164
Carbon dioxide	CO_2	44.01	0.1889	0.846
Carbon monoxide	CO	28.011	0.2968	1.040
Ethane	C_2H_6	30.070	0.2765	1.7662
Ethylene	C_2H_4	28.054	0.2964	1.5482
Helium	He	4.003	2.0769	5.1926
Hydrogen	H_2	2.016	4.1240	14.307
Methane	CH_4	16.043	0.5182	2.2537
Neon	Ne	20.183	0.4119	1.0299
Nitrogen	N_2	28.013	0.2968	1.039
Octane	C_8H_{18}	114.230	0.0729	1.7113
Oxygen	O_2	31.999	0.2598	0.918
Propane	C_3H_8	44.097	0.1885	1.6794
Steam	H_2O	18.015	0.4615	1.8723

APPENDIX D

The following table lists the ideal gas-specific heat at constant pressure at various temperatures.

T(K)	Air kJ/(kg K)	CO$_2$ kJ/(kg K)	CO kJ/(kg K)	H$_2$ kJ/(kg K)	H$_2$O(g) kJ(kmol K)	N$_2$ kJ/(kg K)	O$_2$ kJ/(kg K)
250	1.003	0.791	1.039	14.051	33.324	1.039	0.913
300	1.005	0.846	1.040	14.307	33.669	1.039	0.918
350	1.008	0.895	1.043	14.427	34.051	1.041	0.928
400	1.013	0.939	1.047	14.476	34.467	1.044	0.941
450	1.020	0.978	1.054	14.501	34.914	1.049	0.956
500	1.029	1.014	1.063	14.513	35.390	1.056	0.972
550	1.040	1.046	1.075	14.530	35.891	1.065	0.988
600	1.051	1.075	1.087	14.546	36.415	1.075	1.003
650	1.063	1.102	1.100	14.571	36.960	1.086	1.017
700	1.075	1.126	1.113	14.604	37.523	1.098	1.031
750	1.087	1.148	1.126	14.645	38.100	1.110	1.043
800	1.099	1.169	1.139	14.695	38.690	1.121	1.054
900	1.121	1.204	1.163	14.822	39.895	1.145	1.074
1000	1.142	1.234	1.185	14.983	41.118	1.167	1.090

APPENDIX D

The following table lists the ideal gas-specific heat at constant pressure at various temperatures.

APPENDIX E

The following table lists specific heat at constant pressure for saturated liquid water H_2O.

Temperature (°C)	Specific Heat, Cp kJI(kg K)
0	4.2178
20	4.1818
40	4.1784
60	4.1843
80	4.1964
100	4.2161
120	*4.250*
140	4.283
160	4.342
180	4.417

APPENDIX F

The following tables list thermodynamic data for hydrogen, oxygen, water, carbon monoxide, carbon dioxide, methane, and nitrogen for 200 through 1000 K.

Hydrogen Thermodynamic Data

T(K)	$\hat{g}(T)(kJ/mol)$	$\hat{h}(T)(kJ/mol)$	$\hat{s}(T)(J/mol\ K)$	$C_p(T)(J/mol\ K)$
200	−26.66	−2.77	119.42	27.26
220	−29.07	−2.22	122.05	27.81
240	−31.54	−1.66	124.48	28.21
260	−34.05	−1.09	126.75	28.49
280	−36.61	−0.52	128.87	28.7
298.15	−38.96	0	130.68	28.84
300	−39.20	0.05	130.86	28.85
320	−41.84	0.63	132.72	28.96
340	−44.51	1.21	134.48	29.04
360	−47.22	1.79	136.14	29.1
380	−49.96	2.38	137.72	29.15
400	−52.73	2.96	139.22	29.18
420	−55.53	3.54	140.64	29.21
440	−58.35	4.13	142	29.22
460	−61.21	4.71	143.3	29.24
480	−64.08	5.3	144.54	29.25
500	−66.99	5.88	145.74	29.26
520	−69.91	6.47	146.89	29.27
540	−72.86	7.05	147.99	29.28
560	−75.83	7.64	149.06	29.3
580	−78.82	8.22	150.08	29.31
600	−81.84	8.81	151.08	29.32
620	−84.87	9.4	152.04	29.34
640	−87.92	9.98	152.97	29.36
660	−90.99	10.57	153.87	29.39
680	−94.07	11.16	154.75	29.41
700	−97.18	11.75	155.61	29.44
720	−100.30	12.34	156.44	29.47

T(K)	$\hat{g}(T)(kJ/mol)$	$\hat{h}(T)(kJ/mol)$	$\hat{s}(T)(J/mol\ K)$	$C_p(T)(J/mol\ K)$
740	−103.43	12.93	157.24	29.5
760	−106.59	13.52	158.03	29.54
780	−109.75	14.11	158.8	29.58
800	−112.94	14.7	159.55	29.62
820	−116.14	15.29	160.28	29.67
840	−119.35	15.89	161	2972
860	−122.58	16.48	161.7	29.77
880	−125.82	17.08	162.38	29.83
900	−129.07	17.68	163.05	29.88
920	−132.34	18.27	163.71	29.94
940	−135.62	18.87	164.35	30
960	−138.91	19.47	164.99	30.07
980	−142.22	20.08	165.61	30.14
1000	−145.54	20.68	166.22	30.2

Oxygen Thermodynamic Data

T(K)	$\hat{g}(T)(kJ/mol)$	$\hat{h}(T)(kJ/mol)$	$\hat{s}(T)(J/mol\ K)$	$C_p(T)(J/mol\ K)$
200	−41.54	−2.71	194.16	25.35
220	−45.45	−2.19	196.63	26.41
240	−49.41	−1.66	198.97	27.25
260	−53.41	−1.10	201.18	27.93
280	−57.45	−0.54	203.27	28.48
298.15	−61.12	0.00	205.00	28.91
300	−61.54	0.03	205.25	28.96
320	−65.66	0.62	207.13	29.36
340	−69.82	1.21	208.92	29.71
360	−74.02	1.81	210.63	30.02
380	−78.25	2.41	212.26	30.30
400	−82.51	3.02	213.82	30.56
420	−86.80	3.63	215.32	30.79
440	−91.12	4.25	216.75	31.00
460	−95.47	4.87	218.14	31.20
480	−99.85	5.50	219.47	31.39
500	−104.25	6.13	220.75	31.56
520	−108.68	6.76	221.99	31.73
540	−113.13	7.40	223.20	31.89
560	−117.61	8.04	224.36	32.04
580	−122.10	8.68	225.48	32.19
600	−126.62	9.32	226.58	32.32
620	−131.17	9.97	227.64	32.46
640	−135.73	10.62	228.67	32.59
660	−140.31	11.27	229.68	32.72
680	−144.92	11.93	230.66	32.84
700	−149.54	12.59	231.61	32.96
720	−154.18	13.25	232.54	33.07

T(K)	$\hat{g}(T)(kJ/mol)$	$\hat{h}(T)(kJ/mol)$	$\hat{s}(T)(J/mol\ K)$	$C_p(T)(J/mol\ K)$
740	−158.84	13.91	233.45	33.19
760	−163.52	14.58	234.33	33.30
780	−168.21	15.24	235.20	33.41
800	−172.93	15.91	236.05	33.52
820	−177.66	16.58	236.88	33.62
840	−182.40	17.26	237.69	33.72
860	−187.16	17.93	238.48	33.82
880	−191.94	18.61	239.26	33.92
900	−196.73	19.29	240.02	34.02
920	−201.54	19.97	240.77	34.12
940	−206.36	20.65	241.51	34.21
960	−211.20	21.34	242.23	34.30
980	−216.05	22.03	242.94	34.40
1000	−220.92	22.71	243.63	34.49

$H_2O(l)$ Thermodynamic Data

T(K)	$\hat{g}(T)(kJ/mol)$	$\hat{h}(T)(kJ/mol)$	$\hat{s}(T)(J/mol\ K)$	$C_p(T)(J/mol\ K)$
273	−305.01	−287.73	63.28	76.10
280	−305.46	−287.20	65.21	75.81
298.15	−306.69	−285.83	69.95	75.37
300	−306.82	−285.69	70.42	75.35
320	−308.27	−284.18	75.28	75.27
340	−309.82	−282.68	79.85	75.41
360	−311.46	−281.17	84.16	75.72
373	−312.58	−280.18	86.85	75.99

$H_2O(g)$ Thermodynamic Data

T(K)	$\hat{g}(T)(kJ/mol)$	$\hat{h}(T)(kJ/mol)$	$\hat{s}(T)(J/mol\ K)$	$C_p(T)(J/mol\ K)$
280	−294.72	−242.44	186.73	33.53
298.15	−298.13	−241.83	188.84	33.59
300	−298.48	−241.77	189.04	33.60
320	−302.28	−241.09	191.21	33.69
340	−306.13	−240.42	193.26	33.81
360	−310.01	−239.74	195.20	33.95
380	−313.94	−239.06	197.04	34.10
400	−317.89	−238.38	198.79	34.26
420	−321.89	−237.69	200.47	34.44
440	−325.91	−237.00	202.07	34.62
460	−329.97	−236.31	203.61	34.81
480	−334.06	−235.61	205.10	35.01
500	−338.17	−234.91	206.53	35.22
520	−342.32	−234.20	207.92	35.43
540	−346.49	−233.49	209.26	35.65
560	−350.69	−232.77	210.56	35.87

T(K)	$\hat{g}(T)(kJ/mol)$	$\hat{h}(T)(kJ/mol)$	$\hat{s}(T)(J/mol\ K)$	$C_p(T)(J/mol\ K)$
580	−354.91	−232.05	211.82	36.09
600	−359.16	−231.33	213.05	36.32
620	−363.43	−230.60	214.25	36.55
640	−367.73	−229.87	215.41	36.78
660	−372.05	−229.13	216.54	37.02
680	−376.39	−228.39	217.65	37.26
700	−380.76	−227.64	218.74	37.50
720	−385.14	−226.89	219.80	37.75
740	−389.55	−226.13	220.83	37.99
760	−393.97	−225.37	221.85	38.24
780	−398.42	−224.60	222.85	38.49
800	−402.89	−223.83	223.83	38.74
820	−407.37	−223.05	224.78	38.99
840	−411.88	−222.27	225.73	39.24
860	−416.40	−221.48	226.65	39.49
880	−420.94	−220.69	227.56	39.74
900	−425.51	−219.89	228.46	40.00
920	−430.08	−219.09	229.34	40.25
940	−434.68	−218.28	230.21	40.51
960	−439.29	−217.47	231.07	40.76
980	−443.92	−216.65	231.91	41.01
1000	−448.57	−215.83	232.74	41.27

CO Thermodynamic Data

T(K)	$\hat{g}(T)(kJ/mol)$	$\hat{h}(T)(kJ/mol)$	$\hat{s}(T)(J/mol\ K)$	$C_p(T)(J/mol\ K)$
200	−150.60	−113.42	185.87	30.20
220	−154.34	−112.82	188.73	29.78
240	−158.14	−112.23	191.31	29.50
260	−161.99	−111.64	193.66	29.32
280	−165.89	−111.06	195.83	29.20
298.15	−169.46	−110.53	197.66	29.15
300	−169.83	−110.47	197.84	29.15
320	−173.80	−109.89	199.72	29.13
340	−177.81	−109.31	201.49	29.14
360	−181.86	−108.72	203.16	29.17
380	−185.94	−108.14	204.73	29.23
400	−190.05	−107.56	206.24	29.30
420	−194.19	−106.97	207.67	29.39
440	−198.36	−106.38	209.04	29.48
460	−202.55	−105.79	210.35	29.59
480	−206.77	−105.20	211.61	29.70
500	−211.01	−104.60	212.83	29.82
520	−215.28	−104.00	214.00	29.94
540	−219.57	−103.40	215.13	30.07
560	−223.89	−102.80	216.23	30.20

$T(K)$	$\hat{g}(T)(kJ/mol)$	$\hat{h}(T)(kJ/mol)$	$\hat{s}(T)(J/mol\ K)$	$C_p(T)(J/mol\ K)$
580	−228.22	−102.19	217.29	30.34
600	−232.58	−101.59	218.32	30.47
620	−236.95	−100.98	219.32	30.61
640	−241.35	−100.36	220.29	30.75
660	−245.77	−99.75	221.24	30.89
680	−250.20	−99.13	222.17	31.03
700	−254.65	−98.50	223.07	31.17
720	−259.12	−97.88	223.95	31.31
740	−263.61	−97.25	224.81	31.46
760	−268.12	−96.62	225.65	31.60
780	−272.64	−95.99	226.47	31.74
800	−277.17	−95.35	227.28	31.88
820	−281.73	−94.71	228.07	32.01
840	−286.30	−94.07	228.84	32.15
860	−290.88	−93.43	229.60	32.29
880	−295.48	−92.78	230.34	32.42
900	−300.09	−92.13	231.07	32.55
920	−304.72	−91.48	231.79	32.68
940	−309.37	−90.82	232.49	32.81
960	−314.02	−90.17	233.18	32.94
980	−318.69	−89.51	233.86	33.06
1000	−323.38	−88.84	234.53	33.18

CO_2 Thermodynamic Data

$T(K)$	$\hat{g}(T)(kJ/mol)$	$\hat{h}(T)(kJ/mol)$	$\hat{s}(T)(J/mol\ K)$	$C_p(T)(J/mol\ K)$
200	−436.93	−396.90	200.1	31.33
220	−440.95	396.25	203.16	32.77
240	−445.04	−395.59	206.07	34.04
260	−449.19	−394.89	208.84	35.19
280	−453.39	−394.18	211.48	36.24
300	−457.65	−393.44	214.02	37.22
320	−461.95	−392.69	216.45	38.13
340	−466.31	−391.92	218.79	39
360	−470.71	−391.13	221.04	39.81
380	−475.15	−390.33	223.21	40.59
400	−479.63	−389.51	225.31	41.34
420	−484.16	−388.67	227.35	42.05
440	−488.73	−387.83	229.32	42.73
460	−493.33	−386.96	231.23	43.38
480	−497.98	−386.09	233.09	44.01
500	−502.66	−385.20	234.9	44.61
520	−507.37	−384.31	236.66	45.2
540	512.12	−383.40	238.38	45.76
560	−516.91	−382.48	240.05	46.3
580	−521.72	−381.54	241.69	46.82

T(K)	$\hat{g}(T)(kJ/mol)$	$\hat{h}(T)(kJ/mol)$	$\hat{s}(T)(J/mol\ K)$	$C_p(T)(J/mol\ K)$
600	526.59	−380.60	243.28	47.32
620	−531.46	−379.65	244.84	47.8
640	−536.37	−378.69	246.37	48.27
660	−541.31	−377.72	247.86	48.72
680	−546.28	−376.74	249.32	49.15
700	−551.29	−375.76	250.75	49.57
720	−556.31	−374.76	252.15	49.97
740	−561.37	−373.76	253.53	50.36
760	−566.45	−372.75	254.88	50.73
780	−571.56	−371.73	256.2	51.09
800	−576.71	−370.70	257.5	51.44
820	581.86	−369.67	258.77	51.78
840	−587.05	−368.63	260.02	52.1
860	−592.26	−367.59	261.25	52.41
880	−597.50	−366.54	262.46	52.71
900	−602.76	365.48	263.65	53
920	−608.05	−364.42	264.82	53.28
940	−613.35	−363.35	265.97	53.55
960	−618.68	−362.27	267.1	53.81
980	624.04	−361.19	268.21	54.06
1000	−629.41	−360.11	269.3	54.3

CH_4 Thermodynamic Data

T(K)	$\hat{g}(T)(kJ/mol)$	$\hat{h}(T)(kJ/mol)$	$\hat{s}(T)(J/mol\ K)$	$C_p(T)(J/mol\ K)$
200	−112.69	−78.25	172.23	36.30
220	−116.17	−77.53	175.63	35.19
240	−119.71	−76.83	178.67	34.74
260	−123.32	−76.14	181.45	34.77
280	−126.97	−75.44	184.03	35.12
298.15	−130.33	−74.80	186.25	35.65
300	−130.68	−74.73	186.48	35.71
320	−134.43	−74.01	188.80	36.47
340	−138.23	−73.27	191.04	37.36
360	−142.07	−72.52	193.20	38.35
380	−145.95	−71.74	195.31	39.40
400	−149.88	−70.94	197.35	40.50
420	−153.85	−70.12	199.36	41.64
440	−157.86	−69.27	201.32	42.80
460	−161.90	−68.41	203.25	43.98
480	−165.99	−67.51	205.15	45.16
500	−170.11	−66.60	207.01	46.35
520	−174.27	−65.66	208.86	47.54
540	−178.46	−64.70	210.67	48.73
560	−182.69	−63.71	212.47	49.90
580	−186.96	−62.70	214.24	51.07

T(K)	$\hat{g}(T)(kJ/mol)$	$\hat{h}(T)(kJ/mol)$	$\hat{s}(T)(J/mol\ K)$	$C_p(T)(J/mol\ K)$
600	−191.26	−61.67	215.99	52.23
620	−195.60	−60.61	217.72	53.37
640	−199.97	−59.53	219.43	54.50
660	−204.38	−58.43	221.13	55.61
680	−208.82	−57.31	222.80	56.71
700	−213.29	−56.16	224.46	57.79
720	−217.79	−55.00	226.10	58.85
740	−222.33	−53.81	227.73	59.90
760	−226.90	−52.60	229.34	60.93
780	−231.51	−51.37	230.94	61.94
800	−236.14	−50.13	232.52	62.93
820	−240.81	−48.86	234.08	63.90
840	−245.50	−47.57	235.64	64.85
860	−250.23	−46.26	237.17	65.79
880	−254.99	−44.94	238.70	66.70
900	−259.78	−43.60	240.20	67.60
920	−264.60	−42.23	241.70	68.47
940	−269.45	−40.86	243.18	69.33
960	−274.33	−39.46	244.65	70.17
980	−279.23	−38.05	246.11	70.99
1000	−284.17	−36.62	247.55	71.79

N_2 Thermodynamic Data

T(K)	$\hat{g}(T)(kJ/mol)$	$\hat{h}(T)(kJ/mol)$	$\hat{s}(T)(J/mol\ K)$	$C_p(T)(J/mol\ K)$
200	−38.85	−2.83	180.08	28.77
220	−42.48	−2.26	182.82	28.72
240	−46.16	−1.68	185.31	28.72
260	−49.89	−1.11	187.61	28.76
280	−53.66	−0.53	189.75	28.81
298.15	−57.11	0	191.56	28.87
300	−57.48	0.04	191.74	28.88
320	−61.33	0.62	193.6	28.96
340	−65.22	1.2	195.36	29.05
360	−69.15	1.78	197.02	29.14
380	−73.10	2.37	198.6	29.25
400	−77.09	2.95	200.11	29.35
420	−81.11	3.54	201.54	29.46
440	−85.15	4.13	202.91	29.57
460	−89.22	4.72	204.23	29.68
480	−93.32	5.32	205.5	29.79
500	−97.44	5.92	206.71	29.91
520	−101.59	6.51	207.89	30.02
540	−105.76	7.12	209.02	30.13
560	−109.95	7.72	210.12	30.24
580	−114.16	8.33	211.19	30.36

$T(K)$	$\hat{g}(T)(kJ/mol)$	$\hat{h}(T)(kJ/mol)$	$\hat{s}(T)(J/mol\ K)$	$C_p(T)(J/mol\ K)$
600	−118.40	8.93	212.22	30.47
620	−122.65	9.54	213.22	30.58
640	−126.92	10.16	214.19	30.69
660	−131.22	10.77	215.14	30.8
680	−135.53	11.39	216.06	30.91
700	−139.86	12.01	216.96	31.02
720	−144.21	12.63	217.83	31.13
740	−148.57	13.25	218.69	31.24
760	−152.96	13.88	219.52	31.34
780	−157.35	14.51	220.34	31.45
800	−161.77	15.14	221.13	31.55
820	−166.20	15.77	221.91	31.66
840	−170.64	16.4	222.68	31.76
860	−175.11	17.04	223.43	31.86
880	−179.58	17.68	224.16	31.96
900	−184.07	18.32	224.88	32.06
920	−188.58	18.96	225.58	32.16
940	−193.10	19.61	226.28	32.25
960	−19.63	20.25	226.96	32.35
980	−202.1	20.9	227.63	32.44
1000	−206.3	21.55	228.28	32.54

APPENDIX G

The following table lists binary diffusion coefficients for relevant fuel cell substances at 1 atmosphere.

Substance A	Substance B	T(K)	$D_{AB}(m^2/s)$
Acetone	Air	298	0.11×10^{-4}
Acetone	H_2O	298	0.13×10^{-8}
Ar	N_2	293	0.19×10^{-4}
Benzene	Air	298	0.88×10^{-5}
CO_2	Air	298	0.16×10^{-4}
CO_2	N_2	293	0.16×10^{-4}
CO_2	O_2	273	0.14×10^{-4}
CO_2	H_2O	298	0.20×10^{-8}
Ethanol	H_2O	298	0.12×10^{-8}
G	H_2O	298	0.94×10^{-9}
Glucose	H_2O	298	0.69×10^{-9}
H_2	Air	273	0.41×10^{-4}
H_2	O_2	273	0.70×10^{-4}
H_2	N_2	273	0.68×10^{-4}
H_2	CO_2	273	0.55×10^{-4}
H_2	H_2O	298	0.63×10^{-8}
H_2O	Air	298	0.26×10^{-4}
N_2	H_2O	298	0.26×10^{-8}
Naphthalene	Air	300	0.62×10^{-5}
NH_3	Air	298	0.28×10^{-4}
O_2	Air	298	0.21×10^{-4}
O_2	N_2	273	0.18×10^{-4}
O_2	H_2O	298	0.24×10^{-8}

APPENDIX H

Properties of Saturated Water (Liquid–Vapor): Temperature Table

Temp. °C	Press. Bars	Specific Volume m3/kg		Internal Energy kJ/kg		Enthalpy kJ/kg			Entropy kJ/kg · K		Temp. °C
		Sat. Liquid $v_f \times 10^3$	Sat. Vapor v_g	Sat. Liquid u_f	Sat. Vapor u_g	Sat. Liquid h_t	Evap. h_{fg}	Sat. Vapor h_g	Sat. Liquid s_f	Sat. Vapor s_g	
.01	0.00611	1.0002	206.136	0.00	2375.3	0.01	2501.3	2501.4	0.0000	9.1562	.01
4	0.00813	1.0001	157.232	16.77	2380.9	16.78	2491.9	2508.7	0.0610	9.0514	4
5	0.00872	1.0001	147.120	20.97	2382.3	20.98	2489.6	2510.6	0.0761	9.0257	5
6	0.00935	1.0001	137.734	25.19	2383.6	25.20	2487.2	2512.4	0.0912	9.0003	6
8	0.01072	1.0002	120.917	33.59	2386.4	33.60	2482.5	2516.1	0.1212	8.9501	8
10	0.01228	1.0004	106.379	42.00	2389.2	42.01	2477.7	2519.8	0.1510	8.9008	10
11	0.01312	1.0004	99.857	46.20	2390.5	46.20	2475.4	2521.6	0.1658	8.8765	11
12	0.01402	1.0005	93.784	50.41	2391.9	50.41	2473.0	2523.4	0.1806	8.8524	12
13	0.01497	1.0007	88.124	54.60	2393.3	54.60	2470.7	2525.3	0.1953	8.8285	13
14	0.01598	1.0008	82.848	58.79	2394.7	58.80	2468.3	2527.1	0.2099	8.8048	14
15	0.01705	1.0009	77.926	62.99	2396.1	62.99	2465.9	2528.9	0.2245	8.7814	15
16	0.01818	1.0011	73.333	67.18	2397.4	67.19	2463.6	2530.8	0.2390	8.7582	16
17	0.01938	1.0012	69.044	71.38	2398.8	71.38	2461.2	2532.6	0.2535	8.7351	17
18	0.02064	1.0014	65.038	75.57	2400.2	75.58	2458.8	2534.4	0.2679	8.7123	18
19	0.02198	1.0016	61.293	79.76	2401.6	79.77	2456.5	2536.2	0.2823	8.6897	19
20	0.02339	1.0018	57.791	83.95	2402.9	83.96	2454.1	2538.1	0.2966	8.6672	20
21	0.02487	1.0020	54.514	88.14	2404.3	88.14	2451.8	2539.9	0.3109	8.6450	21
22	0.02645	1.0022	51.447	92.32	2405.7	92.33	2449.4	2541.7	0.3251	8.6229	22
23	0.02810	1.0024	48.574	96.51	2407.0	96.52	2447.0	2543.5	0.3393	8.6011	23
24	0.02985	1.0027	45.883	100.70	2408.4	100.70	2444.7	2545.4	0.3534	8.5794	24
25	0.03169	1.0029	43.360	104.88	2409.8	104.89	2442.3	2547.2	0.3674	8.5580	25
26	0.03363	1.0032	40.994	109.06	2411.1	109.07	2439.9	2549.0	0.3814	8.5367	26
27	0.03567	1.0035	38.774	113.25	2412.5	113.25	2437.6	2550.8	0.3954	8.5156	27
28	0.03782	1.0037	36.690	117.42	2413.9	117.43	2435.2	2552.6	0.4093	8.4946	28
29	0.04008	1.0040	34.733	121.60	2415.2	121.61	2432.8	2554.5	0.4231	8.4739	29
30	0.04246	1.0043	32.894	125.78	2416.6	125.79	2430.5	2556.3	0.4369	4.4533	30
31	0.04496	1.0046	31.165	129.96	2418.0	129.97	2428.1	2558.1	0.4507	8.4329	31
32	0.04759	1.0050	29.540	134.14	2419.3	134.15	2425.7	2559.9	0.4644	8.4127	32
33	0.05034	1.0053	28.011	138.32	2420.7	138.33	2423.4	2561.7	0.4781	8.3927	33
34	0.05324	1.0056	26.571	142.50	2422.0	142.50	2421.0	2563.5	0.4917	8.3728	34
35	0.05628	1.0060	25.216	146.67	2423.4	146.68	2418.6	2565.3	0.5053	8.3531	35
36	0.05947	1.0063	23.940	150.85	2424.7	150.86	2416.2	2567.1	0.5188	8.3336	36
38	0.06632	1.0071	21.602	159.20	2427.4	159.21	2411.5	2570.7	0.5458	8.2950	38
40	0.07384	1.0078	19.523	167.56	2430.1	167.57	2406.7	2574.3	0.5725	8.2570	40
45	0.09593	1.0099	15.258	188.44	2436.8	188.45	2394.8	2583.2	0.6387	8.1648	45

Temp. °C	Press. Bars	Specific Volume m3/kg		Internal Energy kJ/kg		Enthalpy kJ/kg			Entropy kJ/kg · K		Temp °C
		Sat. Liquid $v_f \times 10^3$	Sat. Vapor v_g	Sat. Liquid u_f	Sat. Vapor u_g	Sat. Liquid h_t	Evap. h_{fg}	Sat. Vapor h_g	Sat. Liquid s_f	Sat. Vapor s_g	
50	.1235	1.0121	12.032	209.32	2443.5	209.33	2382.7	2592.1	.7038	8.0763	50
55	.1576	1.0146	9.568	230.21	2450.1	230.23	2370.7	2600.9	.7679	7.9913	55
60	.1994	1.0172	7.671	251.11	2456.6	251.13	2358.5	2609.6	.8312	7.9096	60
65	.2503	1.0199	6.197	272.02	2463.1	272.06	2346.2	2618.3	.8935	7.8310	65
70	.3119	1.0228	5.042	292.95	2469.6	292.98	2333.8	2626.8	.9549	7.7553	70
75	.3858	1.0259	4.131	313.90	2475.9	313.93	2321.4	2635.3	1.0155	7.6824	75
80	.4739	1.0291	3.407	334.86	2482.2	334.91	2308.8	2643.7	1.0753	7.6122	80
85	.5783	1.0325	2.828	355.84	2488.4	355.90	2296.0	2651.9	1.1343	7.5445	85
90	.7014	1.0360	2.361	376.85	2494.5	376.92	2283.2	2660.1	1.1925	7.4791	90
95	.8455	1.0397	1.982	397.88	2500.6	397.96	2270.2	2668.1	1.2500	7.4159	95
100	1.014	1.0435	1.673	418.94	2506.5	419.04	2257.0	2676.1	1.3069	7.3549	100
110	1.433	1.0516	1.210	461.14	2518.1	461.30	2230.2	2691.5	1.4185	7.2387	110
120	1.985	1.0603	0.8919	503.50	2529.3	503.71	2202.6	2706.3	1.5276	7.1296	120
130	2.701	1.0697	0.6685	546.02	2539.9	546.31	2174.2	2720.5	1.6344	7.0269	130
140	3.613	1.0797	0.5089	588.74	2550.0	589.13	2144.7	2733.9	1.7391	6.9299	140

APPENDIX I

Properties of Saturated Water (Liquid–Vapor): Pressure Table

Press. Bars	Temp. °C	Specific Volume m³/kg Sat. Liquid $v_f \times 10^3$	Sat. Vapor v_g	Internal Energy kJ/kg Sat. Liquid u_f	Sat. Vapor u_g	Enthalpy kJ/kg Sat. Liquid h_t	Evap. h_{fg}	Sat. Vapor h_g	Entropy kJ/kg · K Sat. Liquid s_f	Sat. Vapor s_g	Press. Bars
0.04	28.96	1.0040	34.800	121.45	2415.2	121.46	2432.9	2554.4	0.4226	8.4746	0.04
0.06	36.16	1.0064	23.739	151.53	2425.0	151.53	2415.9	2567.4	0.5210	8.3304	0.06
0.08	41.51	1.0084	18.103	173.87	2432.2	173.88	2403.1	2577.0	0.5926	8.2287	0.08
0.10	45.81	1.0102	14.674	191.82	2437.9	191.83	2392.8	2584.7	0.6493	8.1502	0.10
0.20	60.06	1.0172	7.649	251.38	2456.7	251.40	2358.3	2609.7	0.8320	7.9085	0.20
0.30	69.10	1.0223	5.229	289.20	2468.4	289.23	2336.1	2625.3	0.9439	7.7686	0.30
0.40	75.87	1.0265	3.993	317.53	2477.0	317.58	2319.2	2636.8	1.0259	7.6700	0.40
0.50	81.33	1.0300	3.240	340.44	2483.9	340.49	2305.4	2645.9	1.0910	7.5939	0.50
0.60	85.94	1.0331	2.732	359.79	2489.6	359.86	2293.6	2653.5	1.1453	7.5320	0.60
0.70	89.95	1.0360	2.365	376.63	2494.5	376.70	2283.3	2660.0	1.1919	7.4797	0.70
0.80	93.50	1.0380	2.087	391.58	2498.8	391.66	2274.1	2665.8	1.2329	7.4346	0.80
0.90	96.71	1.0410	1.869	405.06	2502.6	405.15	2265.7	2670.9	1.2695	7.3949	0.90
1.00	99.63	1.0432	1.694	417.36	2506.1	417.46	2258.0	2675.5	1.3026	7.3594	1.00
1.50	111.4	1.0528	1.159	466.94	2519.7	467.11	2226.5	2693.6	1.4336	7.2233	1.50
2.00	120.2	1.0605	0.8857	504.49	2529.5	504.70	2201.9	2706.7	1.5301	7.1271	2.00
2.50	127.4	1.0672	0.7187	535.10	2537.2	535.37	2181.5	2716.9	1.6072	7.0527	2.50
3.00	133.6	1.0732	0.6058	561.15	2543.6	561.47	2163.8	2725.3	1.6718	6.9919	3.00
3.50	138.9	1.0786	0.5243	583.95	2546.9	584.33	2148.1	2732.4	1.7275	6.9405	3.50
4.00	143.6	1.0836	0.4625	604.31	2553.6	604.74	2133.8	2738.6	1.7766	6.8959	4.00
4.50	147.9	1.0882	0.4140	622.25	2557.6	623.25	2120.7	2743.9	1.8207	6.8565	4.50
5.00	151.9	1.0926	0.3749	639.68	2561.2	640.23	2108.5	2748.7	1.8607	6.8212	5.00
6.00	158.9	1.1006	0.3157	669.90	2567.4	670.56	2086.3	2756.8	1.9312	6.7600	6.00
7.00	165.0	1.1080	0.2729	696.44	2572.5	697.22	2066.3	2763.5	1.9922	6.7080	7.00
8.00	170.4	1.1148	0.2404	720.22	2576.8	721.11	2048.0	2769.1	2.0462	6.6628	8.00
9.00	175.4	1.1212	0.2150	741.83	2580.5	742.83	2031.1	2773.9	2.0946	6.6226	9.00
10.0	179.9	1.1273	0.1944	761.68	2583.6	762.81	2015.3	2778.1	2.1387	6.5863	10.0

APPENDIX J

Properties of Superheated Water Vapor

$T\,°C$	$v\,m^3/kg$	$u\,kJ/kg$	$h\,kJ/kg$	$s\,kJ/kg\cdot K$	$v\,m^3/kg$	$u\,kJ/kg$	$h\,kJ/kg$	$s\,kJ/kg\cdot K$
	\multicolumn{4}{l}{$p = 0.06\ bar = 0.006\ MPa$}	\multicolumn{4}{l}{$p = 0.35\ bar = 0.035\ MPa$}
	\multicolumn{4}{l}{$(T_{sat} = 36.16\ °C)$}	\multicolumn{4}{l}{$(T_{sat} = 72.69\ °C)$}
Sat.	23.739	2425.0	2567.4	8.3304	4.526	2473.0	2631.4	7.7158
80	27.132	2487.3	2650.1	8.5804	4.625	2483.7	2645.6	7.7564
120	30.219	2544.7	2726.0	8.7840	5.163	2542.4	2723.1	7.9644
160	33.302	2602.7	2802.5	8.9693	5.696	2601.2	2800.6	8.1519
200	36.383	2661.4	2879.7	9.1398	6.228	2660.4	2878.4	8.3237
240	39.462	2721.0	2957.8	9.2982	6.758	2720.3	2956.8	8.4828
280	42.540	2781.5	3036.8	9.4464	7.287	2780.9	3036.0	8.6314
320	45.618	2843.0	3116.7	9.5859	7.815	2842.5	3116.1	8.7712
360	48.696	2905.5	3197.7	9.7180	8.344	2905.1	3197.1	8.9034
400	51.774	2969.0	3279.6	9.8435	8.872	2968.6	3279.2	9.0291
440	54.851	3033.5	3362.6	9.9633	9.400	3033.2	3362.2	9.1490
500	59.467	3132.3	3489.1	10.1336	10.192	3132.1	3488.8	9.3194

$T\,°C$	$v\,m^3/kg$	$u\,kJ/kg$	$h\,kJ/kg$	$s\,kJ/kg\cdot K$	$v\,m^3/kg$	$u\,kJ/kg$	$h\,kJ/kg$	$s\,kJ/kg\cdot K$
	\multicolumn{4}{l}{$p = 0.70\ bar = 0.07\ MPa$}	\multicolumn{4}{l}{$p = 1.0\ bar = 0.10\ MPa$}
	\multicolumn{4}{l}{$(T_{sat} = 89.95\ °C)$}	\multicolumn{4}{l}{$(T_{sat} = 99.63\ °C)$}
Sat.	2.365	2494.5	2660.0	7.4797	1.694	2506.1	2675.5	7.3594
100	2.434	2509.7	2680.0	7.5341	1.696	2506.7	2676.2	7.3614
120	2.571	2539.7	2719.6	7.6375	1.793	2537.3	2716.6	7.4668
160	2.841	2599.4	2798.2	7.8279	1.984	2597.8	2796.2	7.6597
200	3.108	2659.1	2876.7	8.0012	2.172	2658.1	2875.3	7.8343
240	3.374	2719.3	2955.5	8.1611	2.359	2718.5	2954.5	7.9949
280	3.640	2780.2	3035.0	8.3162	2.546	2779.6	3034.2	8.1445
320	3.905	2842.0	3115.3	8.4504	2.732	2841.5	3114.6	8.2849
360	4.170	2904.6	3196.5	8.5828	2.917	2904.2	3195.9	8.4175
400	4.434	2968.2	3278.6	8.7086	3.103	2967.9	3278.2	8.5435
440	4.698	3032.9	3361.8	8.8286	3.288	3032.6	3361.4	8.6636
500	5.095	3131.8	3488.5	8.9991	3.565	3131.6	3488.1	8.8342

T °C	vm³/kg	ukJ/kg	hkJ/kg	skJ/kg·K	vm³/kg	ukJ/kg	hkJ/kg	skJ/kg·K
	p = 1.5 bars = 0.15 MPa				*p = 3.0 bars = 0.30 MPa*			
	(T_{sat} = 111.37 °C)				(T_{sat} = 133.55 °C)			
Sat.	1.159	2519.7	2693.6	7.2233	0.606	2543.6	2725.3	6.9919
120	1.188	2533.3	2711.4	7.2693				
160	1.317	2595.2	2792.8	7.4665	0.651	2587.1	2782.3	7.1276
200	1.444	2656.2	2872.9	7.6433	0.716	2650.7	2865.5	7.3115
240	1.570	2717.2	2952.7	7.8052	0.781	2713.1	2947.3	7.4774
280	1.695	2778.6	3032.8	7.9555	0.844	2775.4	3028.6	7.6299
320	1.819	2840.6	3113.5	8.0964	0.907	2838.1	3110.1	7.7722
360	1.943	2903.5	3195.0	8.2293	0.969	2901.4	3192.2	7.9061
400	2.067	2967.3	3277.4	8.3555	1.032	2965.6	3275.0	8.0330
440	2.191	3032.1	3360.7	8.4757	1.094	3030.6	3358.7	8.1538
500	2.376	3131.2	3487.6	8.6466	1.187	3130.0	3486.0	8.3251
600	2.685	3301.7	3704.3	8.9101	1.341	3300.8	3703.2	8.5892

Index

Page numbers in the form 1*ap* refer to the appendices

Printed and bound by CPI Group (UK) Ltd, Croydon, CR0 4YY

08/05/2025

01864879-0002